Birkhäuser

Modern Birkhäuser Classics

Many of the original research and survey monographs in pure and applied mathematics published by Birkhäuser in recent decades have been groundbreaking and have come to be regarded as foundational to the subject. Through the MBC Series, a select number of these modern classics, entirely uncorrected, are being re-released in paperback (and as eBooks) to ensure that these treasures remain accessible to new generations of students, scholars, and researchers.

Ido Yavetz

From Obscurity to Enigma

The Work of Oliver Heaviside, 1872-1889

Reprint of the 1995 Edition

Birkhäuser

Ido Yavetz
The Cohn Institute for the History
and Philosophy of Science and Ideas
Tel Aviv University
Tel Aviv, 69978
Israel
yavetz@post.tau.ac.il

2010 Mathematics Subject Classification: 01A70, 01A55, 78-03

ISBN 978-3-0348-0176-8 e-ISBN 978-3-0348-0177-5
DOI 10.1007/978-3-0348-0177-5

Library of Congress Control Number: 2011931524

Originally published under the same title as volume 16 in the Science Networks. Historical Studies series by Birkhäuser Verlag, Switzerland, ISBN 978-3-7643-5180-9
Reprint 2011 by Springer Basel AG

Cover design: deblik, Berlin

Printed on acid-free paper

Springer Basel AG is part of Springer Science+Business Media

www.birkhauser-science.com

To my parents

Table of Contents

Acknowledgments

Writing a book about Oliver Heaviside provides one way of appreciating the uniqueness of his work. He wrote his *Electrical Papers* without significant guidance from others, which seems remarkable considering the intellectual debts I have incurred in the course of writing this book. My first gratitude goes to Yehuda Elkana, who supervised my Tel-Aviv University Ph.D. thesis—out of which this book grew—with patience and liberality of mind that are truly rare. Initial development of the thesis into a book was aided by a Post-Doctoral fellowship at Wolfson College, Oxford. I am grateful to Robert Fox for making it possible, and for many useful conversations on physicists and engineers, theoreticians and practicians in 19th century Britain. In the advanced stages of developing the book, I have benefitted greatly from the precise, uncompromising, but always constructive and open-minded criticism of Jed Z. Buchwald. He has also made it possible for me to put the finishing touches on the book in the comfortable and stimulating environment of the Dibner Institute. It is safe to say that without his help this book would not have been published. Roger H. Stuewer went through the manuscript with a fine tooth comb and weeded out many embarrassing errors (all remaining errors are mine). Gerald Holton extended his generous help while I was writing the Ph.D. thesis. Thomas P. Hughes read the thesis and pointed out several important issues. I also had the benefit of Amos Funkenstein's sharp eye and immense knowledge. L. Pearce Williams encouraged me to develop the thesis into a book. Comments from Anna Guagnini and Andy Warwick helped clarify several points of engineering and physics. My work at Swartzrauber–Segan, Inc., with Marc, Sayre, Mike, and Neil has made circuit design into much more than an ideal exercise on paper. Doris Wörner and Annette A'Campo of Birkhäuser Verlag helped in formatting the manuscript into a book. Finally, it gives me special pleasure to thank Lenore Symons, chief archivist of the IEE, for many pleasant weeks of reading through and talking about the rich material held at the IEE's Heaviside Collection.

Chapter I The Enigmatic Legacy of Oliver Heaviside

I felt obliged to give you warning that you are a little obscure for ordinary men.

Hertz to Heaviside, 1889.

1. Introduction

This work traces Oliver Heaviside's electromagnetic investigations from the publication of his first electrical paper in 1872 to the public recognition awarded to him by Lord Kelvin in 1889.[1] By 1891, following Kelvin's unqualified praise, Heaviside became established as a leading authority on electrical matters, particularly on the electromagnetic theory of telegraph and telephone communication.

It should be noted at the outset that Heaviside's work is not an example of great work—such as Da Vinci's Codici—that had practically disappeared from view to be rediscovered later and lend its author the image of one who transcended his time. Physicists, mathematicians, electrical engineers and historians of all three subjects have been referring to Heaviside's work quite regularly since the turn of the century. Lately, his work has proven to be a major primary source for the historical study of several important developments on the British engineering and scientific scenes of the 1880s.[2]

There appears to be, however, something very special about the work of Oliver Heaviside. During his lifetime from 1850 to 1925 he had been a contemporary of great scientists like Maxwell, J. Larmor, H.A. Lorentz, Henri Poincaré, and Einstein. Technically, their work has not been any less challenging than Heaviside's—to modern, as well as to their own contemporary read-

1. William Thomson was knighted in 1866, in recognition of his contributions to the trans-Atlantic cable project. He was raised to the peerage in 1892, under the title of Lord Kelvin of Largs. To avoid the cumbersome multiplicity of names and titles, he will be referred to throughout this book as Kelvin.

2. These studies will be referred to in detail in the course of chapter IV. The general list of references to Heaviside, studies of his work and accounts of his life is much too long for a single footnote. The bibliography lists most of the important works on Oliver Heaviside to date.

ers. Yet, none of these individuals seems to enjoy Heaviside's reputation as the author of almost undecipherable papers. This reputation has been upheld through the years with striking unanimity. Probably the first important scientist who called attention to Heaviside's work was Oliver Lodge, who also set the tone for future commentators:

> ...I must take the opportunity to remark what a singular insight into the intricacies of the subject [of the skin effect], and what a masterly grasp of a most difficult theory, are to be found among the writings of Mr. Oliver Heaviside. I cannot pretend to have done more that skim these writings, however, for I find Lord Rayleigh's papers, in so far as they cover the same ground, so much pleasanter and easier to read; though, indeed, they are none of the easiest.[3]

In 1889, Heaviside received the following friendly warning from one of his greatest admirers, Hertz:

> The fact is that the more things became clearer to myself and the more I then returned to your book, the more I saw that essentially you had already made much earlier the progress I thought to make, and the more the respect for your work was growing in me. But I could not take it immediately from your book, and others told me they could hardly understand your writing at all, so I felt obliged to give you warning that you are a little obscure for ordinary men.[4]

In 1891 Heaviside communicated to the Royal Society a seminal paper on the dynamical structure of Maxwell's theory.[5] The paper presented an unconventional approach to the subject in a highly condensed form, and Rayleigh commented to Heaviside about it in a sterner tone than Hertz's:

> Both our referees, while reporting favourably upon what they could understand, complain of the exceeding stiffness of your paper. One says it is the most difficult he ever tried to read. Do you think you could do anything; viz., illustrations or further explanations to meet this? As it is I should fear that no one would take advantage of your work.[6]

More than forty years later, W.E. Sumpner felt even more strongly about this particular paper:

---

3. Oliver Lodge, *Lightning Conductors and Lightning Guards*, (1892), p. 46.

4. Hertz to Heaviside, 5 May 1889, quoted in J.G. O'Hara and W. Pricha, *Hertz and the Maxwellians*, (1987), p. 65.

5. Oliver Heaviside, "On the Forces, Stresses, and Fluxes of Energy in the Electromagnetic Field," *Electrical Papers*, Vol. II, pp. 521-574.

6. Rayleigh to Heaviside, 31 October 1891, quoted in R. Appleyard, *Pioneers of Electrical Communication*, (1930), pp. 227-228.

> Heaviside summed up his work on Maxwell's theory in a single paper printed by the Royal Society in 1892. This was the most important and the most ambitious paper Heaviside ever wrote. It is fairly safe to say that no one yet born has been able to understand it completely.[7]

The publication of Heaviside's *Electrical Papers* was greeted with a review by his dearest friend, G.F. FitzGerald, in 1893. While FitzGerald described the work as extremely important, he also wrote:

> Oliver Heaviside has the faults of extreme condensation of thought and a peculiar facility for coining technical terms and expression that are extremely puzzling to a reader of his Papers. So much so that there seems very little hope that he will ever attain the clarity of some writers, and write a work that will be easy to read. In his most deliberate attempts at being elementary, he jumps deep double fences and introduces short-cut expressions that are woeful stumbling blocks to the slow-paced mind of the average man.[8]

John Perry, who followed Heaviside's example of treating vectors in terms of their own language,[9] and who introduced Heaviside to the problem of the age of the earth, wrote:

> Now I rank Heaviside with [Kelvin and FitzGerald] but I never pretend to be able to read Heaviside. I wish I could, and so do a lot of people like me.... Somebody will have to write down Heaviside to our level.[10]

Even engineers who had extensive mathematical knowledge found Heaviside's work unwieldy, as the following comment from A.E. Kennelly demonstrates:

---

7. W.E. Sumpner, "The Work of Oliver Heaviside," *Journal of the Institution of Electrical Engineers*, **71** (1932): 841. For a detailed analysis of Heaviside's reasoning in this notorious paper, see J.Z. Buchwald, "Oliver Heaviside, Maxwell's Apostle and Maxwellian Apostate," *Centaurus*, **28** (1985): 288-330.

8. G.F. FitzGerald, "Heaviside's Electrical Papers," reprinted in *The Scientific Writings of the Late George Francis FitzGerald*, edited by Joseph Larmor (1902), p. 293. The above excerpt is quoted in P.J. Nahin, *Oliver Heaviside: Sage in Solitude*, (1988), p. 168. This and other quotations from FitzGerald's review may be found in Michael J. Crowe, *A History of Vector Analysis: The Evolution of the Idea of a Vectorial System*, (1967), pp. 175-176.

9. John Perry, *Applied Mechanics: A Treatise for the use of Students who have Time to Work Experimental, Numerical, and Graphical Exercises Illustrating the Subject*, 2nd ed., (1898), pp. 29-32.

10. Quoted in R. Appleyard, *Pioneers of Electrical Communication*, (1930), p. 244.

> The differential equations of potential and current on a real uniform line, in the steady a.c. state, were given by Heaviside, with their algebraic solutions, in 1887; although the solutions offered were very lengthy and unserviceable.[11]

An obituary notice on Heaviside in *Nature* actually highlighted an important reason for the difficulty as follows:

> [Heaviside] published many papers which gradually became more and more technical and more and more difficult to understand, as it became necessary, in order to avoid repetition, to assume that the reader knew some of the writer's previous work.[12]

But it was Kelvin—to whom Heaviside owed both his initial admission to the Society of Telegraph Engineers in 1873 and the official recognition of the value of his scientific achievements—who gave the most poignant expression to the problem of reading Heaviside's work.[13] In January of 1888, Kelvin sent to J.J. Thomson a paper of Heaviside's to be examined for possible communication to *The Philosophical Magazine*.[14] It seems that Kelvin marked certain parts of the paper, and added the following comment:

> I think O.H. right about X but his + is unintelligible to anyone who had not read all O.H.'s papers, and it and everything else would be unintelligible to anyone who had. No brains would be left.[15]

---

11. A.E. Kennelly, *Artificial Electric Lines: Their Theory, Mode of Construction and Uses*, (1917), p. 24. On p. 163, Kennelly refers to OH as a "competent mathematician."

12. A. Russel, "Mr. Oliver Heaviside, F.R.S.," *Nature*, **115** (1925): 237-238.

13. In 1922, Heaviside described how he obtained membership in the Society of Telegraph Engineers. His brother, Arthur, who initially suggested to Oliver that he should join the Society, informed him that the application fell through because the Society would not accept "telegraph clerks." "What would Edison say if he were here now?" Heaviside wrote. "I was riled. I had already had one of my inventions tried in a rough experimental way by the P.O. with success [probably a particular implementation of duplex telegraphy, in 1873. Appleyard, *Pioneers of Electrical Communication*, (1930), p. 221].... So I went to Prof. W. Thomson & asked him to propose me. He was a real gentleman & agreed at once. But as he had engagements away from London, he got William Siemens to do it. So I got in, in spite of the P.O. snobs" (Heaviside to Highfield, 14 March 1922, Heaviside Collection, IEE, London). In 1876 Heaviside was elected to the Council of the Society, but was not reelected the following year because he did not attend a single meeting. In 1881, having consistently failed to pay his dues, his name was struck off the members' list as well. (Sir George Lee, "Oliver Heaviside—The Man," *The Heaviside Centenary Volume*, [1950], p. 12.)

14. The date of the letter suggests that the paper in questions was "On Electromagnetic Waves, Especially in Relation to the Vorticity of the Impressed Forces; and the Forced Vibrations of Electromagnetic Systems," *Electrical Papers*, Vol. II, pp. 375-467. As we shall see in chapter IV, Heaviside may have made it particularly abstruse on purpose.

Considering this verdict from the greatest scientific authorities of the time, it is hardly surprising that the image persevered and found expression in modern historical work:

> Because the bulk of [Heaviside's] work is extremely technical and difficult, it is unlikely that a full scientific and personal biography of him will be written. ... He was known as a wildly eccentric person and his work was notoriously difficult to understand.[16]

Heaviside's work is certainly not easy. It requires patient work and considerable readiness to adopt his unconventional style to follow his reasoning. However, it is by no means the hopeless maze of unintelligibility that the above quotes seem to illustrate. The question, therefore, is how Heaviside's work acquired its enigmatic legacy. Heaviside's life story yields the first clue to this question, as the following biographical sketch will show.

2. Oliver Heaviside as a Lone Wolf

Oliver Heaviside was born on May 18, 1850, the fourth (and youngest) son of Thomas and Rachel Elizabeth West Heaviside. He spent his early childhood in 55 King Street, Camden Town, London.[17] At the time of his birth, this area of London was bordering on some of the city's poorer sections. In 1897 Heaviside recalled his first home with considerable disdain and complained about the lowly neighborhood, leaving the impression that these living conditions had scarred him for life.[18] At the same time, it would probably be wrong to classify the Heaviside family among the truly poor, and we shall soon see that they had at least one very useful familial connection to the well-to-do. Thomas Heaviside made a precarious living as a skilled wood engraver, in an era that saw the spread of photographic reproduction techniques. In order to supplement his meager and unsteady income, his wife offered elementary schooling for girls, and later worked as a governess. Between them they man-

---

15. Kelvin to J.J. Thomson, 15 January 1888, quoted in Rayleigh (John William Strutt), *The Life of Sir J.J. Thomson*, (1942), p. 33. Thomson's reply to the letter appears to have been lost, but Rayleigh reports that "...like Lord Kelvin [J.J. Thomson] was in general impatient of obscurity, and disinclined to take the trouble to follow authors such as Oliver Heaviside who used unconventional methods in mathematics—it would be easier to do it over again, he said."

16. William Berkson, *Fields of Force: The Development of a World View from Faraday to Einstein*, (1974), pp. 197-198.

aged to earn enough money to raise four boys, move in 1863 to 117 Camden Street, and in 1876 to a somewhat better location on 3 St. Augustine Road, Camden Town. From 1874 to 1889, following a short period of employment in commercial telegraphy, Oliver Heaviside lived with his parents. Under their care he did nearly all the work that will be examined in this book.

Very little seems to be known about Heaviside's life in general until the 1890s, and in particular about his education until he was in his twenties. Most accounts suggest that he acquired his initial education in his mother's girls school.[19] Later, he attended two other schools in his immediate neighborhood. In a 1905 essay on the teaching of mathematics, Heaviside recalled one of his teachers, whom he described as a dedicated if not terribly inspired instructor:

> I feel quite certain that I am right in this question of the teaching of geometry, having gone though it at school where I made the closest observations on the

17. Some details about Heaviside's life may be found in several obituary notices and other short essays about his life and work. There are, however, three main sources of knowledge concerning his childhood and late life. Rollo Appleyard's portrait of Heaviside in *Pioneers of Electrical Communication*, (1930), pp. 211-260, contains most of what we know about Heaviside's early life. In many ways the most remarkable and most revealing account of Heaviside's later life is contained in G.F.C. Searle's *Oliver Heaviside, The Man*, edited by Ivor Catt, (1987). Searle wrote the monograph in 1950 for the centenary celebration of Heaviside's birth, but only a short abstract of it was published in the *Heaviside Centenary Volume*, (1950), pp. 8-9. The full document was published for the first time only in 1987 under somewhat mysterious circumstances (see Catt's introductory note). The only comprehensive biography of Heaviside may be found in P.J. Nahin's painstakingly researched and highly readable *Oliver Heaviside: Sage in Solitude*, (1988). All accounts of Heaviside's life agree that biographical information is scarce and patchy at best. Thus, E.J. Berg, who met Heaviside in person, wrote: "Little is known of [Heaviside's] history, as he was exceedingly reluctant to speak about himself and had evidently requested his brother [probably Charles Heaviside, whom Berg had met] ... not to make public any facts about his career." (*Heaviside's Operational Calculus*, [1936], p. xiv). P.J. Nahin concurs: "Heaviside had remarkably little to say about the personal aspects of life, except for occasional remarks scattered about in letters and his research notebooks." (*Sage in Solitude*, p. 27 [note 23]; see also p. 20.)

18. P.J. Nahin, *Oliver Heaviside: Sage in Solitude*, (1988), pp. 13-14.

19. "[T]here is a legend that [Heaviside] was at an early stage taught by his mother." Rollo Appleyard, *Pioneers of Electrical Communications*, (1930), p. 215. See also P.J. Nahin, *Oliver Heaviside: Sage in Solitude*, (1988), p. 15. E.T. Whittaker ("Oliver Heaviside", Reprinted in Heaviside's *Electromagnetic Theory*, Vol. I, p. xiv) has his own version of this story. He tells how the young boy rebelled against the idea of being alone in a group of girls, until his father dragged him to the nearby boys' school, which apparently was not a very attractive institution, and offered him a simple choice between attending it, or studying with the girls under his mother's care. P.J. Nahin (p. 15) repeats this story, but neither he nor Whittaker supply its origin.

effect of Euclid upon the rest of them. It was a sad farce, though conducted by a conscientious and hard-working teacher.[20]

However, as these remarks were made in the course of one of Heaviside's many excursions into the delights of sarcastic fun, one does not know quite how to take them. At the age of sixteen Heaviside took the College of Preceptors Examination, finished fifth overall out of over five hundred candidates and won the first prize in the Natural Sciences part of the examination. Geometry, on the other hand, seemed to have presented a particular difficulty, and he managed only 15% of the problems in that section. This marked the end of Heaviside's formal schooling, and all biographical accounts suggest that at this point he possessed no more than an elementary knowledge of algebra and trigonometry.[21] It does appear, however, that already in his early schooling days Heaviside took to science and mathematics (save for Euclidean geometry, which he evidently abhorred and failed).

It appears that during the next eight years Heaviside's career was influenced by his illustrious uncle, Sir Charles Wheatstone.[22] The famous telegraph pioneer was married to Rachel Elizabeth West's sister, and it seems that the two families, living not far from each other in London, enjoyed a close relationship.[23] Three of the Heaviside boys ended up in telegraphy. Practically all we know about the eldest, Herbert, is that he was already working as a telegraph operator in Newcastle-on-Tyne when Oliver Heaviside began his six year career in the telegraph service in 1868.[24] Letters to Oliver Heaviside from his two other brothers, Arthur and Charles, reveal a strained relationship between Herbert Heaviside and the rest of the family. If, as some accounts suggest, Herbert Heaviside left home as a result of a row with his father, the letters from Arthur and Charles show that by 1881 their sympathies, as well as Oliver's, were entirely with their parents and not with their older bother.[25] Arthur West Heaviside also started his career as a telegraphist in Newcastle,

20. *Electromagnetic Theory*, Vol. 3, p. 514.

21. See, e.g., E.J. Berg, *Heaviside's Operational Calculus*, (1936), p. xv; P.J. Nahin, *Oliver Heaviside: Sage in Solitude*, (1988), p. 20.

22. For a biography of Wheatstone, see Brian Bowers, *Sir Charles Wheatstone*, (1975).

23. P.J. Nahin, *Oliver Heaviside: Sage in Solitude*, (1988) p. 19.

24. *Ibid.*, p. 20. There is direct evidence of Heaviside's employment with the Great Northern Telegraph Company from 1870-74, in the form of a letter from the company confirming this term of employment (Heaviside Collection, IEE, London).

25. See Ch. Heaviside to O. Heaviside, 28 June 1881 and 27 June 1882, Box 9:3:1. A.W. Heaviside to O. Heaviside, 15 July 1881, Box 9:6:2, Heaviside Collection, IEE, London.

and eventually rose to the respectable position of superintending engineer of the British Post Office telephone department there.[26] He is credited with the design and installation of a novel and highly efficient telephone network in that area. As we shall see, Oliver Heaviside's involvement with this project had far-reaching consequences for his contributions to the theory of telephone and telegraph communication. The third of the brothers, Charles Heaviside, made a career in another of Wheatstone's areas of interest—the music industry. He started out as an instrument maker, eventually owned and managed a music store in Torquay, and took care of his aging parents and his reclusive youngest brother from 1889 on.[27]

At the age of eighteen, two years after he finished his formal education, Oliver Heaviside began working as a telegraph operator in the Danish-Norwegian-English Telegraph Company, which was absorbed in 1870 into the Great Northern Telegraph Company, based in Newcastle-on-Tyne. A description of Oliver Heaviside as a young telegraph operator by a colleague of his, one W. Brown, clearly suggests Wheatstone's influence on this career choice:

> Oliver Heaviside was the principal operator at Newcastle—appointed no doubt by the influence of his uncle, Sir Charles Wheatstone. He was usually on day duty. He was a very gentlemanly-looking young man, always well dressed, of slim build, fair hair and ruddy complexion.[28]

Like most things about Heaviside's life that are not directly related to his scientific work, Wheatstone's influence on his career remains a matter of speculation. On one hand, Appleyard noted in his article on Heaviside for the *Dictionary of National Biography* that no evidence supports the allegation that Wheatstone actively shaped Heaviside's career,[29] and the speculative remark quoted above is no exception. On the other hand, all four Heaviside brothers ended up in businesses in which Wheatstone had a direct stake, and this could suggest more than mere coincidence.

---

26. q.v. "Heaviside, Arthur West", in *Who was Who, 1916-1926*, (London: Adam & Charles Black, 1967).

27. R. Appleyard, q.v. "Heaviside, Oliver," in *The Dictionary of National Biography, 1922-1930*, p. 413.

28. Oliver Lodge, Obituary notice, *Journal of the Institution of Electrical Engineers*, **63** (1925): 1154.

29. R. Appleyard, q.v. "Heaviside, Oliver," in *The Dictionary of National Biography, 1922-1930*, p. 413.

All we know of Heaviside's activities during the two years between his graduation in 1866 and the beginning of his employment in 1868 is that he used them to further his education on his own. Owing to lack of direct evidence, we can only speculate about what this privately pursued course of studies involved. Heaviside's first published work, and the surviving manuscript records from 1870 to 1871 do not reveal any of the mathematical sophistication that so strongly characterized his work from 1874 on.[30] What the early work and manuscripts do reveal is intimate familiarity with the details of circuit design pertaining to all aspects of telegraphy, and that at least for a while, Heaviside was stationed in Denmark. This seems to support Sir George Lee's suggestion that Heaviside used the two years between his graduation and the beginning of his employment to acquire a knowledge of Danish, Morse Code, perhaps some German, and probably an elementary acquaintance with the electrical circuitry used in telegraphy.[31] It does not appear likely that he used these two years to considerably further his knowledge of mathematics and physics.

Little more is known about Heaviside's personal life during the six years of his employment with the telegraph industry. At the end of this period Heaviside already served as chief operator, a position he was promoted to in 1871 with an increase in salary from £150 to £175 per annum. It is noteworthy that his duties included the location of faults in telegraph cables. In the case of a long submerged cable, the procedure involved a working knowledge of Kirchhoff's circuit laws, and the ability to manipulate them algebraically. William Edward Ayrton, who from the late 1870s emerged as a leading figure in British technical education, followed a similar track in his own early career in the telegraph industry. He too was responsible for fault location,[32] and it

30. Heaviside's notebook 1a:83-118 (Heaviside Collection, IEE, London) contains diary entries from 26 December 1870 to 6 July 1881. These entries appear in the middle of the notebook, following descriptions of experimental work carried out in 1886. On one occasion (pp. 113-118), the diary entries do not follow each other chronologically. It is almost certain, therefore, that this is not the original diary, and that Heaviside must have copied these entries into the notebook from an earlier manuscript that did not survive. This is a recurring feature in almost all of Heaviside's note books. For the most part, they contain copies of previously worked problems, and Heaviside did not always record when the original work was done. This often makes it very difficult to date the original work on the basis of the notebook entries.

31. Sir George Lee, "Oliver Heaviside—The Man," *The Heaviside Centenary Volume*, (1950), p. 11.

seems that this activity was the mark of particularly able telegraphists who mastered the basic mathematical theory of telegraphic circuitry.

Heaviside stayed with commercial telegraphy until 1874. He then left his job with the Great Northern Telegraph Company, and returned to London to live with his parents. Several reasons have been suggested for this, and the paucity of information makes it difficult to assess their relative weight. In 1873 Heaviside wrote a paper on duplex telegraphy, in which he ridiculed the conservative, short-sighted attitudes of certain superiors regarding this rapidly developing technique. Heaviside explicitly referred to R.S. Culley's authoritative *Handbook of Practical Telegraphy*,[33] and quoted the work of another unnamed authority. The unnamed writer may well have been W.H. Preece, to whom Culley gave special thanks in the introduction to his book. Culley was the engineer-in-chief of the nationalized telegraph service under the control of the British Post Office. Preece served under him as chief engineer of the Southern Telegraph Division, and was already well under way to becoming the most influential telegraph and telephone official of the 1880s and 1890s.[34] Correspondence between Preece and Culley shows that both were expressly unhappy with Heaviside's comments—so much so that Culley wrote, "we will try to pot Oliver somehow" (more on that in chapter IV). We have no evidence that this directly led to Heaviside's resignation, but the incident certainly could not have helped further his career. In fact, just prior to his resignation, Heaviside seems to have applied for a salary raise, but his request was denied. In addition to these difficulties, Heaviside suffered from partial deafness since before he began his work. Some accounts point to this as the main reason which compelled him to leave his job.[35] It should be noted that telegraph receivers in Britain, unlike their American counterparts, depended more on visual than auditory cues.[36] Therefore, it does not seem likely that Heaviside's partial deafness could have seriously interfered with his telegraphic work. His

32. Philip Joseph Hartog, q.v. "Ayrton," *The Dictionary of National Biography*, Supplement, January 1901 - December 1911, p. 73.

33. Heaviside quoted from R.S. Culley, *A Handbook of Practical Telegraphy*, 5th edition, (1871), p. 223. The 6th edition, published in 1874, already contains (pp. 387-404) a long contribution on duplex telegraphy by Stearns, who designed the implementation most commonly used in the U.S. and in England.

34. E.C. Baker, *Sir William Henry Preece, F.R.S., Victorian Engineer Extraordinary*, (1976), p. 94.

35. See A. Russel, "Mr. Oliver Heaviside, F.R.S.," *Nature*, **115** (1925): 237-238.

36. R.S. Culley, *A Handbook of Practical Telegraphy*, 5th edition, (1871), pp. 200-201.

deafness, however, might have made personal relations somewhat awkward and uncomfortable for him. Indeed, E.C. Baker quotes Arthur West Heaviside as having sadly acknowledged his younger brother's growing isolationist attitudes that made him unsuitable for coordinated teamwork.[37] Considering that Heaviside did live most of his life in seclusion, this personality trait must have contributed to the termination of his telegraphic career. Baker's book, however, is devoted to the life of William Henry Preece, who later became the British Post Office's Chief Engineer. In the sharp dispute that erupted between Preece and Heaviside in 1887, Baker's sympathies lie squarely with Preece.[38] Considering that Preece and Heaviside clashed as early as 1873, Baker's account of Heaviside's departure from the telegraph service may be somewhat colored by his desire to exonerate Preece. Still, Heaviside's general demeanor does seem to have helped bring his telegraphic career to an end. Commenting on Heaviside's resignation from the Company, a supervisor described him as a very capable operator, but a rather insubordinate one, with a very high opinion of himself. All things considered, this particular supervisor felt that Heaviside's departure would not be a great loss.[39] The general picture that emerges from all of this is one of a bright, capable, but somewhat cocky and socially awkward young operator at the beginning of his professional career. Add to the above that he embarrassed his higher-ups, and that his request for a salary raise had been denied, and a resignation seems just about inevitable.

From 1874 to 1889 Heaviside lived with his parents in London, and continued to educate himself while publishing papers of growing scientific sophistication. Everything we know suggests that he never obtained another job in the commercial sector. There is evidence to suggest that there was at least one job offer in 1881. The Western Union Company acquired a number of Wheatstone telegraph recorders, and was looking for technical experts to maintain them, and perhaps instruct others in their use. Arthur West Heaviside called his younger brother's attention to the job, which offered a salary of £250 per annum. Apparently Preece was also involved, and ready to help Oliver Heaviside get the job, but it came to naught in the end. Perhaps Heaviside found Preece's involvement reason enough to stay away from the job; perhaps he simply decided that staying at home suited his research plans better; and

37. E.C. Baker, *Sir William Henry Preece, F.R.S., Victorian Engineer Extraordinary*, (1976), pp. 208-209.

38. *Ibid*. p. 206.

39. P.J. Nahin, *Oliver Heaviside: Sage in Solitude*, (1988), p. 22

perhaps Arthur Heaviside was right about his younger brother's isolationist tendencies. For one reason or another, Heaviside never became a Western Union employee and remained at his parents' house in London.[40] Despite that, there is some evidence to suggest that Heaviside did not live at his parents' expense, and that he actually contributed to the household's income. Letters from Arthur and Charles show that they, as well as Herbert and Oliver, were regularly extending financial help to their parents in London, although Herbert seems to have made his contributions in an insulting manner.[41] Surviving fragments of some letters from W.E. Ayrton to Heaviside dating from 1878 to 1881 reveal that he offered Heaviside an opportunity to write abstracts of various scientific papers for the *Journal of the Society of Telegraph Engineers*:

> Last year you published a very interesting paper in the phil. mag. in connection with signalling through faulty cables. Could you let us have a short abstract of this for our journal if possible during the next two or three days.
>
> We are endeavouring to organize a regular system of abstracting....[42]

The rest of the letter did not survive, but other fragments clearly show that Heaviside did accept the offer, and received material for abstracting from Ayrton. Another partially preserved note from Ayrton indicates that the abstracting work was sparse, especially since Heaviside did not want to abstract articles written in foreign languages:

> Will you kindly prepare an abstract of the accompanying of about three quarters of a page *small* print of our journal.
>
> I am sorry I have not been able to send you more abstracting, but the majority of what we have published is from the French and German, and I think you mentioned you did not care to translate from ....[43]

In view of this, it does not seem likely that the abstracting work could have amounted to very much financially.

In 1892, however, Heaviside wrote to Oliver Lodge that the editor of *The Electrician* paid him £40 per annum for the articles he contributed to this journal from 1882 to 1887. According to Appleyard, the rent for the house on 3 St. Augustine Road was £45 per annum.[44] It would seem, therefore, that while

---

40. R. Appleyard, *Pioneers of Electrical Communication*, (1930), p. 222; E.C. Baker, *Sir William Henry Preece, Victorian Engineer Extraordinary*, (1976), pp. 208-209.

41. Arthur West Heaviside to Oliver Heaviside, 15 July 1881, Box 9:6:2, Heaviside Collection, IEE, London.

42. Ayrton to Heaviside, 16 February (no year), Box 9:6:2, The Heaviside collection, IEE, London.

Heaviside's earnings were by no means large, he could actually contribute significantly to his parents' expenses. Some letters to Oliver Heaviside from his older brother Arthur indicate that there were also hopes for other sources of income. On June 29, 1881, Arthur wrote what appears to suggest that the two were in the advanced stages of patenting, and perhaps even selling, an invention of theirs:

> I think the agreement right except to which way to read against Reid's or against you—"the sum of eleven shillings per mile" and "for pair of wires" and initial addition.
>
> The £500 clause means that if the Royalties don't amount to that Reid's must pay the £100 stamp duty and if they do you must pay the stamp duty so I would let that pass.
>
> Sign the blooming agreement and take your copy signed by Reid's and the cheque for £100. Signature must be witnessed.
>
> Yours in haste,
> A W Heaviside

We have no evidence regarding the outcome of this effort, but we do know that Oliver Heaviside's was never a story of "rags to riches." At best, the two brothers reaped only a small reward from whatever they were working on.[45]

Another question to consider is why Heaviside did not pursue some form of higher education, either before or after his short period of employment. Probably the simplest and most persuasive reason is financial. In 1874 additional arguments against attending a university may have been his age and his

---

43. Ayrton to Heaviside, 15 March (no year), Box 9:6:2, Heaviside Collection, IEE, London. Ayrton did not specify the year of most of his letters to Heaviside. One letter, however, is dated 1881. Further estimates of the period from which this correspondence dates are aided by the knowledge that Ayrton came back to England from Japan in 1878. Heaviside's paper on signalling through faulty cables that Ayrton referred to in the first quotation was published in 1879 ("On the Theory of Faults in Cables", *Electrical Papers*, Vol. I, pp. 71-95). One other letter specifying instructions for Heaviside on abstracting the third part of John Perry's "On the Contact Theory of Voltaic Action" further indicates that the relevant period is 1878-81: "Will you, in accordance with the proposed arrangement for abstracting, make an abstract of about two pages small print of it, at your earlier convenience, for the April number of the journal. I enclose ... the abstract which I wrote some time ago for the Proceedings of the Royal Society. Your abstract should differ from this as this abstract has already appeared in the Electrician and elsewhere. It might be well if you glanced at paper N° 1 Proc. Roy. Soc. No 86 1878...."

44. R. Appleyard, *Pioneers of Electrical Communication*, (1930), p. 215.

45. According to R. Appleyard, the invention concerned means of "neutralizing disturbances in cables." See *Pioneers of Electrical Communication*, (1930), p. 221.

reclusive tendencies. There may have been other, less obvious motivations as well. By 1874 Heaviside had six years of experience in practical telegraphy, and extended his knowledge of mathematical electric-circuit theory to the point of publishing several papers on the subject. We may therefore consider with Nahin that there was little incentive for Heaviside to pursue a university education. He may have felt that there was little for him to learn there, and some—though by no means all—practical telegraphists at the time did not have a high opinion of university education anyway.[46] However, the opinions of other practical telegraphists should not have bothered Heaviside too much, considering his interests and independent turn of mind. Furthermore, by 1874 Heaviside discovered Maxwell's treatise, and a course of studies in Cambridge may not have been as unattractive to him as Nahin suggests. Heaviside did write later that he exceedingly regretted not to have had the benefit of a Cambridge education; but since he placed the comment in the context of poking fun at Cambridge mathematics, one cannot make too much of the remark.[47] Cambridge, however, was not the only framework within which Heaviside could have furthered his scientific knowledge. E.C. Baker describes a training program designed specifically for promising young telegraphists, which might have suited Heaviside even better than a university degree.[48] Tyndall and Kelvin were responsible for the theoretical sections of the course, while Preece taught aspects of practical telegraphy. Heaviside already met Kelvin in person, and obtained his help in becoming a member of the Society of Telegraph Engineers. Tyndall was also no stranger to Heaviside; in fact, his treatise, *Heat as a Mode of Motion*, considerably influenced Heaviside's scientific thought.[49] Thus, good reasons could be offered for Heaviside to have had more than casual interest in this program. But Preece's involvement may have disposed Heaviside against taking any part in it. In the end, we are left once again with uncertainty. We know that Heaviside pursued his studies on his own; but we cannot assess the various reasons that may have prompted him to do so with a very high degree of confidence.

It appears that until 1888, Oliver Heaviside's sole scientific collaborator was his brother Arthur. Heaviside's experimental notebooks show that from

---

46. P.J. Nahin, *Oliver Heaviside: Sage in Solitude*, (1988), p. 24.

47. *Electromagnetic Theory*, Vol. II, p. 10.

48. E.C. Baker, *Sir William Henry Preece, F.R.S., Victorian Engineer Extraordinary*, (1976), pp. 83-87.

49. John Tyndall, *Heat as a Mode of Motion*, 4th edition, (1870).

1880 to 1887 he worked very closely with his older brother. There can be little doubt that the two corresponded quite regularly, and discussed scientific as well as personal matters. A few letters and fragments of letters from Arthur to Oliver are currently in the possession of the Institution of Electrical Engineers (IEE). They have been preserved accidentally, because Oliver used their blank sides for his notes and calculations. This may also explain why some of the letters are incomplete; having no use for sheets with written text on both sides, he probably discarded them. It is most unfortunate that nearly all of what must have been a voluminous correspondence did not survive. The nature of the remaining letters clearly suggests that the correspondence contained precisely the sort of data out of which a living picture of Oliver Heaviside from 1874 to 1890 could be reconstructed. From the surviving letters we can learn that Arthur kept his younger brother up to date on technical developments, and used him as a sort of technical adviser on theoretical and technical matters:

> I think you told me when last in London that Bridge's algebra was better than Todhunter? Which is the best book on magnetism. Next time I write I will give you all the news as to what we are doing electrically in the post office and some facts about insulation that will make you stare.[50]

And on another occasion:

> My dear O.
> Can you suggest an experiment for comparison of E.F's [electromotive forces] of Leclanche and Dan[iel] by means of condensers.[51]

Arthur also supplied Oliver with electrical equipment for experimenting, as the following note from January 22, 1881, reveals:

> Please receive 3 Gower Bell Telephones, for export to India in a/c of A.W. Heaviside of Newcastle.[52]

It appears that Arthur visited London quite frequently and occasionally the two met and dined together. On other occasions Arthur provided his reclusive brother with first-hand accounts of meetings at the Society of Telegraph Engineers, complete with personal observations like: "... how Ayrton speaks when he opens his mouth but does not give all the truth...."[53]

---

50. A.W. Heaviside to O. Heaviside, 12 October 1881, Heaviside Collection, IEE, London, 9:6:2.

51. A.W. Heaviside to O. Heaviside, 25 (month missing) 1880, Box 9:6:2, Heaviside Collection, IEE, London.

52. *Ibid.*

Further on, we shall see that Heaviside had an exceptional ability to keep the most sophisticated mathematical investigations in close association with intuitive physical notions. However, a letter that Arthur wrote on April 29, 1881, just prior to going abroad for an unspecified reason, suggests that this practical bent of Oliver Heaviside's mind did not extend to the more mundane aspects of life. The letter also demonstrates Arthur's ability to humor his younger brother and the open and amicable relationship that the two seem to have enjoyed:

> ... [you] are an amusing cuss. Your calculations sometimes, like the officer's sword when it gets between his legs, upset you. If I examine my salary in the way you suggest, I should not be, but as I am, that is sufficient answer. I have often given you the key to my rate of living at home and if you make allowance for my life abroad and cost of life insurance you will find a different result. ... I don't wish to send my boys to an expensive school but to a good school.
>
> Good bye old man, take care of yourself,
>
> Yours affectionately,
>
> A W Heaviside.[54]

Scientifically, Heaviside's London years were his most productive, and he seems to have been of that opinion himself later on. On more than one occasion he wrote that he did practically all his original work by 1887, that it was collected in the two-volume *Electrical Papers*, and that the three-volume *Electromagnetic Theory* was "simply developmental."[55] Until 1886 Heaviside worked quietly at home, publishing his papers in *The Electrician*, the

---

53. A.W. Heaviside to O. Heaviside, undated fragment, Box 9:6:2, Heaviside Collection, IEE, London.

54. A.W. Heaviside to O. Heaviside, 25 (month missing) 1880, Box 9:6:2, Heaviside Collection, IEE, London.

55. Heaviside wrote this in his official statement of acceptance of the Faraday Medal awarded to him by the Institution of Electrical Engineers: "I wish to say that practically all my original work was done before 1887, and is contained in my Electrical Papers. The Electromagnetic Theory work is simply developmental and had to be forced upon the wooden headed Royal Society mathematicians first" (Heaviside to Highfield, Box 9:1:8, Heaviside Collection, IEE, London). At about the same time, Heaviside repeated his judgement on the relative merit of his two great publications to E.J. Berg: "Pray do not forget that my *Electrical Papers* are actually my Great Work ... out of which my E.M.T. grew." (quoted in P.J. Nahin, *Oliver Heaviside: Sage in Solitude*, [1988], p. 294). Heaviside wrote that in reply to Berg's report that the *Electromagnetic Theory* was selling well in the U.S. Heaviside's strong emphasis on the *Electrical Papers* may have been partially motivated by an attempt to drum up demand for them, but he did not distort the truth. The main themes that guided all his scientific work are already fully defined in the *Electrical Papers*.

Philosophical Magazine and the *Journal of the Society of Telegraph Engineers* while taking no part in the long, often heated debates over the nature of electricity that continuously raged over the pages of the *Electrician* since 1878. It was only between 1886 and 1889 that Heaviside finally emerged from this self-imposed obscurity by making significant contributions to several hotly contested electrical issues. As the events of 1886 to 1889 will be closely examined in chapter IV, a rough outline will suffice for the purposes of this short biographical sketch. Heaviside's first significant involvement in scientific controversy was occasioned by certain experiments of David E. Hughes, which were published in February of 1886. Heaviside's comments on the experiments brought a quick, though not very favorable, reaction from Hughes. This was the first time that Heaviside's work was remarked upon in public, on the occasion of a well publicized scientific event. It was, however, owing to the events of 1887 that he eventually catapulted onto the public scientific stage. During 1886 and 1887, Arthur and Oliver Heaviside collaborated on the design of an innovative telephone network for Newcastle. As usual, Arthur did the practical work, and Oliver supplied theoretical guidance. The result of this collaboration was a joint paper, in which Arthur described the circuit design, while Oliver wrote three appendices that provided its theoretical underpinnings. This paper brought about a bitter dispute with Preece, who was by then the senior electrician of the Post Office, and who objected to the paper's conclusions regarding the theory and design of long-distance telephone lines. The dispute with Preece lasted through 1888, and culminated in 1889 when Kelvin publicly supported Heaviside's position.

The year 1889 was something of a watershed in Heaviside's career. In January Kelvin became President of the Institution of Electrical Engineers. In his presidential address he called attention to Heaviside's work on telegraph and telephone communication, describing it as the best available analysis of the subject. This was by far the most influential official recognition of the importance of Heaviside's work. Six months after this dramatic moment in Heaviside's life, he left London with his parents to live in Paignton with their third son, Charles. Thus, at the moment of his greatest triumph, with official recognition and widening contacts with the British scientific community, Heaviside removed himself from the scene. The move suggests an irresistible, though not necessarily intentional, gesture: it looks as though Heaviside emerged from his obscurity only to certify his reluctance to join the lively activities of Britain's scientific capital. As Oliver Lodge later wrote:

> ... as soon as he began to be recognized he fled to Devonshire, and thence emerged no more—never, so far as I know, attending the Royal Society or the Electrical Engineers, or coming to hear the congratulations which might—late in life—have been showered on him, and living to the end the life of a recluse.[56]

During the first ten years in the Torquay area, Heaviside appears to have progressed with his scientific work and career. In 1891 he was elected Fellow of the Royal Society. In 1892, his publications from 1872 to 1891 were collected and printed in the two-volume *Electrical Papers*. Also in 1892, he began to communicate a new series on the operational calculus to the Royal Society. The first two parts of this series were published in 1893 in the Society's *Proceedings*.[57] In 1894 the first volume of *Electromagnetic Theory* was published, and was followed by the second volume in 1899. In 1896 he was awarded a Civil List pension of £120 per annum. He finally had a steady, if modest, income. However, all was not as well as this suggests. In 1893 the Royal Society refused to publish the third part of Heaviside's operational calculus series, following a negative review by an anonymous referee.[58] The very appointment of a referee was highly uncharacteristic, which suggests that feelings ran high against Heaviside's investigations. In fact, J.L.B. Cooper later suggested that the wonder is not the rejection of the third part, but the publication of the first two. Others saw the whole affair as an expression of the inability of the mathematical establishment to come to terms with a particularly innovative work.[59] This time, however, there was no official vindication of Heaviside's position, and the Royal Society did not publish the contents of the rejected paper in any form. Heaviside ended up publishing the substance of the paper in the second volume of *Electromagnetic Theory*, amidst many witty, sarcastic, and bitter remarks on the woes of rigorous Cambridge mathematics. By the second decade of the twentieth century, some Cambridge mathematicians did find interest in Heaviside's operators. Most prominent among them

---

56. Oliver Lodge, "Oliver Heaviside, F.R.S.," *Electrical World*, (21 February 1925): 403-405, esp. 403.

57. O. Heaviside, "On Operators in Physical Mathematics", *Proceedings of the Royal Society of London*, **LII** (Feb. 1893): 504-529; **LIV** (June 1893): 105- 143.

58. Bruce J. Hunt discovered that the referee was William Burnside, professor of mathematics at the Royal Naval College in Greenwich. See B.J. Hunt, "Rigorous Discipline: Oliver Heaviside Versus the Mathematicians," in Peter Dear (ed.), *The Literary Structure of Scientific Argument*, (1991), pp. 72-95. Heaviside published a condensed version of the rejected paper in *Electromagnetic Theory*, Vol. II, pp. 457-482.

59. See appendix 4.2 for further remarks.

was T.J.I'A. Bromwich, who corresponded with Heaviside on the subject, and attempted to establish Heaviside's operational procedures on the basis of complex integrals. Heaviside undoubtedly enjoyed Bromwich's attention, but his comments on the margins of Bromwich's paper manifestly show that he did not like Bromwich's mathematical approach.[60]

Heaviside's disappointment over the reception of his mathematical work by the Royal Society was dwarfed by the fate of his recommendations for the elimination of distortion from long-distance telephony. These recommendations stood at the heart of Kelvin's praise in 1889, and Heaviside could realistically hope to see them implemented with some profit to himself. In the early 1890s, S.P. Thompson tried to implement Heaviside's scheme without success; but there were other displays of interest in the subject which could have given Heaviside cause for hope. Thus, in 1891, John Stone Stone of the Bell Telephone Company wrote to Heaviside for some advice regarding the elimination of distortion from telephone lines.[61] However, after these promising beginnings, the first U.S. patent for distortionless telephone lines designed according to Heaviside's general theory was awarded in 1901 to Professor Michael I. Pupin of Columbia University. The invention turned out to be a financial gold mine, and brought Pupin hundreds of thousands of dollars in royalties by the mid-1910s.[62] Heaviside never recovered from the shock of being deprived of the rights to an invention he considered his own. For the rest of his life, he was haunted by Pupin and the thought of the financial rewards he undeservedly reaped.

In 1894 Heaviside's mother died, to be followed by his father in 1896. Until that time, Heaviside lived with his parents in an apartment above Charles Heaviside's music store in Paignton. In 1897, after what he described as a long

60. T.J.I'A. Bromwich, "Normal Coordinates in Dynamical Systems," *Proceedings of the London Mathematical Society*, **15** (1916): 401-448, (Heaviside Collection, IEE, London).

61. There are four letters from Stone to Heaviside in the Heaviside Collection at the IEE in London. The letters date from 1891 to 1894. In the first of these letters, Stone sought Heaviside's help in formulating a specific version of the latter's general transmission-line theory to fit the metallic circuits used by the Bell Telephone Company. Stone also wrote of the intense interest aroused in him by Heaviside's 1887 work on long-distance telephony. He stated that it was this interest that prompted him to join the Bell Telephone Company following his graduation from the Johns Hopkins University.

62. For details of this remarkable turn of events, see James E. Brittain, "The Introduction of the Loading Coil: George A. Campbell and Michael I. Pupin," *Technology and Culture*, **11** (1972): 36-57 (esp. pp. 36-38).

and frustrating search, he purchased a house in the neighboring village of Newton Abbot. He dwelled in "Bradley View," Newton Abbot, until 1908. Heaviside's letters to G.F. FitzGerald and to G.F.C. Searle reveal that at first he was very excited by this development. For the first time in his life he felt truly independent, and he savored the image of himself as the owner of property. His enthusiasm was short lived, however. The house itself was old and run-down. Heaviside's initial investment of £30 in interior decorating was apparently insufficient to lift the house from its decaying state. Gardening without proper tools also proved to be more frustration than relaxation. As his letters often reveal, Heaviside quickly discovered that the coveted life of a country squire is liberating and rewarding only when supported by the proper bank account. His neighbors did not help the situation either. They seem to have been relatively uneducated, hard-working people who scraped a living from small farms and related services. To them, the newcomer must have appeared more than a trifle odd: living on a government pension, incapable of properly taking care of his garden and spending most of his time among his books, with an occasional break for a bicycling spree around the hills of Devon.[63] To Heaviside, who must have grown accustomed to a sheltered routine in his parents' home, these people appeared crass, vulgar and noisy. Before long, local children discovered the joys of pestering their awkward, introverted neighbor. Towards the end of his stay at Bradley View, Heaviside often complained in his letters and notebooks of local hooligans flinging stones at his house and breaking his windows. Furthermore, while Heaviside's neighbors could not possibly have understood much about his work, they still managed to follow the more sensational side of their neighbor's career. They were certainly clever enough to know that chanting "poop, poop, poopin" outside Heaviside's window would rile the strange man who kept claiming that professor Pupin of Columbia University stole the glory that was rightfully his own. The hardships of living alone, with insufficient heat and what may well have been an inadequate diet, finally affected Heaviside's health. In 1907 he fell seriously ill. Searle, who saw Heaviside almost every Christmas since the late 1890s, recalled in 1950 that the illness put a permanent stop to Heaviside's en-

63. Heaviside was an avid cyclist—a passion he shared with FitzGerald. Searle told of several bicycle trips he had taken with Heaviside, and described Heaviside's impish habit of cycling to the top of a hill, then putting both feet up on the handlebars, and allowing himself to accelerate uncontrollably downhill, leaving Searle far behind. See G.F.C. Searle, *Oliver Heaviside, The Man*, edited by Ivor Catt, (1987), p. 10.

thusiastic and vigorous bicycling career. In 1908 he left Newton Abbot and his failed attempt at independence. Miss Mary Way, sister of Charles Heaviside's wife, was living by herself in "Homefield," Lower Warberry Road, in nearby Torquay.[64] The house is situated a stone's throw away from Torwood Street, where Charles Heaviside's home and music store were located. It seemed convenient for Miss Way to take Oliver Heaviside as a paying lodger, and it was undoubtedly a relief for Heaviside to have his meals cooked for him, and his living quarters taken care of once again. In a letter to Lodge, he expressed mixed feelings regarding the new turn in his life-style:

> I should have left even the first year I was there, finding the people to be so savage (not *all* of them) except for the impossibility of finding a house to suit my purpose and other things. I remember I rejoiced to find that house at all, it seemed the only one in a large area, after a long hunt. ... At last, however, I have no house; I am only a lodger; I have lost my independence...[65]

Miss Way's presence seems to have helped Heaviside regain some of his strength. However, what he grew to call his "Torquay marriage" of convenience, quickly became a casualty of incompatible personalities. Most of the blame for the two's inability to forge a peaceful coexistence has been put on Heaviside, and probably rightly so. From 1874 until his mother's death, he lived in a protected environment all his own, in which he could cultivate both his highly individualistic scientific style, and his equally individualistic day-to-day habits. His letters to Searle from Miss Way's home reveal that he found it very hard to part with these cherished habits, which accompanied his years of greatest scientific productivity. It is worthwhile to quote one of these letters at length, if only to show what Miss Way had to endure (Heaviside referred to her as "the baby," even though she was three years his senior):

> The great lentil Question cropped up today (not the first time). Shall I when I want Pork and Pease pudding hot, this being the proper time for that wholesome and vulgar fare, to make the system able to resist the cold, shall I be diddled into eating lentils instead on the plea that they are much nicer, and so nutritious? Never! I had enough of it before. I was introduced to lentils at Paignton, by a niece who took charge when my mother became too feeble; it was substituted for

64. "Homefield" is as of this writing "The El-Marino Hotel." The interested visitor will find a short glass-encased article on Heaviside by the main door, and a light blue plaque that was embedded by the Institution of Electrical Engineers in the stone wall around the house.

65. Heaviside to Lodge, 10 December 1908, quoted from G.F.C. Searle, *Oliver Heaviside, the Man*, (1987), pp. 23-24.

my mother's pease pudding, most unwarrantably and without any consideration for our feelings or wishes, but merely because this new cook was a vegetarian, and vegetarians seem to have a spite against pease and always preach lentils. Why? I hardly know, probably because they have been proved by chemical analysis to contain a little more nitrogen than pease. This learned girl (a woman now) had nuts for breakfast, because they were recommended by some idiotic vegetarian journal, and contained more nitrogen than anything else. Save me from nitrogen! It's a mad world. I preferred the pease, but never had 'em again. It was always that sloppy lentil soup.

But why does the Baby do it? She isn't a vegetarian, eating nuts for breakfast, with vegetarian butter (a fraud), and vegetarian cheese (another fraud) at other meals, all very nutritious and nitrogenous, no doubt. Because she once was strongly under the vegetarian niece's influence, and so imbibed a lot of her nonsense, and it hasn't gone off yet. I have, however, got rid of cabbage stalk soup, and some ot'er wretched frauds. She eats real good cheese no Cheddar and St. Ivel, and all sorts of non-vegetarian food. (Perhaps too much).

Having asked for the seasonable dish (a change from chopped up steak and potatoes—half black), I got the pork because there was some in the house, rather stale, and not the right sort, but wouldn't have the lentils or their nutritiousness. (Several times same thing before). She wasn't amenable to my very civil remonstrances that I knew lentils very well; I wanted pease. 'Oh! You know everything!' She is going to buy some, if procurable. To keep her from forgetting I drop down a note periodically. No.1 (new series) informed her that the Jews ate lentils in the Bible, but there is no mention of pease pudding. No.2 (in preparation) there was a plague of lentils in Egypt in the time of Moses. Also there was one case of living for forty days on lentils and wild honey, or else honey and wild lentils, they were so nutritious, No.3 (ready tomorrow) mentioned in Magna Carta. Felony to rob the villein [sic] of his pease pudding. No.4 (soon) Act of George IV. Fine 40/- or one month on grocers and *others* for substituting lentils for pease pudding. And so on. I shall get my pease pudding in time, as I did my Brawn. That's another story.[66]

Despite Heaviside's dissatisfaction with Miss Way's menu, he seems to have been dependent on her presence, and protested in his unique style when she left the house to do her errands without informing him. Thus, upon returning from such departures, she would sometimes find him in the garden with a lit candle, looking for her dead body.[67] In view of this, it seems ironic and astonishing to find that Heaviside projected his own rigid attachment to a previ-

66. G.F.C. Searle, *Oliver Heaviside, The Man*, (1987), pp. 26-27.
67. *Ibid.*, p. 25.

ous way of life onto his caretaker; he actually claimed she had been a spoiled child who got used to having everything her own way.[68] Still, having said all that, Heaviside's uncompromising insistence on his old habits is no proof that Miss Way was a model of flexibility. Perhaps Heaviside was not the only old dog who refused to learn new tricks in Homefield. At any rate, the "Torquay marriage" did not last long. In 1913, following what may have been a nervous breakdown, Miss Way moved to the house of her sister and brother-in-law, and Oliver Heaviside remained on his own once again. Whatever ailment she suffered did not cause permanent damage. Indeed, she outlived Heaviside, and Searle who saw her several times at her new home, remembered her as a perfectly sane and cheerful, though somewhat overweight, old lady.

In 1912 the third and last volume of *Electromagnetic Theory* appeared. There are several indications that at one time or another, Heaviside was considering a fourth volume. However, Searle's recollections supported by a letter Heaviside wrote in 1912 suggest that by that time Heaviside's analytical powers were waning. "... I fear my mental activity is gone for good," Heaviside wrote. "I cannot concentrate upon anything now save for a short time. Of course the constant thinking about money matters is contributing to this."[69] Science, during Heaviside's final years, was part of a past that he vividly remembered and often thought of. His present consisted of worries regarding unpaid bills and the hardships of old age without a family he critically depended on. At the same time, the honors continued to accumulate. In 1904 he was offered the Hughes Medal by the IEE, but turned it down. In 1905 he accepted an honorary Ph.D. from the University of Göttingen. In 1908 he became an honorary member of the IEE. In 1910 Heaviside was nominated for the Nobel Prize in physics. In 1918 he accepted, after considerable persuasion,[70] an honorary membership in the American Institute of Electrical Engineers, and in

68. *Ibid.*, p. 31.

69. Heaviside to Searle, 23 January 1912, quoted in G.F.C. Searle, *Oliver Heaviside, The Man*, (1987), p. 31.

70. B.A. Behrend apparently took it upon himself to make up for the wrong done to Heaviside by conferring upon him the official recognition of the AIEE, of which Pupin himself was a member. He tried to explain to Heaviside that he should accept the honorary membership even though the financial rewards went to another, "... or else the powers of darkness will exult and acclaim their man the discoverer of what [we] know is your work and, while the 'rude boys of Newton Abbot' bellow forth his name, evil will once more hold its sway. I shall be glad to hear from you and bespeak your kindly consideration at a [time] when America stretches [out] her hands across the sea." (Undated fragment, Box 9:6:2, Heaviside Collection, IEE, London).

1922 he became the first recipient of the Faraday Medal.[71] But Heaviside did not pursue these honors. What he wanted in his final years was "justice," as he put it in his own words. He wanted to be recognized as the rightful inventor of distortionless telephone lines, and was sorely disappointed when his honorary membership in the AIEE was not accompanied by an official denunciation of Pupin's claim to the invention.[72] He caustically summed up his unfulfilled wish for justice in 1918:

> An engineer writing in the [*Electrician*] once said that my description & directions only wanted to be put into the conventional language of patents to make a patent. But there is a rather funny notion prevalent that an invention is not an invention unless it is patented, and then it is the patentee's invention.[73]

Claims to the effect that during the last thirteen years of his life Heaviside practically lost all his intellectual ability and was teetering on the verge of insanity cannot be substantiated. Regarding his intellectual ability, both his letters and those of the people who met him during this period clearly show that he maintained active interest in current scientific developments. At the same time, it seems equally clear that he was no longer up to more original research, or to composing another text of four hundred pages. As to his sanity, Heaviside was always highly eccentric, probably since early childhood. He was also brilliant and prolific during most of his adult life. When his days of scientific productivity came to an end, what remained was an eccentric old man. There is a powerful contrast between the eccentric, but scientifically brilliant and productive Heaviside of 1872 to 1891 and the still eccentric, but now unsheltered, aging and scientifically unproductive Heaviside of 1912 to 1925. Perhaps this contrast proved too hard for some observers to contend with in mea-

---

71. Sir George Lee, "Oliver Heaviside—The Man", in *The Heaviside Centenary Volume*, (1950), pp. 13-15.

72. "Now I had correspondence with a Boston U.S.A. gentleman connected with the A.I.E.E. He was very friendly, too much so. The A.I.E.E. wanted to 'recognize' me. I didn't want their recognition. I had excellent recognition from the best men in Britain, and from some first rate men in U.S.A. I wanted some justice. And they didn't even mention that I had invented the telegraphic and telephonic loading system which Pupin had made his fortune by." (Heaviside to Highfield, 4 January 1922, Box 9:1:8, Heaviside Collection, IEE, London).

73. Heaviside to Behrend, 24 June 1918, Heaviside Collection, IEE London.

sured terms.[74] However, Searle's final verdict, emphasizing a friendship that lasted thirty-three years, seems by far the most plausible:

> On 21 December [1924] he wrote me a lively and humorous letter describing his recent fall from a ladder, and showing that he was still the same old, odd and impish Oliver... Many legends grew up about Oliver. I believe I do right to record the conviction that he was never a 'mental' invalid. Of course he was a first rate oddity—he was Oliver. I had been his friend for 33 years.[75]

Heaviside spent his final days in the "Mount Stewart Nursing Home" easily visible on the opposite hill from his residence in Torquay, following the above-mentioned fall from a ladder that he suffered in an attempt to help a local workman mend leaks in the roof of his house.[76] According to Searle, who saw him there just days before he died, Heaviside was in full command of his senses, and "won the affections of the nurses and others in the Home." He died on February 3, 1925, at the age of 75.

As the preceding pages demonstrate, the available biographical information about Heaviside sketches the fascinating life of a remarkable personality. Many details may be added to this short sketch of Heaviside's life. Appleyard reproduced two wood engravings that Heaviside produced as a young boy. Appleyard also described Heaviside's one attempt at creative writing—an essay entitled "Muscular Characters," in which Heaviside recorded his impressions of the youths who frequented a London gymnasium known as "the Pimple." Searle provided a most striking personal portrait of Heaviside as an eccentric old man. Most of the information obtainable from obituaries and other short personal sketches has been collected and convincingly put together by Nahin. However, we have also seen that many basic questions regarding Heaviside's life and character remain open to speculation owing to lack of evidence. In particular, what stood behind Heaviside's odd personality remains shrouded in deep fog, and a satisfactory portrait of Heaviside the man may keep eluding historical investigators more stubbornly than his scientific contributions. One may speculate that his childhood under the Dickensian condi-

74. See B.A. Behrend, "The Work of Oliver Heaviside," in E.J. Berg, *Heaviside's Operational Calculus*, (1936), p. 208; and B.R. Gossick, "Where is Heaviside's Manuscript for Volume 4 of his 'Electromagnetic Theory'?" *Annals of Science*, **34** (1977): 601-606.

75. G.F.C. Searle, *Oliver Heaviside, The Man*, (1987), p. 72.

76. "Mount Stewart" is no longer a nursing home, but a regular apartment building. It is a light green three-story building, overlooking the bay and affording a view of Lower, Middle, and Higher Warberry roads. Heaviside's home can be glimpsed among the trees on Lower Warberry road.

tions of life in poverty-ridden London deformed Heaviside's character. However, both Arthur and Charles Heaviside endured the same conditions, but showed no sign of sharing any of their younger brother's eccentric traits. Perhaps it was the specter of his father's outbreaks of bad temper, aggravated by life on the brink of bankruptcy that made Oliver Heaviside introverted and reclusive.[77] In *Electromagnetic Theory*, he wrote:

> The following story is true. There was a little boy, and his father said, "Do try to be like other people. Don't frown." And he tried and tried, but could not. So his father beat him with a strap; and then he was eaten up by lions.
>
> Reader, if young, take warning by this sad life and death. For though it may be an honour to be different from other people, if Carlyle's dictum about the 30 millions [of British citizens being mostly fools] be still true, yet other people do not like it. So, if you are different, you had better hide it, and pretend to be solemn and wooden-headed. Until you make your fortune. For most wooden-headed people worship money; and really, I do not see what else they can do. In particular, if you are going to write a book, remember the wooden-headed. So be rigorous; that will cover a multitude of sins.[78]

Perhaps this was Oliver Heaviside's way of remembering his father as a short-tempered, insensitive man; but then, should the seemingly autobiographical first paragraph be taken at face value considering that its whole purpose is to poke sarcasm at mathematical rigorists? There may have been tensions between the financially stressed Thomas Heaviside and his sons. Herbert's departure from the household may attest to that. However, we have seen some evidence to suggest that this may have had more to do with Herbert than with his father. Whatever tensions did exist in the Heaviside household, they were neither sufficiently disruptive to undermine the care and affection manifested in Arthur's letters, nor to prevent Charles Heaviside from personally tending to the needs of his parents and youngest brother from 1889 to 1896. Finally it should always be remembered that Thomas and Rachel Heaviside gave their youngest son a home during his lifelong career of partial unemployment. If Oliver Heaviside's parents were somehow responsible for his odd personality, they were also sensitive and caring enough to support him with more than average devotion from 1874 to 1889. If nothing else, they gave him the basis for the intellectual freedom that so strongly characterizes all of his work.

---

77. P.J. Nahin, *Oliver Heaviside: Sage in Solitude*, (1988), pp. 4, 15.

78. *Electromagnetic Theory*, Vol. 3, p. 1.

Any attempt to trace the development of Heaviside's ideas must always take into account that he did not obtain his more advanced knowledge in the usual way of registering in an institution for higher education. Likewise, in following his later career it must be considered that while he was officially a member of the Society of Telegraph Engineers from 1873 to 1881, and of the Royal Society from 1891 on, he never participated in either institution's social or political life. All the known records suggest that he never once appeared in person in either place. In general, Heaviside seems to have bypassed the powerful constraints and incentives that an institution so often imposes on the professional careers of its fellows. His dependence on his family for the needs of everyday life was more than counterbalanced by the fierce intellectual independence that it made possible.[79]

The freedom to work independently, without a teacher to guide him, and unhampered by the pressures of a professional career, provides the first clue to the enigmatic legacy of Heaviside's scientific work. The isolation in which he developed his ideas helped to enhance and preserve their unique character. One cannot help recalling in this connection the words of another lone wolf, France's great entomologist, Jean Henri Fabre:

> I was denied the privilege of learning with a master. I should be wrong to complain. Solitary study has its advantages: it does not cast you in the official mould; it leaves you all your originality. Wild fruit when it ripens, has a different taste from hothouse produce: it leaves on a discriminating palate a bitter-sweet flavour whose virtue is all the greater for the contrast.[80]

---

79. Searle's account of how Heaviside obtained his fellowship in the Royal Society crisply illustrates his fierce independence and bears further testimony to his prickly character (*Oliver Heaviside, the Man*, [1987], pp. 76-78). Fellowship in the Royal Society usually followed upon recommendation by existing Fellows. There was a waiting list, and candidacy of the proposed new Fellow would then be examined relative to that of others on the list. As a result, candidates were not always successful on their first attempt, and had to wait several years before being admitted to the Society. Heaviside, however, would have none of this. When Lodge asked for his consent to be put on the candidates' list, Heaviside replied that he would agree only if given explicit guarantee of election "on the first go." He would rather not be proposed at all than have the dubious honor of being considered for rejection. Lodge tried to assure him that he had good chances of being elected immediately, but Heaviside was not satisfied. As he saw it, if he was good enough to be recommended, he was also good enough to be elected. He wanted a guarantee of election, and solemnly promised Lodge that if he were proposed and then rejected, he would make a public row over it, and drag the image-conscious Society into a controversy it could surely do without. Heaviside was elected as he wanted, "on the first go," while Silvanus Thompson and Joseph Larmor, both candidates at the same time, patiently awaited another chance.

Heaviside's prose does not possess the poetic flare of Fabre's. Yet, he must have shared the sentiment. In 1895 he wrote the following as an introduction to a discussion of the relationship between Fourier-series analysis and his own operational techniques:

> The virtues of the academical system of rigorous mathematical training are well known. But it has its faults. A very serious one (perhaps a necessary one) is that it checks instead of stimulating any originality the student may possess, by keeping him in regular grooves. Outsiders may find that there are other grooves just as good, and perhaps a great deal better, for their purposes. Now, as my grooves are not the conventional ones, there is no need for any formal treatment.[81]

The bitter-sweet flavor of Heaviside's unconventional grooves presents itself in virtually every section of his books. It is the one immediately perceived characteristic that permeates his entire work. Without a doubt, the unconventional nature of his papers must have contributed greatly to his mystifying image.

3. The Character of Heaviside's Work

By themselves, the biographical details above cannot explain what predisposed Heaviside's readers to perceive his work as so difficult to master; nor can the unique character of his work, reflected in novel mathematical methods and revisions of nomenclature, fully account for this perception. After all, Heaviside was by no means the first physicist to invent new mathematical methods and conventions for the purpose of rendering his work more coherent. Moreover, the charge of abstruseness is not the only characteristic that various appraisals of his work have in common. The unanimous view of Heaviside's publications as hopelessly unintelligible is nicely counterbalanced by sharply divergent opinions regarding his professional classification. To G. Doetch, who formulated a version of the operational calculus on the basis of the Laplace transform, Heaviside was merely "an English Engineer," whose methods were, from the mathematical point of view, "very inadequate."[82] B. Van Der Pol and H. Bremmer, who extended and generalized the Laplace transform approach to the operational calculus, disagreed; despite having their

---

80. Jean Henri Fabre, Tr. by A.T. de Mattos, *Life of the Fly*, (1919), p. 277.

81. *Electromagnetic Theory*, Vol. 2, p. 32.

82. G. Doetch, *Theorie und Anwendung der Laplace-Transformation*, (1937), pp. 337, 421.

own misgivings about the unrigorous character of Heaviside's mathematics, they considered him more than an engineer.[83] Writing about the rejection of "On Operators in Physical Mathematics" sixty years after the event, J.L.B. Cooper considered that:

> [Heaviside] was primarily a physicist—though he had an intense interest in some parts of pure mathematics—and was not very widely read in mathematics.[84]

By contrast, just one year prior to this observation by Cooper, Ernst Weber wrote in the preface to a reprint of Heaviside's *Electromagnetic Theory* that:

> Oliver Heaviside, one of the most unusual characters among great modern scientists, could probably be classified best as an outstanding applied mathematician. He was truly a pioneer in this new branch of science.[85]

When J.A. Fleming discussed Heaviside's contributions to long-distance telephony, he did not even bother with the adjective "applied":

> Nevertheless, there is a further remedy for distortion, which was strongly urged by an eminent mathematician, Mr. Oliver Heaviside.[86]

Rollo Appleyard, however, wrote:

> [Heaviside] was proud to have been at one time a 'practitioner' himself, and his correspondence shows that when practical men approached him in a way of which he approved he was ever ready to assist them, as well as men of science, with their problems.[87]

Finally, the following often quoted classification was produced in 1932 by one of Heaviside's contemporaries, W.E. Sumpner:

> [Heaviside] regarded all theoretical work as subsidiary. He was a mathematician at one moment and a physicist at another, but first and last, and all the time, he was a telegraphist.[88]

---

83. B. Van Der Pol and H. Bremmer, *Operational Calculus, Based on the Two-Sided Laplace Integral*, (1950), p. 2.

84. J.B.L. Cooper, "Heaviside and the Operational Calculus," *The Mathematical Gazette*, **36** (1952): 13.

85. Ernst Weber, "Oliver Heaviside" preface to O. Heaviside, *Electromagnetic Theory*, (1951), p. xv.

86. J.A. Fleming, *Fifty Years of Electricity: The Memories of an Electrical Engineer*, (1921), pp. 104-105.

87. R. Appleyard, *Pioneers of Electrical Communication*, (1930), p. 230.

88. W.E. Sumpner, "The Work of Oliver Heaviside," (23rd Kelvin Lecture), *Journal of the Institution of Electrical Engineers*, **71** (1932): 837.

In conclusion, we have seen that the circumstances under which Heaviside worked enabled him to develop the unconventional scientific style so often remarked upon by his contemporaries. We have also seen that one general way in which the uniqueness of his work manifests itself is through the difficulties others had in classifying it among the established specialized domains of engineering, mathematics, and physics. This, however, is as far as we can get by examining biographical details and reactions to his books and papers. Only by closely consulting his work will we be able to understand and extract the particular characteristics that gave rise to the reception documented in the previous pages.

Heaviside made positive contributions to three fields of knowledge. These fields form the basis of three distinct professional doctrines, namely, mathematics, physics and electrical engineering. The coherence of Heaviside's papers stems primarily from the inextricable interdependence of mathematical, physical and engineering themes. The manner in which he presented his mathematical innovations cannot be readily understood if the physical problems that motivated them are not clearly perceived. His exposition of Maxwell's theory appears like a rather disordered, fragmented series of papers unless one perceives the engineering theme that directs the presentation. When investigated on their own, many of Heaviside's papers give the impression that he could not make up his mind as to whether he was using a circuit problem as a means of presenting Maxwell's electromagnetic field theory; or using it as a means of infusing a mathematical problem with physical meaning; or applying Maxwell's theory in a mathematically novel way as a means of resolving a basic engineering problem. However, once certain themes are explicitly exposed and firmly kept in mind, Heaviside's work will be seen to possess a degree of thematic coherence that far exceeds the special flavor of an individualistic style. Unfortunately, the difficulty of distilling this coherence out of hundreds of pages of electrical papers proved to be a stumbling block for prospective readers ever since the initial publication of his work. Without clearly perceiving the unifying themes, one would often be perplexed by what must have seemed like a most awkward path Heaviside followed to the resolution of a particular question. The problem was further exacerbated by certain events that affected the publication of his work so as to effectively disguise its unifying themes. It should be noted that these problems are mostly formal in nature. Removing them helps bring out the essential themes that guided Heaviside's work; but it does not remove the difficulties that Heaviside's work presented to his readers over the years. Once elucidated, the guiding ideas in Heaviside's work indicate that the problem of classifying Heaviside as a scientist is rooted in his own difficulty of finding a proper scientific niche for himself. As we shall see in the conclusion of this book, Heaviside himself had something to say about his classification as a scientist. However, only when seen in the light of a close examination of his work from 1872 to 1891 does it become apparent that his humorous remarks actually provide the deepest insight into the reasons behind the enigmatic legacy of his work.

4. Outline of this Book

The remaining three chapters of this book are organized primarily along chronological lines. The second chapter deals with Heaviside's work from 1872 to 1882. During these years he published various investigations pertaining to linear circuit theory. They hint on certain occasions that his basic electromagnetic outlook had been undergoing fundamental changes as early as 1876. But it is only in hindsight, keeping in mind his work from 1882 to 1885, that these hints can be identified as reflections of a newly acquired Maxwellian view.

In 1882 the first sharp discontinuity appeared in Heaviside's work. He abandoned the analysis of telegraph circuits in favor of a different theme. By 1884 he produced four long papers that were broken into many short installments for publication in *The Electrician.* The main part of the third chapter is dedicated to a detailed analysis of these four papers, which may be regarded collectively as Heaviside's introduction to field thinking for the highly motivated non-mathematical electrician. The chapter begins with a discussion of the apparent break in continuity that this different topic entails, and suggests that in many ways the discontinuity is more apparent than real.

A thorough understanding of the concepts, methods and problems Heaviside introduced in his papers from 1882 to 1885 is practically indispensable for a reading of his work from 1885 to 1891, which forms the subject of the fourth chapter. During the latter period Heaviside published some of his most important, and sometimes most difficult papers. However, if the lessons of the 1882 to 1885 publications are well-understood, one should have little difficulty discerning at least the general gist of these later works, as well as the unifying themes that permeate them. The basic approach in this chapter is to examine Heaviside's scientific ideas in comparison with some of the prevalent scientific ideas of his time. A central aim is to show that with certain themes in mind, Heaviside's work appears to possess a high degree of internal coherence. A secondary goal is to show how certain events between 1886 and 1889 influenced the publication of his work so as to impede the perception of these unifying themes and render an inherently unconventional work even more prone to be regarded as baffling and incomprehensible. Finally, it will be shown that by considering Heaviside's work on the background of the scientific trends of its day and the particular circumstances of its publication, we can understand some of the main reasons for its enigmatic legacy.

Steven Weinberg cautioned against confusing physics with history, or history with physics.[89] While this book contains a fair amount of physics, it is first and foremost a historical account of the evolution of Heaviside's ideas. Maintaining a continuous historical narrative comes at the expense of consolidating the discussion of specific technical topics. Some repetition, for which I can only beg the reader's indulgence, is therefore unavoidable. Two topics in particular will be discussed several times. The basic aspects of Heaviside's particular brand of electrodynamics will be developed in chapter III. They will be encountered again and further elaborated in chapter IV, in connection with their effect on the reception of his contributions to physics and electrical engineering. The second topic that will be discussed several times is transmission-line analysis. This subject occupied Heaviside as a student of telegraphy during most of the 1870s. His transmission-line work from this period will be described in chapter II. Chapter III will show how transmission-line analysis both influenced and was affected by Heaviside's reformulation of Maxwell's field theory during the early 1880s. Transmission-line analysis will be discussed yet again in chapter IV, with the difficulties Heaviside encountered during the late 1880s in making known his novel theory of distortionless telephony. Bound up with Heaviside's electrodynamics and transmission-line work are his contributions to mathematics. These too will therefore be discussed in conjunction with the evolution of Heaviside's contributions to field and circuit theory.

Construction of a narrative conforming to the above outline depends to a large extent on tracing the development of Heaviside's original conventions

---

89. Steven Weinberg, *Gravitation and Cosmology: Principles and Applications of the General Theory of Relativity*, (New York: John Wiley & Sons, 1972), p. 1.

and ideas. It should be noted that some of the most valuable sources for such an endeavor are sadly deficient. One of the best ways to examine the evolution of a scientist's ideas is through his correspondence with trusted scientific friends during his scientifically formative years. In Heaviside's case, such correspondence is almost entirely non-existent. The vast majority of the letters in the Heaviside collection at the IEE date back to 1888. Everything seems to indicate that he did not begin a routine correspondence with scientific colleagues and friends like Lodge, FitzGerald, Hertz, Larmor and Searle any earlier. The Lodge collection at University College, London, contains a single letter from Heaviside to Lodge dating from January 1885. Continuous correspondence between the two (comprising well over 100 letters) began in June 1888, and it was only in 1889 that Heaviside began to open his letters with "My Dear Lodge," as opposed to "Dear Professor Lodge," or "Dear Sir." Thus, it appears that Heaviside began to correspond with those who became his closest scientific friends after his main scientific ideas had already been formed and set. Indeed, Heaviside's own observation that the *Electrical Papers* contain practically all his original work implies that by mid-1888 most of his original work was already in print. For the most part, therefore, Heaviside's correspondence reveals the established ideas of a mature scientist, not his struggles to develop them during the formative stages of his career.

We have already seen that Heaviside's closest scientific correspondent and collaborator from 1874 to 1887 was his older brother Arthur. The surviving letters and notebooks show that the two designed many experiments together. Oliver provided the theoretical input, while Arthur either supplied him with equipment, or with information on full-scale tests on working telephone lines and networks. We have also seen, however, that very little remains of this correspondence, which could have provided valuable information both about Heaviside's life and about the development of his scientific thought. Some ideas about the conceptual origins of Heaviside's scientific thought may be gathered from the preserved notebooks, which contain short summaries of a few of the books Heaviside read. We know that Heaviside used several well known texts of the period. In one of his manuscripts we find references to Peacock's treatise on the calculus, and to Todhunter's text on the same subject. In "On Operators in Physical Mathematics" Heaviside showed familiarity with Boole's work on divergent series. For various topics in analytical mechanics Heaviside referred to the classic treatise by Thomson and Tait. For keeping in touch with recent developments, he appears to have closely followed *The*

Electrician. Later on, probably in the 1890s, he added *Nature* to his possession, and several copies of this journal, annotated by his hand still survive in the IEE Collection.

By Heaviside's own account, the three texts that had the greatest influence on his scientific thinking were Maxwell's *Treatise on Electricity and Magnetism*, Tyndall's *Heat as a Mode of Motion*, and Fourier's *Theory of Heat.* Maxwell's *Treatise* undoubtedly exerted the most decisive influence on Heaviside's scientific career. However, if he took reading notes and worked through parts of this often difficult treatise on paper, all such material did not survive. Therefore, save for the remarks in his published work, we have no direct basis for reconstructing the manner in which Heaviside formulated his initial impressions, interpretations, and queries regarding Maxwell's work. From Tyndall, Heaviside extracted a notion of "dynamics" that, in a somewhat changed form, permeates his entire electromagnetic work. From Fourier, he abstracted a model of what he considered the proper use of mathematics in physics and the correct relationship between the two. Once again, however, we do not know whether he encountered these two works before he read Maxwell so that they conditioned his understanding of the *Treatise*, or whether he read them following an earlier exposure to Maxwell's work. We do not know whether Heaviside originally derived his views of "physical mathematics" from Fourier's *Theory of Heat*, or whether he found support in it after he had already developed his own notions along a different path.

Despite these difficulties, the conceptual development of Heaviside's scientific ideas can be traced in considerable detail from a careful examination of his published work. Indeed, to a large extent his published work makes up for the gaps in the manuscript records. Heaviside actually regarded publication as a scientist's high moral duty. In an introductory note in *Electromagnetic Theory*, he discussed Cavendish's scientific secretiveness in the harshest terms:

> I can see only one good excuse for abstaining from publication when no obstacle presents itself. You may grow your plant yourself, nurse it carefully in a hothouse, and send it into the world full-grown. But it cannot often occur that it is worth the trouble taken. As for the secretiveness of a Cavendish, that is utterly inexcusable; it is a sin. ... [T]o make valuable discoveries, and to hoard them up as Cavendish did, without any valid reason, seems one of the most criminal acts such a man could be guilty of.[90]

90. *Electromagnetic Theory*, Vol. 1, p. 3.

This emphasis on publication was probably associated with Heaviside's idea of scientific progress and how it comes about:

> Original research teams with error, because it is on the borderland. It gets corrected by one investigator or another, and the result of its elimination is scientific progress.[91]

Comparison of Heaviside's surviving manuscripts and correspondence with his published work shows that he published just about any notion or idea he considered potentially useful. While the manuscripts and correspondence often support and further elucidate Heaviside's published work, they reveal very little about his ideas that cannot be gathered with equal or greater coherence from his published work. In particular, Heaviside's published work from 1872 to 1891 marks quite clearly the general lines along which his scientific ideas developed.

91. Notebook 10:160, Heaviside Collection, IEE, London.

Chapter II Outlining the Way

One must learn by doing the thing; for though you think you know it, you have no certainty until you try.

Sophocles

In 1872 Oliver Heaviside embarked on what turned into a fascinating career of scientific publication. It was to last until the early years of the twentieth century. The beginning, however, seems to have been quite humble. The papers he published between 1872 and 1881 bear little, if any, indication of the radical revision of telegraph and telephone practices that he provided in 1887. They hint at neither his formulation of vector algebra, nor at the powerful, innovative and controversial version of the operational calculus that he developed and used; and only in hindsight do they reveal elements of his insightful interpretation of Maxwell's electromagnetic theory. Perhaps for these reasons his early papers have been largely ignored in previous examinations of his work. Yet, they provide some key insights into his view of the relationship between mathematics and physics, his style of analysis, and the slow emergence of a research program that guided him in most of his subsequent work. The purpose of this chapter is to outline these fundamental aspects of Heaviside's work as they appear in his papers from 1872 to 1881.

1. Early Lessons: Electrical and Mathematical

Heaviside's earliest papers were devoted to the design and analysis of electrical circuits that form the basic building blocks of telegraphic receivers and transmitters. These papers share two fundamental aspects: they are guided by linear circuit theory based on Kirchhoff's laws and they display a very conscious policy of applying mathematical reasoning to the settlement of various problems. Most of this section will be devoted to demonstrating the manner in which Heaviside employed mathematics in these early investigations. However, one particular aspect of linear circuit theory must be clearly kept in mind throughout the discussion, for it will figure prominently later on, especially in chapter IV. The word "linear" above reflects the assumption that for all practical purposes the current in the wire is homogeneously distributed over the wire's cross section. Thus, a thick wire is equivalent to a thin wire characterized by electrical properties per unit length identical to those of the thick wire, and carrying the same integral current as the thick wire. Under these assumptions the wire may be considered as a geometrical line, having no thickness at all and supporting an integral current. All other circuit elements are either discrete or linearly distributed along the wire. This is the significance of the word "linear" as Heaviside used it in this context.[1] A remarkably wide range of practical applications is served by this simple view, together with Ohm's and Kirchhoff's laws, and two relations that define capacitance and inductance. In particular, it will be seen later on that Heaviside required no further knowledge in order to derive his famous condition for distortionless telephone communications.

1.1 Electrical and Mathematical Manipulation

In June of 1873 Heaviside published the first of two papers dedicated to duplex telegraphy. The distinguishing mark of this technology is the ability to transmit one message while simultaneously receiving another one, both messages being sent via the same line. The attraction of duplex systems is obvious; they can handle twice the amount of information without requiring the considerable expense of laying another telegraph cable between the communicating stations. It appears that in the early 1870s the demand for telegraph services in England had grown to such an extent that practical duplex telegraphy became a hotly pursued goal. As an enthusiastic young telegraphist, Oliver Heaviside set forth to make his own contribution to this new technology.[2]

In order to effect duplex telegraphy one must isolate receiver from transmitter within each of the stations connected by the line. With such isolation effectively implemented, any message sent by a particular station, say S_1, will not register on its own receiver. As it turns out, there exist many ways of achieving the desired isolation. In his paper on the subject, Heaviside introduced two duplex methods of his own design. However, for the present purposes, it is not his particular contribution to telegraphic technology that is of interest, but rather the following general characteristic of such design work.

1. "As regards the interpretation of ... results, showing departure from the linear theory, by which I mean the theory that ignores differences in the current-density in wires, I have before made the following remarks..." (*Electrical Papers*, Vol. II, p. 170).

2. *Electrical Papers*, Vol. I, p. 18.

There exists, it seems, a certain kind of similarity between circuit design and algebraic manipulation. In algebraic work we begin with a more or less complicated expression, and then manipulate it while keeping it equal to the original. In the end, we obtain an equivalent expression, which differs from the original only in form. The advantage of such manipulation is that within the framework of a particular problem one form of the expression may have far greater significance than the others. Keeping this in mind, consider Heaviside's description of the favored duplex system of the time.[3] The galvanometer in a balanced Wheatstone Bridge (wherein $a/b = c/d$) will not register a reading (see figure 2.1*A*).[4],

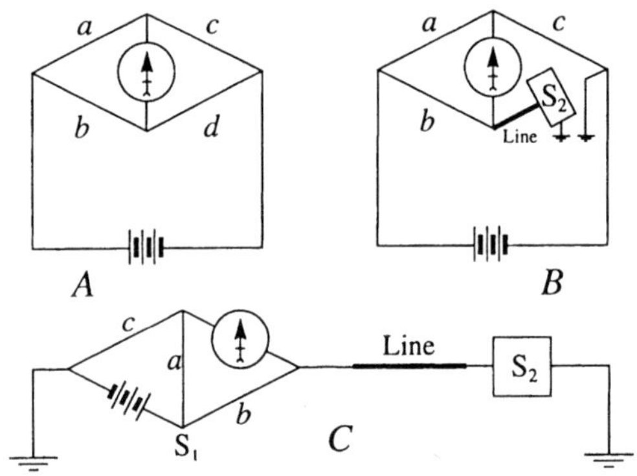

Figure 2.1: Duplex telegraphy based on the Wheatstone Bridge

Obviously, this is precisely the effect wanted for duplex telegraphy; all one has to do is put the telegraph receiver in place of the galvanometer. To convert this into an explicit telegraph circuit, one of the resistances, say *d*, should represent the total resistance of the line and the connected apparatus of the second station, say S_2. This is shown in figure 2.1*B*; but the circuit still does not look like a telegraph circuit, with two stations connected to the ends of a long telegraph wire. However, the same circuit can be drawn differently, as in figure 2.1*C*, explicitly displaying the basic circuitry of the duplex system. To understand why the diagram in 2.1*C* represents an effective duplex system, simply revert back to the equivalent 2.1*B*. Since $a/b = c/d$, making contact with the

3. *Electrical Papers*, Vol. 1, p. 21.

4. For further details on the Wheatstone Bridge, see the next section.

battery at S_1 will send a current through S_2, without registering in the galvanometer (receiver) of S_1. If a similar arrangement is made in S_2, then its own signals will register only in S_1. Thus both stations may receive and transmit at the same time.

By itself, this particular aspect of circuit design says little beyond the suggestion that good circuit designers are not all that different from clever punsters who excel at manipulating words. It would be wrong to conclude from this that clever punsters make good circuit designers or vice versa. Similarly, a skilled mathematician may not make a good electrician, and a good electrician may still be a very mediocre algebraist. Indeed, even when an individual possesses more than a fair share of skills in both circuit design and algebra, the two abilities do not necessarily reflect some deeper talent, of which they are merely two different expressions.

The two skills can, however, be intimately related when an individual versed in both also excels at expressing electrical ideas mathematically, and at interpreting mathematical expressions electrically. A generalized form of this ability manifests itself throughout Heaviside's work. He appreciated mathematics not merely as a calculating tool, but as a way of reasoning about concepts that transcend the symbols and manipulation rules of mathematics. He believed that when devoid of meaning beyond the formal rules of manipulation, mathematics is of little use. At the same time, it appears that he had little use for those subjects that he could not bring into the realm of mathematical discussion. It will be seen in chapter III that he actually expressed these ideas quite plainly. However, they are clearly discernible already in his earliest work as the following examples will show.

1.2 Three Examples of Electro-Mathematical Reasoning

The differential galvanometer is a measuring instrument for the determination of unknown resistances. Like the Wheatstone Bridge, it can also be used as a telegraph receiver. The simplest version of a differential galvanometer is composed of a magnetic needle placed in the middle of two equal coils, such that its plane of rotation is perpendicular to the plane around which the coils are wound. Current is made to flow in the coils in opposite directions. When the two currents are equal, there is no resultant torque on the needle. Otherwise, it is proportional to the difference between the two currents (after the earth's magnetic field has been corrected for), hence the instrument's name—differential galvanometer.

Heaviside showed two arrangements for the differential galvanometer used as a resistance meter. Figure 2.2*A* is the common one, figure 2.2*B* is Heaviside's new arrangement. At this point one sees the competent circuit

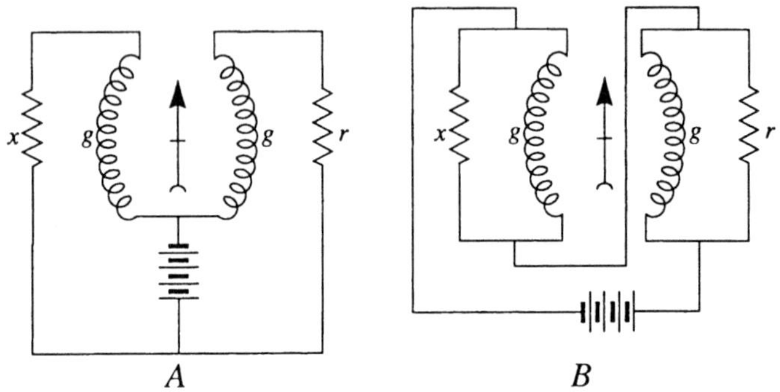

Figure 2.2: Complementary arrangements of the differential galvanometer for effective measurements of large and small resistances (g is the resistance of each of the coils, r is a known resistance and x is the resistance to be determined).

designer once more. The arrangements are not the same, but they do the same job. In both cases the magnetic needle will not move when the reference resistance r is equal to the unknown resistance x. The question is then, why bother with the new arrangement. The answer is that while both arrangements register zero when $r = x$, they differ in their sensitivity to deviations from equality. In order to go further, we must carefully define the meaning of sensitivity in this situation. It can be done as follows: let D_1 be the differential current in arrangement 2.2*A*, and let D_2 be the differential current in 2.2*B*. If the ratio D_1/D_2 is greater than unity, than 2.2*A* responds more strongly to the difference between r and x. Therefore it will detect deviations from equality more sensitively and register the equality of r and x more accurately. If D_1/D_2 is less than unity, the reverse holds. Let E be the battery's E.M.F., and g the resistance of the galvanometer's coils. Then the current differences D_1 and D_2 are:[5]

$$D_1 = \frac{E(r-x)}{b(x+r+2g) + (x+g)(r+g)},$$

$$D_2 = \frac{Eg(r-x)}{b(x+g)(r+g) + gx(r+g) + gr(x+g)},$$

$$\therefore \lim_{x \to r} \frac{D_2}{D_1} = \frac{(2b+r+g)g}{b(r+g)+2gr}. \qquad (2\text{-}1)$$

Inspection of equation (2-1) will quickly disclose that D_2/D_1 is less than unity when g is smaller than r, and that it is greater than unity when g is greater than r. Thus, the sensitivity of Heaviside's new design to deviations from equality of r and x is greater than the sensitivity of the traditional arrangement when one measures resistances smaller than the coil resistance. There is a remarkable resemblance between these two complementary methods of estimating resistances with the differential galvanometer, and the approximation of the value of a mathematical expression by series expansion. Take as a simple example the expression $1/(1+S)$. It may be expanded as $1 - S + S^2 - S^3 + \ldots$. This series will provide a good approximation of the expression for values of S that are smaller than unity. When S is greater than unity, this series will not be a good practical estimator. However, for such cases we may write the original expression as $1/S(1 + 1/S)^{-1}$. This will yield a different series, of the form $1/S - (1/S)^2 + (1/S)^3 - (1/S)^4 + \ldots$. For values of S that are greater than unity, this expansion will provide the better practical approximation of $1/(1+S)$. Thus, just as in the case of the two differential galvanometer arrangements, a

5. There is an error in Heaviside's expression for D_2; he gives it as the negative of the above. In the ratio D_2/D_1, the negative sign disappears without explanation. This may be due to a printer's error; but there exist other simple mathematical errors like this in Heaviside's work. Compare for example the expression for the galvanometer current in Wheatstone's bridge from his 1873 paper on the subject, to the expression of the same from the 1879 revision of the problem. My own calculations show the 1879 expression to be correct. The fascinating thing is that the analysis of the electrical question Heaviside was considering is quite unaffected by the error. It will be seen in chapter III that such incidents occur periodically in Heaviside's work. It seems he stopped checking his mathematics once the physical investigation was satisfactorily settled, and rarely gave erroneous answers to the physical questions he was studying. The combination of apparent mathematical sloppiness with physical precision further highlight the manner in which Heaviside constantly guided his mathematical investigations by the physical ideas they represented.

judicious choice must be made for the estimator that best fits the requirements of the problem. In his investigations of operational solutions to differential equations during the 1890's, Heaviside often made use of precisely such complementary series (see also the concluding remarks in appendix 4.2).[6]

A seasoned electrician may have sufficient practical experience with resistances connected in series and in parallel to perceive the properties of the two arrangements without going through the mathematics.

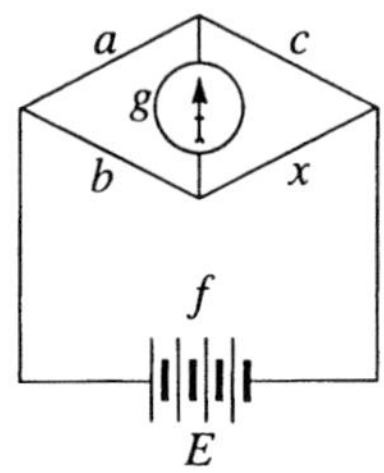

Figure 2.3: The Wheatstone Bridge

The experienced tinkerer may therefore frown at all the mathematical rigmarole necessary to arrive at the above conclusion. However, no amount of practical experience will ever guide one to the most sensitive arrangement of the Wheatstone Bridge. The balance condition for the bridge is very easily discerned without calculating the galvanometer current explicitly. For there to be no current in the galvanometer branch, Kirchhoff's laws require that the current through branch c must be the same as the current through branch a, and the current through branch x must be the same as the current through branch b (see figure 2.3). In addition, there cannot be any voltage drop across the galvanometer, or else current will flow there in accordance with Ohm's law. Hence, the voltage drop across c must be the same as the voltage drop across x, and the same must hold for a and b. Let the current through c and a be I_1, and the current through x and b be I_2. Then, writing the second condition mathematically, using Ohm's law, we have:

$$\left.\begin{array}{l} cI_1 = xI_2 \\ aI_1 = bI_2 \end{array}\right] \Rightarrow \frac{a}{b} = \frac{c}{x}.$$

6. For an example of typical use by Heaviside of these two complementary forms, see E.T. Whittaker, "Oliver Heaviside" in Heaviside's *Electromagnetic Theory*, Vol. I, pp. xviii-xxix.

This is simple enough if we merely wish to determine the unknown resistance, say *x*, when the other three are known. Consider, however, that for every value of *c* there exists an infinite number of associated pairs *a* and *b* that will satisfy the balance condition. Since *c* itself is variable, we have a doubly infinite number of balance combinations, as Heaviside says, but only one will be the most sensitive. It should be plainly evident that trial and error is not the practical way to find that most sensitive arrangement. To be truly practical, the electrician must turn to mathematical theory:

> Some difference of opinion prevails amongst electricians as to what constitutes the most sensitive arrangement of Wheatstone's Bridge for comparing electrical resistances. Now, were Wheatstone's Bridge little used, this would be of no importance; but as it has, on the other hand, most extensive employment, it is certainly desirable that the matter should be thoroughly threshed out. When it is considered that Wheatstone's Bridge is by no means a complicated electrical arrangement, and that the laws regulating the currents in the different branches, and the proportions in which they are divided when division takes place, are extremely simple, and their accuracy as well established as that of the law of gravitation, the wonder is that there should be any doubt respecting a question which can be brought under mathematical reasoning without any hypothetical assumptions whatever.[7]

Heaviside's first *Philosophical Magazine* paper was devoted to this problem.

Before proceeding to the paper's significance regarding Heaviside's approach to electrical problems, it should be noted that he was rather proud of this paper, as his private comments show. These comments also reveal that he had met Kelvin in person as early as 1873, while he was still an employed telegraph operator:

> 'On the Best Arrangement of Wheatstone's Bridge', [Phil. Mag. Feb. 1873 ...] My first Philosophical Magazine paper. A very short time after it appeared, saw Sir W. Thomson at Newcastle, who mentioned it, so I gave him a copy, which no doubt he didn't read. They say he never reads papers. Cuff told me Sir W. said he had tried to work it out, but found the algebra too heavy. S.E. Phillips also congratulated me upon it, as he had tried at it. So paper was a good beginning. Sent Maxwell a copy, & he noted it in his 2nd Edn.[8]

7. *Electrical Papers*, Vol. I, p. 8.

8. Second entry, Notebook 3A, Heaviside Collection, IEE, London. Maxwell's reference to Heaviside's paper may be found in J.C. Maxwell, *A Treatise on Electricity and Magnetism*, 3rd Edition (New York: Dover Publications, Inc., 1954), Vol. I, Art. 350, p. 482.

The comment on the algebra being too heavy cannot be taken seriously. The problem is technically taxing, for it requires a considerable amount of manipulation simply to solve for the galvanometer current from a set of six simultaneous equations. However, it is simply unthinkable that Kelvin could not handle such a problem. It seems more likely that he found it too time consuming to go through every detail, while his remark to Heaviside seems more like a way of humoring an enthusiastic young beginner. It is far more important to note that Kelvin was aware of the paper in the first place, especially considering that Heaviside's work has a reputation of never having been read. In one sense, Kelvin's uncritical reading of the paper is regrettable, for Heaviside seems to have erred in his algebra. In 1879, when he again considered the problem, he put forth the correct expression for the galvanometer current. Curiously enough, he did not refer to the difference between the 1879 expression and the earlier one from 1873.

The paper sets out to do the work in a very businesslike fashion, stating the problem and proceeding directly to the solution. Later on, when Heaviside returned to the problem in a somewhat more contemplative mood, he pinpointed the heart of the matter explicitly:

> It becomes, in the first place, necessary to give a precise meaning to the word sensitiveness. If as nearly perfect a balance as possible be obtained, and then any one of a, b, c, x be altered by a given small fraction of itself—as, for instance, x changed to $x(1+\delta)$, where δ is a small fraction—then a current will appear in G of the same strength whichever one of the four be altered, though of opposite direction for b and c as compared with a and x. Obviously one arrangement will be more sensitive than another if in the first the change of x to $x(1+\delta)$, or corresponding changes in a, b, or c, causes a greater current through the galvanometer than in the second. And the importance of an error is to be reckoned by the ratio it bears to the quantity measured; thus $\delta x/x = \delta$. Whence the balance of greatest sensitiveness is that one in which a given small change from x to $x(1+\delta)$ causes the greatest current through the galvanometer. For the greater this current the nearer can approximation to accuracy be made by adjustment, and if it is inappreciable, as in a coarse balance, no further accuracy can be reached.[9]

If this is well understood, it becomes clear that the variation in the galvanometer current should be examined for values of *x* near balance, with the goal of maximizing this variation. With the problem thus defined, application of the calculus to find maxima of the variation is obviously suggested.

9. *Electrical Papers*, 1, p. 9.

In both of the above problems the calculating power of mathematics recommends its use. But mathematics actually provides a different way of reasoning about a situation, and this is somewhat obscured by its use as a calculating tool. There exists, however, one particularly striking example in Heaviside's early work which clearly demonstrates his use of mathematics as a general reasoning tool. While he was still working in Newcastle, Heaviside and some of his colleagues were perplexed by a curious phenomenon: they could signal from England to Denmark more quickly than the other way around; this, despite the identity of the sending and receiving instruments on both sides of the line.

In a telegraph system such as the above, signalling speed is limited by the rate at which the current rises and subsides at the receiving end. If signals are sent too quickly, the current associated with one pulse may still be rising while the next signal begins to register its own current. This will result in such a blending of signals that the pattern of discrete pulses that makes up a telegraph message will no longer be discernible. The dominant cause for the slow rise of current in the receiving end is the electrostatic capacity of the submerged cable, compared to which the capacity of the land lines is negligible. Even if the line were perfectly insulated, the current at the receiving end would not attain its maximum value until the submerged cable has been charged to its full capacity.

The telegraph line in the particular problem above consisted of land lines on both the British and Danish sides, connected by a submerged cable. That the two land lines were unequal in length and hence had different resistances, cannot by itself account for the difference in signalling speeds. If only a source of electromotive force were connected at one end and a simple ground at the other, the rise of current at the grounded end will be the same regardless of which end the E.M.F. is applied to. Thus, the asymmetry must have something to do with the interaction between the end apparatus—which has a finite resistance—and wire system. For the purpose of analyzing this particular problem, it suffices to consider the wire system as composed of five discrete parts (see figure 2.4): resistance a representing the land line on the British side, two resistances $c/2$, each representing half the resistance of the submerged cable, a capacitor S representing the electrostatic capacity of the cable, and another resistance b representing the land line on the Danish side.

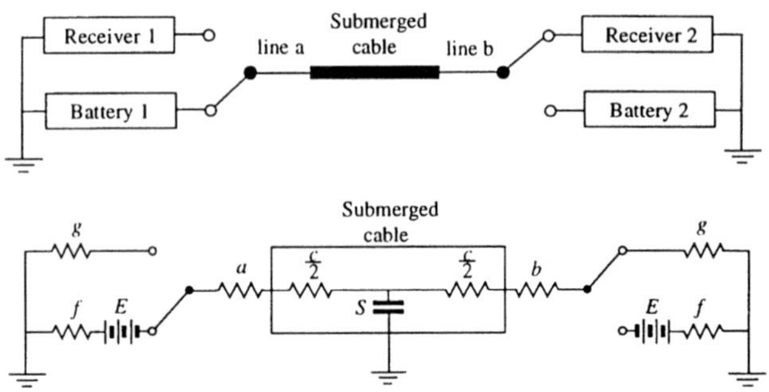

Figure 2.4: Elements of the English-Danish telegraph system that account for its signalling asymmetry

In addition, one has a receiving instrument with resistance g on each side, and a battery of E.M.F. E and internal resistance f on each side. Now connect the battery to a, and study the rise of current in the receiving instrument connected to b. It can be readily shown by application of Kirchhoff's laws together with the definition of capacitance, that the current at b is:

$$I_b = \frac{E}{R}(1 - e^{-t/T}).$$

Similarly, when transmitting in the opposite direction the current in the receiver at a will be:

$$I_a = \frac{E}{R}(1 - e^{-t/T'}),$$

$$\text{where } R = c + a + b + f + g,\ T = \frac{S}{R}\left(\frac{c}{2} + a + f\right)\left(\frac{c}{2} + b + g\right),$$

$$\text{and } T' = \frac{S}{R}\left(\frac{c}{2} + a + g\right)\left(\frac{c}{2} + b + f\right).$$

The first thing to observe is that after the battery has been connected for a very long time, I_a and I_b reach the same maximum value of E/R, for the exponent rapidly approaches zero as t becomes larger than T or T'. Next, note that I_a

comes to within (1 - 1/*e*) of this maximum value after T' units of time, while I_b arrives at the same value after T units of time. Hence, to investigate the rate of current rise at the receiving instruments, one must obviously compare the expressions for T and T'. It should be clear that if $g = f$, T and T' are identical regardless of the values of a and b. In particular, this holds when both g and f are zero, which agrees with the previous assertion that the wire system by itself cannot account for the difference in the speed of signalling. Further reflection will reveal that when a is less than b and f is less than g, then T will be less than T'; while if a is less than b but f is greater than g, then T will be greater than T'.[10] In this manner Heaviside resolved the puzzle.

This simple example exhibits Heaviside as more than a clever manipulator of mathematical syntax. He translated an electrical problem into a mathematical one, and directed the mathematical manipulation by constantly reverting to the physical problem. Heaviside considered mere formal manipulations, or pure mathematical concerns of rigor to be quite useless. In physical questions, he believed, "...the physical nature of a particular problem will usually suggest, step by step, the necessary procedure to render the solution complete."[11] Thus, Heaviside strongly advocated the use of mathematical reasoning in physical questions, but he strongly objected to turning mathematical physics into something resembling the work of the pure mathematician.

Many years later, after the Royal Society rejected his paper "On Operators in Physical Mathematics, Part III", Heaviside wrote a short essay entitled "Rigorous Mathematics is Narrow, Physical Mathematics Bold and Broad." In it, he quoted Lord Rayleigh's observation that to the physicist, whose mind is "exercised in a different order of ideas, the more severe procedure of the pure mathematician may appear not more but less demonstrative." To this observation, Heaviside added his own:

> The best result of mathematics is to be able to do without it. To show the truth of this paradox by an example, I would remark that nothing is more satisfactory to a physicist than to get rid of a formal demonstration of an analytical theorem and to substitute a quasi-physical one, or a geometrical one freed from co-ordinate symbols, which will enable him to *see* the necessary truth of the theorem, and

10. To see this, simply subtract T from T'. The result is $(b - a)(g - f)$. Clearly, this product is greater than zero if b is greater than a and g is greater than f, or when b is less than a and g is less than f. Conversely, the expression is less than zero when b is greater than a and g is less than f, or when b is less than a and g is greater than f.

11. *Electrical Papers*, Vol. I, p. 94.

> make it be practically axiomatic. Contrast the purely analytical proof of the Theorem of Version well known to electrical theorists, with the common-sense method of proof by means of the addition of circuitations. The first is very tedious, and quite devoid of luminousness. The latter makes the theorem be obviously true, and in any kind of coordinates. When seen to be true, symbols may be dispensed with, and the truth becomes an integral part of one's mental constitution, like the persistence of energy. [12]

One should not make the mistake of writing these comments off as propagandistic remarks in the wake of Heaviside's clash with the mathematicians of the Royal Society. The clash only prompted him to state explicitly the manner in which he had worked since his very first papers in 1873, as the discussion in these pages shows. It will become evident in chapter III that this physical manner of doing mathematics is one of the most distinctive marks of Heaviside's work.

2. At the Crossroads: Two Ways of Looking at a Transmission Line

In 1874 Heaviside introduced a subject that lends coherence to his entire work all the way to 1891. In a paper entitled "On Telegraph Signalling with Condensers" he began to analyze the transmission line. The basic theory underlying the discussion is again linear circuit theory. The current through the wire is assumed homogeneously distributed over its cross section at all times, so that the thickness of the wire may be effectively ignored, and only the integral current through the wire needs to be considered. The distinguishing mark of the analysis is that here Heaviside no longer found it satisfactory to consider the wire as a simple circuit element. Instead, he viewed the wire as having distributed electrical properties, namely, resistance, capacitance and leak resistance, all reckoned per unit length of wire. This approach to the problem enabled Heaviside to study the development of current and voltage at each and every point of the line. It will soon be observed that Kirchhoff's laws, which Heaviside referred to in his first published paper, suffice for putting down the equations of the line's state. At the same time, it will be seen that toward 1880 the discussion began to reveal that there may be more to consider besides the circuit laws of the linear theory.

---

12. *Electromagnetic Theory*, Vol. 2, pp. 7-8.

2.1 *"On Induction Between Parallel Wires"*

In 1881, Heaviside published a paper entitled "On Induction Between Parallel Wires." It marks a watershed of sorts with regard to his work, for while it appears to provide a natural extension of issues he dealt with before, it also contains a clear indication that something fundamental may be about to change in his general work plan. Throughout the ensuing discussion, Heaviside sought to determine the current and potential, $I(x,t)$ and $V(x,t)$, of a transmission line at any moment in time and at any point along the line. The basic theme was not new. As stated above, he began to consider it in 1874. In this particular study he introduced one new element: he solved for the voltage and current under the assumption that there are other conducting wires in the vicinity.

Consider first the case of a single, isolated transmission line. For simplicity's sake, the line examined can be regarded as a single conductor, say a solid copper wire, surrounded by an insulating sleeve and suspended in the air. This would correspond to the simplest case of a land telegraph cable. Much of Heaviside's work was done with something like that in mind. As usual, the line has four basic electrical characteristics:

1) Ohmic resistance per unit length, denoted by R.
2) Electro*static* capacitance, or simply capacitance, per unit length, denoted by C.
3) Electro*magnetic* capacitance, or self-inductance, per unit length, denoted by L.
4) Leak resistance, which unlike the first three *diminishes* as the line gets longer (there is more insulating surface to leak through). It is measured in Ohm-miles, and denoted by r.

Heaviside did not take the leak resistance into account until the paper's very end. Only then did he add a short remark pointing in general terms to the manner in which the basic equations must be altered to account for the fourth parameter. The explicit solutions for the voltage and current throughout the paper are all based on the limiting case that the leak resistance is infinite, or that the wire is perfectly insulated.

2.2 *Reconsidering the Problem In Light of Kirchhoff's Circuit Laws*

Before examining Heaviside's approach to the problem in this paper, review it in the way he had dealt with similar ones before, namely, using Kirchhoff's circuit laws without any further observations concerning the nature of electrical conduction. There are two such laws:

1) The sum of voltages around any closed circuit path must be zero.
2) The sum of currents entering an "intersection" must be the same as the sum of currents leaving the intersection.

These rules suggest the view of each infinitesimal segment of the transmission line as made up of four discrete components (see figure 2.5). Throughout this discussion use the convention that current flows *out* of the positive pole of a voltage source and *into* the negative pole of a voltage source.

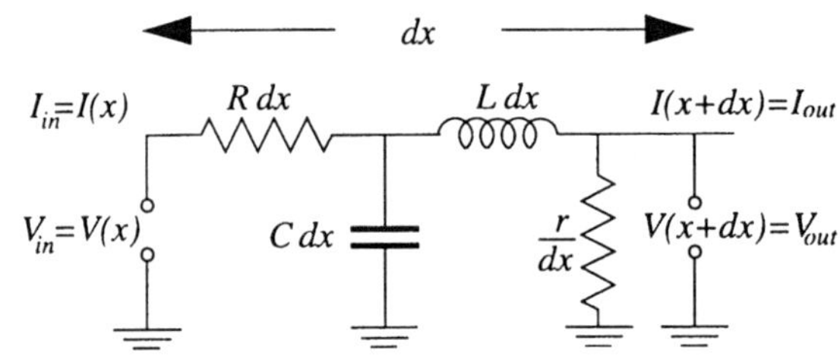

Figure 2.5: Discrete-component representation of an infinitesimal segment of telegraph line.

Current in the circuit is assumed to flow from left to right, hence, all the loops around this circuit will be in the clockwise direction. First, sum the voltages around the outermost loop with V_R and V_L the voltage drops over the resistance and coil respectively. This yields:

$$V_{in} + V_R + V_L + V_{out} = 0.$$

However, according to the convention, the current flows against V_R, V_L and V_{out}. Consequently, using Ohm's and Faraday's laws, the voltage relation turns into:

$$V_{in} - V_{out} = RdxI + Ldx\frac{\partial I}{\partial t}.$$

As for the current, it enters the segment in question from the left, and exits it from the right. However, part of the current entering the segment goes into the capacitor, and another part leaks to ground through r. Therefore, to conform with Kirchhoff's current law, we must conclude: $I_{in} = I_{out} + I_C + I_r$ where I_C and I_r represent the current into the capacitor and through the leak resistance respectively. It must be remembered that in the infinitesimal segment in question, the infinitesimal voltage drops across the resistance Rdx and inductance Ldx are negligible when directly compared with $V(x,t)$. This justifies the assumption that the voltages that drive current into the capacitor and into the leak resistance are all equal to V_{in}. Hence, using Ohm's law and the definition of capacitance, the current equation above becomes:

$$I_{in} - I_{out} = Cdx\frac{\partial V}{\partial t} + \frac{dx}{r}V.$$

Considering that:

$$I_{in} - I_{out} = I(x, t) - I(x + dx, t) = -dI,$$

$$V_{in} - V_{out} = V(x, t) - V(x + dx, t) = -dV,$$

the two equations transform into two partial differential equations as follows:

$$-dV = -\frac{\partial V}{\partial x}dx = RIdx + L\frac{\partial I}{\partial t}dx,$$

$$-dI = \frac{\partial I}{\partial x}dx = \frac{V}{r}dx + C\frac{\partial V}{\partial t}dx,$$

$$\Rightarrow \quad \left.\begin{aligned} -\frac{\partial V}{\partial x} &= RI + L\frac{\partial I}{\partial t} \\ -\frac{\partial I}{\partial x} &= C\frac{\partial V}{\partial t} + \frac{V}{r} \end{aligned}\right\}. \tag{2-2}$$

The last two are the basic equations of the transmission line. Since in his paper "On Induction between Parallel Wires" Heaviside worked under the assumption that r is infinite, it can easily be seen that the second term on the right hand side of the second equation drops out.

2.3 From Electromagnetism to Electrodynamics

However elegant, the above is not the way Heaviside treated the problem in his paper "On Induction Between Parallel Wires." The previous analysis presents the connection between current and voltage in the circuit on a purely empirical basis, and it seems Heaviside was not satisfied with that. Already his paper on electromagnets seems to indicate that by 1879 he had begun to adopt a quasi-dynamical view of the relationship between current and electromotive force. Furthermore, in the paper "On Induction Between Parallel Wires" he set up a problem more general than the one involving a single wire, and attempted to determine the currents and voltages in a system of interacting parallel wires.

How then, does one deal with the mutual interaction of several conductors?

> During the establishment of the current in the first wire (supposed to take place uniformly all along its length), a current in the *opposite* direction is [electromagnetically] *induced* in the second (also uniform all along), which ceases when the current in the first reaches its steady strength. And on the cessation of the current in the first wire a current is *induced* in the second in the same direction. But this, though sufficient for many, is but a very rudimentary statement of the case.
>
> According to Thomson and Maxwell's theory, the electric current is a kinetic phenomenon, involving matter in motion, and the motion is not confined to the wire alone, but is to be found wherever the magnetic force of the current extends. As matter has to be set in motion when a current is in the course of establishment, inertia has to be overcome[,] real inertia of moving matter having the negative property of remaining in the state of motion it may have. ... Now respecting the currents induced in neighbouring conductors. The momentum [of the initial current] exists in all parts of the field, and on the removal of the E.M.F. becomes visible in all of them, the energy becoming degraded into heat in all. Granting this, the currents induced must be all in the same direction, viz, as that in the primary wire; and it follows immediately that on setting up a current the opposite occurs, currents in the opposite direction to that set up being caused in all the wires. In the secondary wires it is evident as such; in the primary it is evident as retarding the rise of the current.[13]

13. *Electrical Papers*, Vol. 1, p. 120.

With the observation that magnetic induction is to E.M.F. as mass is to mechanical force, the parallel wires in the problem can be viewed as analogous to a mechanical system that behaves under the influence of an impressed force according to the laws of Newtonian dynamics. Thus, electromagnetism has just turned into electrodynamics; or so it seems. All one requires now is the explicit mechanism underlying it all. As it turns out, the last requirement is not fulfilled, and the dynamics is only a dynamics of sorts. Immediately following the paragraph above, Heaviside carefully noted that self-inductance is not mass, and potential differences are not forces in the true Newtonian sense. It appears that Heaviside did not require of a theory to state what is *really* out there, but rather to provide a useful description of phenomena that correspond to, and are measured by certain theoretical parameters:

> Not knowing the actual mechanism of the current and of the magnetic force, we cannot know what the actual amount of real momentum is, although the amount of energy, the connecting link between all forces, may be calculated. But, in a dynamical system, it is not at all necessary that the mechanism should be known completely. If the state of the system is completely defined by the values of a certain number of variables [then] the relations between forces, momenta, etc. corresponding to these variables may be calculated on strictly dynamical principles. Thus Maxwell's electromagnetic momentum of a circuit bears the same relation to the impressed E.M.F. in the circuit that momentum does to force in ordinary dynamics. Ohm's law, however, remains an experimental fact, and is taken as such alone.[14]

What we have then, is not a theory of the real nature of electromagnetism, but rather a remarkably useful analogy.

Guided by this view, the electric current (whatever it really is) may be regarded as something analogous to a material current, which must therefore satisfy the following requirements:

1) Ohm's law suggests that the current is *opposed* by a resistive "force" proportional to the "speed" of flow, which corresponds to the electric current's intensity.
2) The "mass" associated with every infinitesimal segment of the flow resists the driving "force" in accordance with Newton's second law of motion.
3) If there is any difference between the amount of electricity (whatever that really means) entering an infinitesimal line segment and the

14. *Ibid.*

amount leaving it per unit time (that is to say, if the current into the segment is different from the current exiting it at any given moment in time), then the difference must be accounted for by a change in the electrification of the segment plus any current that happens to leak out of the segment altogether in addition to the current exiting it.

Stated mathematically, the first two requirements yield directly:

$$\frac{\partial V}{\partial x} = -RI - \frac{\partial p}{\partial t} .$$

In the above expression, $\partial V/\partial x$ is the total "force" per unit distance that opposes the "flow" I along the line, and p is the electromagnetic "momentum" of the flow. Stated explicitly for the first wire, the "momentum" is:

$$p_1 = L_1 I_1 + M_{1,2} I_2 + M_{1,3} I_3 + \ldots,$$

where L_1 is the self-inductance of the first wire, $M_{1,2}$ the mutual inductance between the first and second wires and I_2 the current in the second wire, etc. Thus, the dynamical statement for the current and voltage along the first conductor becomes:

$$\frac{\partial V_1}{\partial x} = -R_1 I_1 - \frac{\partial}{\partial t}(L_1 I_1 + M_{1,2} I_2 + M_{1,3} I_3 + \ldots) .$$

The mathematical statement of the third requirement is just as straight forward:

$$\frac{\partial I}{\partial x} + \frac{\partial \mu}{\partial t} + I_g = 0.$$

Here $\mu(x,t)$ stands for the density of electricity (again, whatever electricity may really be) per unit length, and I_g is the leak current per unit length. The equation is an exact analogue of the continuity equation for fluid flow. We may use Ohm's law again to express the leak current as a function of the leak resistance and the voltage at the particular point and time in question. As for the density, it is proportional to the voltage and capacitance at the various parts of the system. Hence, for the first wire:

$$\mu_1 = C_1 V_1 + C_{1,2} V_2 + C_{1,3} V_3 + \ldots,$$

where μ_1 is the density along the wire, C_1 and V_1 stand for its capacitance and voltage respectively. $C_{1,2}$ is the mutual capacitance of the first and second wires, etc. The general continuity requirement above can now be stated for the first conductor as:

$$\frac{\partial I_1}{\partial x} = -\frac{V_1}{r} - \frac{\partial}{\partial t}(C_1 V_1 + C_{1,2} V_2 + C_{1,3} V_3 + \ldots)\,.$$

Consequently, each of the wires in the system is described by two equations, expressing a force law and a continuity requirement. Mathematically, this requires the solution of a set of simultaneous partial differential equations for the voltage and current along the wires. In the particular case of a single, isolated wire, all the coefficients of mutual inductance—electromagnetic and electrostatic—are zero, and the equations describing the system become:

$$\left.\begin{aligned} \frac{\partial V}{\partial x} &= -RI - L\frac{\partial I}{\partial t} \\ \frac{\partial I}{\partial x} &= -\frac{V}{r} - C\frac{\partial V}{\partial t} \end{aligned}\right\}. \qquad (2\text{-}3)$$

Clearly, these equations are identical to equations (2-2) above.

Thus, we arrive at the same mathematical description of the transmission line in two very different, but not necessarily conflicting ways. As far as the transmission line taken as a linear circuit is concerned, the empirical derivation and the quasi-dynamical one actually complement one another. The dynamical view may be regarded as enhancing the first derivation by the addition of a useful analogy that guides the analysis. This leaves Heaviside's reference to "the field" in the quotation on page 52 as an inessential ornament. It seems one could proceed to do the problem while totally ignoring this remark; after all, none of the crucial steps in the argument's development seems to depend on it. However, the significance of the apparently useless remark is a very important one: Heaviside was undoubtedly looking beyond the analogy between electric current and massive fluid flow. Having pointed to the equations that bring the effects of self-induction into the picture, he made it quite clear that as he saw it, they do not represent the energy and momentum of the current, Γ, in the wire:

> These equations are exactly similar to those used in the waterpipe analogy. $L\Gamma$ is the electromagnetic momentum of the circuit containing the current Γ, corresponding to MV, the momentum of the water. Also $1/2MV^2$ is the kinetic energy

> of the fluid, and $1/2L\Gamma^2$ is the electrokinetic energy of the current, which, however, does not reside merely in the wire, as the kinetic energy of the water is confined in the pipe, but in the surrounding space as well.[15]

However, it must be clearly understood that only in hindsight, with Heaviside's later work in full view, does the Maxwellian origin of this statement become clear. At this stage, considering only Heaviside's work up to this point, the quotation above is no more than a puzzling curiosity. The analysis may proceed quite unhampered by simply taking the analogous connections between mechanical force and fluid velocity on one side, and E.M.F. and integral current on the other. An attentive reader who had not been exposed to Maxwell's *Treatise* could only note that Oliver Heaviside may have far more to say about electricity and magnetism than he is actually expressing at the moment. The same attentive reader would follow the analysis every step of the way without once having to assimilate the radically innovative views of Maxwell's electrodynamic field theory. Only after the full Maxwellian view of electromagnetism had been explicitly developed, would it become evident that the linear theory of the conducting wire was only a first approximation of the situation (see chapter III).

As for the mutual interaction of several wires, the equations will follow from the linear theory with the additional assumption that individual current-bearing conductors interact with one another electromagnetically and electrostatically across space. The beauty of Maxwell's view is that having regarded the current's energy and momentum as something that permeates all of space, one can dispense with such additional assumptions of action at a distance. Naturally, the discussion of a single, isolated transmission line is quite transparent to all of this. Only when electromagnetism in more than one dimension is studied, do the deep conceptual differences between the Maxwellian view and its various rivals become apparent.

2.4 Playing Both Sides of the Court

It is fascinating to witness the effortless agility with which Heaviside switched back and forth between the two views outlined above. Indeed, he switched among many more than two variations, as the following quotation demonstrates:

---

15. *Electrical Papers*, Vol. 1, p. 96.

> As the first wire is being charged, a positive current flows in from the battery to do it. The negative charge on the earth and second wire may be considered as resulting from a negative current from the latter to earth and the second wire. Or we may say, using old-fashioned language, that the [+] electricity on the first wire attracts [-] from the earth to the second wire. Or that the [+] charge on the first wire induces a [-] charge on the earth and second wire. Or that the potential of the second wire due to the [+] charge on the first is [+], therefore a [+] current must flow from the second wire to earth until its potential is brought to zero, leaving it negatively charged. Or, more accurately, *because more comprehensively,* we may consider all the elementary circuits, as partly conductive and partly inductive, from one pole of the battery to the first wire, and from the latter to earth direct, and also via the second wire to the other pole of the battery, in every one of which circuits a [+] current flows, producing electrical polarization of the dielectric, whose residual polarization appears as a [+] charge on the first wire, and a [-] charge on the second wire and the earth. *But whatever mode of expression be used the result is the same.*[16] (All emphases mine)

The discussion that follows these remarks is devoted to the study of voltage and current in a telegraph line. All one needs to comprehend it is a firm grasp of linear circuit theory. However, after such comments it can no longer be presumed that Heaviside was satisfied with presenting this theory either on the empirical grounds sketched above, or under the analogy of fluid flow. If nothing else, the paragraph above suggests that at this point he regarded current and voltage as complicated notions, requiring to be embedded in a larger theoretical framework. The multitude of view points he mentioned indicates that he became keenly aware of several such frameworks; it suggests further that this awareness may have come about as a result of an intense search for a comprehensive theory of the electric current and its associated effects.

16. *Electrical Papers*, Vol. I, p. 117. Note how the concept of electrical charge has changed in the Maxwellian view as understood by Heaviside. There is no such thing as an independent charge anymore. Charge is simply the end of a polarized line stretching through the dielectric and resting at both ends on conducting materials. This fascinating paragraph clearly indicates that at this stage Heaviside had already selected Maxwell's as the way to go; but not because it was "true" and the others "false". In fact, considerations of truth and falsehood, and therefore proof and refutation, seem to be conspicuously missing from the entire discussion. It appears that Heaviside considered various equivalent descriptions of the same electrical phenomena, and elected to prefer one to all of the others on account of its comprehensiveness.

3. The Solution of the Non-Leaking Transmission Line, a General Comment on Leakage, and a Nagging Puzzle

Thus far into the paper, Heaviside defined the problem to be considered and introduced a dynamical view of voltage and current with which to guide the work. This primarily physical part of the paper ends with the mathematical formulation of the ideas it advances in the form of the transmission-line equations as outlined above in section 2.3. At this point Heaviside began the business of solving the equations. He did that in several stages:

1) He set up and solved the equation for one wire, having assumed both the self-induction and the leakage to be zero.

2) Again without self-induction and leakage, he solved the problem of two electrostatically interacting wires.

3) Having outlined the noninductive solution for the two-wire system, Heaviside reconsidered it with magnetic induction included. The leakage is still zero. He provided an explicit solution for the case of a single wire under specific initial and boundary conditions (namely, specifying the voltage and current along the wire at $t = 0$, as well as the electrical connections at the wire's ends).

With an attitude which had now become a matter of course, Heaviside was not satisfied with the explicit mathematical solution of the problem. The problem at hand is essentially a physical one, and once its explicit mathematical solution has been reached, its physical significance must be interpreted and elucidated. What follows is an enlightening example of Heaviside's physical interpretation of mathematical expressions, but even more important, the particular conclusion reached is of crucial importance to his subsequent work.

To fix ideas, consider the particular solution for the voltage along an isolated wire, the receiving end of which is insulated while a battery of constant voltage V is connected to its sending end at $t = 0$. It is also assumed that initially both the voltage and current along the line are zero. This is not the particular problem Heaviside solved in his paper "On Induction Between Parallel Wires." In this paper he considered the solution under the assumption that the receiving end is put directly to earth, namely, that $V(l,t) = 0$. Heaviside's purpose was to study transmission lines under various conditions, and therefore, understandably enough, he varied the problem from paper to paper. The purpose here, however, is to examine the development of his general outlook.

This would be better served by examining how the analysis of a line under the same initial and boundary conditions changes with the changing outlook. Since the insulated receiving end in a pulsed wire was the case he studied in his first general analysis of transmission lines ("On signalling With Condensers"), we shall stay with it for the entire discussion.

The solution would probably scare away the uninitiated. It looks like this:

$$V(x,t) = V - \frac{2V}{l} e^{-at} \sum_{n=0}^{\infty} \left(\frac{a}{b_n} \sin(b_n t) + \cos(b_n t) \right) \frac{\sin \sqrt{g_n} x}{\sqrt{g_n}}, \quad (2\text{-}4)$$

where:

$$a = \frac{R}{2L};\ b_n = \sqrt{\frac{g_n}{CL} - \frac{R^2}{4L^2}};\ \sqrt{g_n} = \frac{(2n+1)}{2l}.$$

In the above, R is the resistance per mile, C is the capacitance per mile, L is the self-inductance per mile, and l is the cable's total length.

Having overcome an initial reluctance to examine the expression, one may note that because of the exponential e^{-at} that multiplies the summation, this entire part will grow smaller in value as time progresses. When t tends to infinity, the sum will tend to zero, and the solution will reduce to $V(x,t) = V$. After a very long time then, the voltage along the entire line will be uniformly equal to the battery's. This makes perfect intuitive sense if one stops to think about it. The entire wire is perfectly insulated, so that no current can flow out of it. This means that current will flow into the wire until it was charged to the value of the battery's potential, and then all current will cease.

The next conclusion cannot be anticipated by such intuitive considerations. Using the trigonometric identities:

$$\cos(\alpha - \beta) = \cos\alpha\cos\beta + \sin\alpha\sin\beta;$$

$$\cos(\alpha + \beta) = \cos\alpha\cos\beta - \sin\alpha\sin\beta;$$

$$\sin(\alpha - \beta) = \sin\alpha\cos\beta - \sin\beta\cos\alpha;$$

$$\sin(\alpha + \beta) = \sin\alpha\cos\beta + \sin\beta\cos\alpha;$$

equation (2-4) may be rewritten as follows:

$$V(x,t) = V - \frac{V}{l}e^{-at}\sum_{n=0}^{\infty}\frac{a}{b_n\sqrt{g_n}}\cos\sqrt{g_n}\left(x - \frac{b_n}{\sqrt{g_n}}t\right) - \frac{1}{\sqrt{g_n}}\sin\sqrt{g_n}\left(x - \frac{b_n}{\sqrt{g_n}}t\right)$$

$$+\frac{V}{l}e^{-at}\sum_{n=0}^{\infty}\frac{a}{b_n\sqrt{g_n}}\cos\sqrt{g_n}\left(x + \frac{b_n}{\sqrt{g_n}}t\right) - \frac{1}{\sqrt{g_n}}\sin\sqrt{g_n}\left(x + \frac{b_n}{\sqrt{g_n}}t\right).$$

In other words:

$$V(x,t) = V - e^{-at}\sum_{n=0}^{\infty} W_n(x - v_n t) - W_n(x + v_n t)\ ;\ \text{where } v_n = \frac{b_n}{\sqrt{g_n}}.$$

The functions $W_n(x,t) = W_n(x-v_nt)$ have the following interesting property:

$$\begin{aligned} W_n(x + v_n t, t) &= W_n(x + v_n t - v_n t) \\ &= W_n(x, 0). \end{aligned}$$

Thus, after a time t the function's value at $x = x + v_nt$ is the same as what its value used to be at x when t was zero. $W_n(x-v_nt)$ then describes a trace that moves to the right of the origin with a uniform velocity v_n. In a similar vein, $W_n(x+v_nt)$ moves to the left with the same velocity. Now, the superposed wave train in equation (2-4) describes a sharp step from zero to V at $t = 0$. However, as n grows larger, v_n grows larger, and thus the superposed waves will gradually spread out, with the higher frequencies travelling faster than the lower ones. As a result, the sharp voltage step that was initiated at $t = 0$ will become progressively distorted as it moves forward along the wire.

All of this does not happen in one special case, namely, when $R = 0$. In this case, v_n reduces to:

$$v = \frac{1}{\sqrt{CL}}.$$

Thus, the dependence on n is lost, and all frequencies travel at an equal speed that is determined solely by the capacitance and self-inductance of the wire. Under these conditions the initial form of the voltage pulse will be preserved no matter how long it travels.

With one crucial exception, Heaviside carefully put down all these conclusions at the end of the paper and illustrated them by a number of graphs that sketch the pulse's advance. He did not, however, provide the detailed explanation of the wave-front's behavior as outlined above. Instead, he only noted that the resistance has the effect of "rounding off" the edges of the initial pulse.

This is how distortionless transmission is encountered for the first time in Heaviside's work. Of course, such a transmission line was not practical. All conductors had finite resistances (superconductivity was not a practical concept at the time), and hence it appeared that all transmission lines would distort the signals travelling in them. However, Heaviside did not complete the analysis: he left out the effects of leakage. At the very end of the paper Heaviside outlined in general terms how to alter the basic equations of the line to account for leakage. He made no attempt to reformulate the line equations explicitly, let alone to discuss the propagation of signals in a leaky line.

A mathematically educated, devoted reader of Heaviside would possibly begin to anticipate an approaching climax. For the past three or four years Heaviside had been studying transmission lines in growing complexity. Slowly but surely he was providing an analysis of unprecedented sophistication. Now, after his paper "On Induction Between Parallel Wires," the completion of the analysis would appear close at hand; but it was not to be. Only in 1887 did Heaviside finally consider the combined effects of all four line coefficients. What took him so long? He had already formulated the solution's beginning in the present paper. He needed nothing more by way of mathematical tools and theoretical outlook. The problem falls squarely within the scope of the linear circuit theory that underscores all of his work to this point. The same techniques he used to solve the problem of the non-leaking wire would smoothly solve the problem of the leaking transmission line. It should have taken no more than a few hours of work to obtain the explicit expression for the voltage under the same initial and boundary conditions. The effort would have been handsomely rewarded by a most fascinating conclusion, which derives from the changed form of b_n:

$$b_n = \sqrt{\frac{h^2}{LC} + \frac{g_n}{LC} - a^2}\text{; where } a = \frac{R}{2L} + \frac{1}{2Cr}\text{, and } h^2 = \frac{R}{r}\text{,}$$

r being the leak resistance. Now:

$$\frac{h^2}{LC} - a^2 = \frac{h^2}{LC} - \left(\frac{R}{2L} + \frac{1}{2Cr}\right)^2$$

$$= -\frac{1}{4}\left(\frac{R}{L} - \frac{1}{Cr}\right)^2 = -K^2,$$

$$\Rightarrow v_n = \frac{b_n}{\sqrt{g_n}} = \frac{\sqrt{\frac{g_n}{LC} - K^2}}{\sqrt{g_n}} = \sqrt{\frac{1}{LC} - \frac{K^2}{g_n}}.$$

When $R/L = 1/Cr$, then $K = 0$; v_n is $1/(LC)^{1/2}$ and equal for all wave frequencies. Thus, the distortionless transmission line reappears, but no longer in the form of an unattainable ideal. One may construct this new ideal line by using perfectly practical values for the four line parameters. Of course, in this case the decay coefficient a is different from zero, so the signal will decay as it moves along the wire. But the decay will develop without distortion. A telephone conversation for example, will become fainter as the line's length increases, but a sensitive detector can correct for that to a large degree.[17]

---

17. Note that when $g_n/LC < K^2$, the oscillating trigonometric functions $\sin(b_nt)$ and $\cos(b_nt)$ turn into the non-oscillating exponential functions $\sinh(b_nt)$ and $\cosh(b_nt)$ and all wave motion ceases. In order to have waves, then, we must have $b_n > 0$. Under this condition, as n becomes predominantly large, the wave velocity in the conductor reaches a maximum value of $1/(LC)^{1/2}$. With typical values for L and C the speed turns out to be close to the speed of light. Already in 1857, Kirchhoff analyzed the non-leaking wire and wrote: "The velocity of propagation of an electric wave is here found to be ... very nearly equal to the velocity of light in vacuo." (G. Kirchhoff, "On the Motion of Electricity in Wires," *The Philosophical Magazine*, ser. 4, **13** (June 1857): 393-412, esp. 404-406.) Heaviside learned of this paper only in 1888. In one sense this simply attests to the power of hindsight. With Maxwell's theory firmly in mind we may regard this as foreshadowing the general observation that all electromagnetic waves, including light of course, cannot exceed a certain velocity. But why should the non-Maxwellian even consider this remarkable observation? Without the image of electromagnetic fields spreading by oscillation through a dielectric medium, the statement remains confined to the current and voltage waves inside a conducting wire. The fascinating general implications of the Maxwellian view will most likely remain invisible to anyone who interprets the mathematics in a non-Maxwellian way. The distortionless condition, on the other hand, is totally independent of the above consideration and equally evident to both Maxwellians and non-Maxwellians.

It seems that an electrical engineer, studying the development of voltage and current in a wire, could have obtained all of the above by simply analyzing the leaking transmission line in the spirit of Heaviside's paper "On Induction Between Parallel Wires." Moreover, he would not have needed to adopt Heaviside's dynamical outlook, for as we have seen, the equations for a single line can be obtained from the circuit laws of Kirchhoff taken as purely empirical statements. Thus, the nagging puzzle remains: What took Heaviside so long?

4. Summary, and a First Hint of the Puzzle's Solution

As already mentioned, "On Induction Between Parallel Wires" may be regarded as continuing an already well-established Heavisidean theme, namely, the analysis of transmission lines. In it Heaviside concluded that with self-induction included, the current and voltage may be regarded as evolving along the wire in the manner of propagating waves. He also realized that the wire's resistance has the effect of gradually "rounding off" the sharp edge of the initial voltage pulse. Looked at from this point of view, the paper furnishes further clarifications concerning the development of current and voltage in a line under specific conditions. In particular, it seems to indicate the possibility of distortionless transmission. Considering especially the last point, it seems hard to understand why Heaviside did not proceed directly with the next logical step of examining the leaky transmission line and uncovering the condition for distortionless transmission.

Tempting as it may appear, the above viewpoint neglects the significance of Heaviside's switches between various conceptions of current and voltage as exemplified by the quotations on pages 52, 55 and 57. They may very well indicate that "On Induction Between Parallel Wires" was not the simple continuation of an old theme as depicted above. During several years of an increasingly penetrating study, Heaviside often came to challenge views advanced by leading authorities like W.H. Preece and Cromwell Varley. However, throughout his published work until 1881, voltage and current remain simple, unambiguous basic concepts that need no explanation and no analysis. "On Induction Between Parallel Wires" changes that. Suddenly, basic electrical concepts like charge, current and E.M.F. appear to be heavily theory-laden. Furthermore, if we stop to think about it, Heaviside's excursion into a critical examination of basic concepts is totally unnecessary for the results obtained at the end of the paper. He could have proceeded as he did in the past by introducing the problem of multiple wires and solving the equations that can be set up by considering the well-accepted laws of linear circuit theory. Yet, certain remarks Heaviside made in the course of the discussion suggest that he had been looking beyond the framework of linear circuit theory.

It appears then, that Heaviside reached a crossroads of sorts. On one the hand, he was poised to deliver the coup-de-grace of transmission-line analysis; on the other, he had come to feel that the entire subject might be conceived much more comprehensively in terms of a new fundamental theory of electromagnetism. What should he have done then? Continue the analysis in terms of an older view that he found wanting, or delay the continuation to first set up a new framework? As the next chapter will reveal, Heaviside decided in favor of the latter option. The student of telegraph and telephone lines became a student of electricity and magnetism. Between 1881 and 1886, prior to publishing his most comprehensive studies of linear circuit theory, he wrote a long series of articles, methodically exposing the basic elements of Maxwell's theory as he understood it.

Still, one wonders whether it is plausible to suggest that Heaviside actually delayed by six years the publication of an important, concrete discovery merely for the sake of rewriting Maxwell's theory in his own words. Under normal circumstances, it seems more plausible that he would not have waited, especially if the discovery was a simple matter of generalizing existing results to include the case of leakage. However, it may not have been that simple a matter. It has already been noted in the previous section that Heaviside did not furnish a detailed explanation of distortion in his paper "On Induction Between Parallel Wires." All he did was to observe that resistance has the effect of rounding off the sharp edges of the square pulse. In fact, he did not use the word "distortion" in this context until 1887.

Reviewing the situation with all of the above in mind, the picture changes quite dramatically. There can be no doubt that Heaviside possessed the necessary mathematical skills to derive the solution for a pulse's propagation in a leaking transmission line. But it is not at all clear that he formulated the concept of distortion necessary for leading the solution to the distortionless condition. One first needs to fix in mind the notion that the signal's analog shape is the key to clear reception. Heaviside did know how to describe distortion in terms of the harmonic content of a signal.[18] However, the extraction of the distortionless condition from this point of view requires analysis of the full solution in the form of a Fourier series. There is no a priori reason to transform this series solution using the trigonometric relations as shown in section 2.4 above, unless one knows in advance what to look for. It will be seen in chapter IV that once Heaviside clearly defined the concept of distortion he derived the distortionless condition in a far simpler way, completely avoiding the cumbersome Fourier expansion. Indeed, he did not even need to solve the telegraph equation in order to derive the condition.

Heaviside was indeed poised to deliver his transmission line coup-de-grace, but it is not so clear that he was aware of this in 1881. All we can say with certainty is that in 1881 he was perfectly capable of solving the full telegraph equation. However, we must ask what Heaviside could have expected out of the solution. Without discerning a new special meaning behind it, all he could anticipate out of adding leakage was just one more variation on an already well-known theme.

Now consider again what would have been the more plausible course for Heaviside to follow: continue a seemingly straight forward, uneventful analysis of transmission lines in terms of an old, inferior view, or delay the above to first develop a new comprehensive framework? At this point it seems more plausible that he would have opted for the latter. After all, this way he could teach electrical engineers a whole new way of dealing with their subject. As for leakage, what could possibly be so important about it?

18. *Electrical Papers*, Vol. 1, p. 99.

Chapter III The Maxwellian Outlook

> [T]here are some matters which no mind, however gifted, can present in such a way as to be understood in a cursory reading. There is need of meditation, and a close thinking through of what is said.
>
> Johannes Kepler, 1609.

1. A New Theme and a New Approach

In his analysis of signal propagation in telegraph cables Heaviside demonstrated a masterful ability to manipulate the mathematics of partial differential equations. Maxwell's theory, to which he turned his full attention at this point, carried overtones of Hamilton's powerful, if somewhat abstruse, algebra of quaternions. Having read Heaviside's early work, one could expect him to follow suit, expand the analysis, and apply his mathematical skills to the detailed study of specific, illustrative cases just as he did with the telegraph cable. However, the remarks with which he introduced his new undertaking suggest that he had a very different task in mind:

> Every one knows that electric currents give rise to magnetic force, and has a general notion of the nature of distribution of the force in certain practical cases, as within a galvanometer coil, for example. Further than this few go. The subject is eminently a mathematical one, and few are mathematicians. There are, however, certain higher conceptions, created mainly by the labours of eminent mathematical scientists, from Ampère down to Maxwell, which are usually supposed to be within the reach of none but mathematicians, but which I have thought could be to a great extent stripped of their usual symbolical dress, and in their naked simplicity made to appeal to the sympathies of the many. Let not, however, the reader (if he belong to the many) imagine that thinking can be dispensed with; there is no royal road to knowledge, and hard thinking and rigid fixation of ideas are required. ... But earnest students, if they will not or cannot learn the mathematical methods, need not therefore be discouraged, for the name of Faraday will shine forth to the end of time as a beacon of hope and encouragement to them. He was no mathematician, yet achieved results apparently only attainable by such methods. It need not be supposed that he had the peculiar brains of a calculating boy, able to do long sums 'in his head' by special methods of his own. The work was of quite a different kind, and probably Faraday could never have made an ordinary mathematician, with the best of training. In fact, mathematical reasoning does not necessarily involve any calculating in the usual sense, though it

is, of course, greatly assisted thereby sometimes; and as for the use of symbols, they are merely a sort of shorthand to assist the memory, which even those who openly contemn mathematical methods are glad to use so far as they can make them out—in the expression of Ohm's law for instance, to avoid spinning a long yarn.[1]

These remarks are intriguing not merely because they state Heaviside's intention to write about electromagnetism for the intelligent and highly motivated non-mathematical electrician. They also amount to a first explicit indication that he harbored certain ideas about mathematics that he, at least, considered quite different from the generally accepted ones. In particular, the quotation seems to imply that there is more to mathematics than the art of explicit calculation, although it does not explain the essence of this additional aspect.

More generally, the quotation above is taken from the first of a series of papers that Heaviside published from 1882-1884. The entire discussion in the series is essentially a methodological and philosophical introduction to Maxwell's theory as Heaviside understood it. It also outlines the most important general aspects of Heaviside's work. The papers clearly express his view of the force field as the generalized mechanism for the transference of energy, and provide the most balanced account of his views concerning the relationship between physics and mathematics. Accordingly, the rest of this chapter is devoted to a detailed analysis of these papers, using several case studies to demonstrate Heaviside's remarkable style of reasoning along with his particular idea of what electromagnetic field theory is really all about.

2. Magnetic Field of a Straight Wire and a First Generalization

It is known, Heaviside says, that the magnetic field at a distance r from the axis of a long, straight wire of radius a carrying a current C is:

$$B = \begin{cases} \dfrac{2C}{r}\theta \text{ for } r > a \\ \dfrac{2Cr}{a^2}\theta \text{ for } r \leq a \end{cases} \tag{3-1}$$

1. *Electrical Papers*, Vol. I, pp. 195-196.

In the above, θ is a unit vector perpendicular to both r and the current axis. It defines a clockwise rotation when viewed along the direction of the current, that is to say, if the current flows into the face of the watch, then the magnetic field follows the direction of rotation of the watch hands. The so called "right hand rule" can help clarify this a great deal. Use the same convention for current "flow" that Heaviside used throughout his work, namely, that current goes from the positive pole to the negative pole of a battery. If the extended thumb of the right hand points along the current's direction, then the circulation indicated by curling the other four fingers around the thumb is the direction of the magnetic field around the current axis.[2]

An interesting observation emerges upon inquiring how much work it would take to move a unit magnetic pole once around the conducting wire. Two cases must be examined, namely, when the path lies outside the wire, and when it lies inside it. Any other path may be broken into combinations of these two. Heaviside begins with the first case. Since the magnetic field is always perpendicular to the radius r, any radial motion costs no energy. This leaves only pure rotation to consider. At a distance r from the current's axis an infinitesimal rotation $d\theta$ produces an infinitesimal arc $rd\theta$. Recalling that the magnetic force acts along the arc, the work done against it per unit magnetic pole in covering that distance is simply $Brd\theta$. In other words, substituting the magnitude of **B** from the first case of eq. (3-1), the energy required to move through an angle $d\theta$ is $2Cd\theta$. One full rotation around the current axis implies summing all the infinitesimal contributions of 2π radians. Therefore, the total work per unit magnetic pole is $4\pi C$. Thus, regardless of the path chosen around the conductor, the energy required to move a unit magnetic pole once around the conductor is 4π times the total current enclosed by the path. Instead of talking about energy per unit magnetic pole, it is customary to speak of the line-integral of the force field around the conductor. This is actually quite an appropriate mode of speech, since one really sums up, or integrates all the little contributions of $\mathbf{B} \cdot \mathbf{dl}$ once around. The infinitesimal vector **dl** stands for an element of the path.

Using similar reasoning Heaviside proceeds to show that inside the conductor the line integral of a circular path centered on the current axis is

2. This is by no means a simple fact, revealed by trivial observation, and Heaviside did not present it as such. He wrote: "The magnetic force is *known* to be of intensity 2c/r in electromagnetic measure at distance r from the axis...." (my emphasis). It is also 'known' that two masses gravitate according to $F = GMmR^{-2}$. No one, however, would pretend that this is a simple fact.

$4\pi R^2 C/a^2$, a being the conductor's radius. However, during the whole discussion Heaviside assumes that the current is steady and equally distributed over the conductor's cross-section. Therefore CR^2/a^2 measures the total current passing through the circle of radius R. So again, it comes down to the same thing: the line integral is 4π multiplied by the total current enclosed by the path.

Of course, no general property has been proven by these cases. The entire analysis is based only on the case of a straight conductor and its second part is further limited to the particulars of a circular integration path. But while proving nothing in general, the above does suggest the possibility of a general rule: the line integral of the magnetic field associated with a current distribution around an arbitrary closed curve is equal to 4π times the total current enclosed by the curve.

This generalization, known as Ampère's law after its original formulator, leads to a very special relationship between current and associated magnetic field because it is independent of the shape and size of the closed curve of integration. By allowing the curve to shrink down to infinitesimal dimensions, the rule actually describes any current distribution given its associated magnetic field. With this consequence of Ampère's law in mind, Heaviside introduced a conceptual tool of primary importance:

> When one vector or directed quantity, **B**, is related to another vector, **C**, so that the line-integral of **B** *round* any closed curve equals the integral of **C** *through* the curve, the vector **C** is called the curl of the vector **B**.[3]

Using this definition, Heaviside restated the relationship between current and magnetic field as follows. The original generalization states that the line integral of the magnetic field **B** around any closed curve equals 4π times the total current through any surface bounded by the curve. Therefore, by the definition of the curl, Heaviside wrote: $\text{curl}\,\mathbf{B} = 4\pi\mathbf{C}$, where **C** defines the direction and strength of the current per unit area everywhere in space. It is usually known as the current-density.

To get a better understanding of the relationship between current strength I and current density **J**, think of the current as a flowing fluid, and orient a tiny surface, da, perpendicularly to the flow. The ratio dI/da between the total current flowing through da and the area of da is the strength of the current density **J** at the point, so that **J** is measured in units of current strength per unit

3. *Ibid.*, p. 199.

area. The direction of **J** coincides with the normal to *da*. Thus, it is **J**, not *I* that fully describes the flow. Note that unlike the current-density **J**, the current *I* is not a vector function. *I* is the total current that passes through a given surface *A*, arbitrarily oriented in the flow field defined by **J**. Now the surface need not be plane, nor must **J** be everywhere perpendicular to it. Obviously, the component of **J** that is tangent to the surface at a given point does not contribute to the current that passes through the surface at that point. Therefore, the total current through the surface is the contribution over the entire surface of only the components of **J** that are locally perpendicular to the surface. If **n** is a unit vector perpendicular to the infinitesimal area element *da*, then one may express all of the above symbolically as:

$$I = \int_A (\mathbf{J} \cdot \mathbf{n})\, da,$$

where the subscript *A* indicates that the integration must be carried out over the entire surface. Heaviside deplored the use of the word "intensity" to denote the strength of a current:

> I would ... like to see the word "intensity," as applied to the electric current, wholly abolished. It was formerly very commonly used, and there was an equally common vagueness of ideas prevalent. It is sufficient to speak of the current in a wire (total) as "the current," or "the strength of current," and when referred to unit area, the current-density.[4]

In his papers from 1872 to 1881, Heaviside did not use a consistent notation for the conduction current, and symbolized it on different occasions by g, G, C, Γ, and γ. Beginning with his 1882 paper on "The Relations between Magnetic Force and Electric Current", however, he always symbolized current strength by a capital *C*. Heaviside did not use different letters to distinguish between the current *I* and its density **J**. He always used **C** for current density, and the reader must let the context decide whether *C* stands for the magnitude of **C** or for the full current in the sense of the integral above. This may sound terribly misleading, but in practice the context does make the distinction quite clear. It may be noted in passing that the word intensity has indeed been all but replaced by "the current," just as Heaviside had wished; but curiously enough, the symbol for current strength remained *I*. In order to keep as close

4. *Electrical Papers*, Vol. II, p. 23.

as possible to Heaviside's discussions, his notation for the current will be adhered to from this point on.

A number of observations are in order thus far. In the first place, although the curl has been precisely defined, the definition does not provide a recipe for calculation, or a formula that could be substituted wherever the word "curl" appears. Furthermore, Heaviside's intended reader was not expected to be able to translate the notion of integration used to define the curl into an explicit calculation. Practically speaking, the analysis is qualitative, and one may wonder what use it may serve under the circumstances.[5]

3. A Breach of Continuity?

A more general observation emerges from consideration of the basic theme Heaviside introduced in this essay. The papers he published between 1872 and 1881 deal with various aspects of telegraphy. The earlier ones concentrate mainly on the analysis of receiving and transmitting instruments, while the later ones, reviewed in some detail in the previous chapter, address the problem of signal propagation in telegraph cables. In 1882 Heaviside wrote one more paper on signal propagation. It is a direct continuation of his discussion in his paper "On Induction Between Parallel Wires," in the sense that here the telegraph equation is generalized (but not solved) for the first time to include the case of variable resistance, capacitance, and self-inductance. Then follows a solution of the very same problem that was solved in his paper "On Induction Between Parallel Wires," except that it now introduces the case of a semi-infinite line and the mathematical tools necessary for dealing with it. In June of 1882 Heaviside wrote a short comment for *The Electrician*, entitled "Dimensions of a Magnetic Pole," in reply to criticism leveled by Rudolf Clausius at Maxwell's choice of units for a magnetic pole. Heaviside rejected the criticism, and pointed out that it was based on a failure

5. One may note in passing that even in textbooks that generally excel in breeding familiarity with these ideas the explication of the curl is inextricably linked to its Cartesian expression and the proof of Stokes's theorem (see for example E.M. Purcell, *Electricity and Magnetism*, [New York, McGraw-Hill Book Company, 1963], pp. 64-65; or H.M. Schey, *Div, Grad, Curl, And All That*, [New York: W.W. Norton & Company, 1973], pp. 75- 80.) Heaviside's qualitative introduction and the informal manner in which he applies the concept without any calculation to a variety of specific problems, is, to the best of my knowledge, unique.

to correctly relate the Maxwellian concepts of magnetic force and magnetic induction to electric currents and magnetic poles. Then, in November 1882, without any apparent warning (save for the tendency in the later telegraphic papers to describe electrical conductance from several different, sometimes conflicting, points of view), Heaviside published the first in a series of papers on "The Relations Between Magnetic Force and Electric Current."[6] Five more papers followed in quick succession between November 1882 and January 1883.

As a unit, the six papers form a coherent, well-planned monograph that introduces the fundamental connection between a steady electric current and its magnetic field. From January to March 1883, Heaviside published a second monograph in four sections, on "The Energy of the Electric Current." A third, "Some Electrostatic and Magnetic Relations," followed suit between April and June. Finally, between June 1883 and March 1884 he published a fourth series, "The Energy of the Electric Current," which continues the second in terms of the relationships developed in the third. (It is evident that the fourth part was intended as a direct continuation of the second since its first section is numbered V, while the last section of the second is IV). Taken together, the four series constitute a short treatise that might well have been entitled "An Introduction to Field Thinking for the Intelligent Non-Mathematical Electrician." Thus, from the elaborate and highly technical analysis of practical telegraphic problems, Heaviside turned his attention to a mathematically nontechnical introduction of basic concepts in electromagnetic theory. It will soon become evident that nontechnical does not mean unsophisticated, nevertheless, it would seem natural to conclude that the decision Heaviside faced at the crossroads of the previous chapter led to a breach of continuity in his work.

Further light may be shed on this apparent breach of continuity by a comparison of Heaviside's introduction to electromagnetic theory and the general organization of his greatest source of inspiration—Maxwell's *Treatise on Electricity and Magnetism.* The latter opens with a mathematical introduction that Maxwell began with the observation that every measurable physical property has a twofold character: qualitative or dimensional—represented by a standard unit, and quantitative—represented by a number that measures how much of the standard unit in question is involved in a particular case. He proceeded to discuss units and dimensional analysis briefly, and immediately thereafter

6. *Electrical Papers,* Vol. I, pp. 195-231.

introduced the basic mathematical and conceptual tools necessary for the ensuing quantitative discussion. The mathematical introduction is followed by a summary of the known electric and magnetic phenomena relevant to Maxwell's investigation. Having thus presented his general terminology and prepared the required phenomenological groundwork, Maxwell engaged in a most exhaustive analysis of the electrostatic field. Only some 500 pages later, in volume II of his *Treatise*, did he begin the study of the magnetic field in relation to the electric current.[7] Finally, in the last quarter of his work, Maxwell put together all of the elements he had developed in the preceding chapters, and sums up in his famous equations the electrodynamics of force fields as stresses, and inductions as related strained states in a dielectric medium.

A casual perusal through the physics section in any scientific library will quickly reveal a multitude of texts on classical electromagnetic theory. The approach these texts take to the subject is by no means uniform, but by and large, most of them follow a common general outline. Much like Maxwell, they begin with Coulomb's law of electrostatic interaction between two charges, then proceed to study the electrostatic field, electrostatic potential, Gauss's law and finally Gauss's theorem, otherwise known as the divergence theorem. Sometimes a brief mathematical preliminary precedes the above. Discussion of the magnetic field usually appears only several chapters later, often after Stokes's theorem has been established on independent grounds. This organization of material roughly corresponds to the history of the subject in that electrostatics was the first to be discovered and analyzed in great detail. Besides that however, the treatment follows the inner logic of the theory presented rather than the lines of its historical development.

The modern reader of Heaviside's *Electrical Papers* may find it somewhat curious that Heaviside elected to introduce the basic elements of electromagnetism through a discussion of the magnetic field with its inherently more complicated mathematics. Indeed, throughout the presentation, which rapidly becomes more sophisticated as it progresses, the electrostatic field and its properties receive attention mostly as an afterthought, or sometimes as a preparatory example serving to train the mind for some particularly involved investigation of a magnetic property. Keeping in mind that Maxwell's *Treatise* constituted the main theoretical motivation behind Heaviside's work, his un-

7. James C. Maxwell, *A Treatise on Electricity and Magnetism*, 1st Edition, (Oxford: At the Clarendon Press, 1873).

common order of presentation becomes quite puzzling. Why would anyone choose to introduce the subject through the magnetic field and its connection with the electric current rather than begin with the mathematically simpler study of the electrostatic field?

A possible answer to this question resides in another question, which involves the apparent breach of continuity discussed above. In the preface of *Electrical Papers*, Vol. I, while recounting how the papers he published between 1872 and 1891 came to be collected in book form, Heaviside wrote:

> ... it had been represented to me that I should rather boil the matter down to a connected treatise than republish in the form of detached papers. But a careful examination and consideration of the material showed that it already possessed, on the whole, sufficient continuity of subject-matter and treatment and even regularity of notation, to justify its presentation in the original form. For, instead of being like most scientific reprints, a collection of short papers on various subjects, having little coherence from the treatise point of view, my material was all upon one subject (though with many branches), and consisted mostly of long articles, professedly written in a connected manner, with uniformity of ideas and notation. And there was so much comparatively elementary matter (especially in what has made the first volume) that the work might be regarded not merely as a collection of papers for reference purposes, but also as an educational work for students of theoretical electricity.[8]

True, Heaviside's *Electrical Papers* deal only with electrical matters. By itself, however, this does not warrant describing them as having "sufficient *continuity* of subject matter and treatment" to the extent of being regarded "not merely as a collection of papers for reference purposes, but also as an educational work for students of theoretical electricity." It may, of course, be the case that these introductory remarks, written many years after the work was initially undertaken, constitute little more than an elegant evasion of the time-consuming and often frustrating task of reformulating what had already been written. More likely, however, they were not intended as mere window-dressing of this kind. Indeed, upon further reflection, it turns out that in one very important respect the apparent breach of continuity mentioned above is rather illusory. There exists a definite element of continuity in the *Electrical Papers*, which, once understood, makes Heaviside's decision to begin with magnetism and steady currents quite natural.

8. *Electrical Papers*, Vol. I, p. vi.

The first experimental phenomena that the modern university student encounters in a first course on electromagnetism are usually of the electrostatic kind. They include charged spheres attracting or repelling one another, pieces of paper drawn to combs, and various sparking discharges. The first measuring instrument discussed in some detail is often a modern version of the gold leaf electrometer. This is hardly surprising considering the organization of the material in most introductory textbooks as noted above: the order of the presentation of theory necessarily determines the selection of phenomena with which one must become acquainted first. In contrast to the student of theoretical electromagnetism, an electronic technician would probably make his acquaintance with the subject through the elements of the electric circuit: resistors, capacitors and coils. The main measuring instrument would most probably boil down to a galvanometer of one kind or another. Ohm's law, not Coulomb's, would undoubtedly be the first theoretical statement the practical electrician would study in detail. In short, for the practical electrician the central subject of interest is the electric current and its effects. Actually, this is how most of us get to know electricity nowadays. The electric current and its diverse household uses is the side of electricity we constantly come into contact with. Electrostatics is usually restricted to the occasional unfriendly sparking door handle in a well heated and well carpeted house during cold winter days. Recall now, that Heaviside began his electromagnetic career as a telegraph operator. As such he was constantly surrounded by conducting circuits, batteries, resistors and Ohm's law. The vast majority of detecting instruments were current sensors, all operating on the principle of the galvanometer. Heaviside himself pointed this out: "It so happens that my first acquaintance with electricity was with the dynamic phenomena" (meaning the phenomena associated with the transitions of energy through the electric current).[9] If such an operator ever began to have serious theoretical thoughts, they would probably revolve around the nature of the electric current. Furthermore, his thoughts would be more likely to end up in concrete questions like "what do I mean by 'strength of an electric current'"? rather than the vague "what is electricity"? even though the latter may very well have been the one to have started him thinking in the first place. With such a question in mind it would quickly become clear that the electric current is never measured directly. It reveals itself through its magnetic effects. One actually detects the latter and postulates

9. *Electrical Papers*, Vol I, p. 435.

the former. One measures the current by measuring the force on another magnet or on another current-carrying circuit of known configuration. All galvanometers work by applying this principle in one way or another. Thus, the natural starting point for a practically motivated theoretical dissertation should be the relationship between magnetic force and electric current.

From this point of view, Heaviside seems to have made the natural choice in beginning with the magnetic field and not with the electrostatic field. Furthermore, regarded in this light the breach of continuity suggested above seems significantly less drastic. In hindsight it appears that from 1872 to 1881 Heaviside produced a glorified version of Maxwell's short phenomenological chapter. The phenomena he presented and investigated in great detail during this period are the ones relevant to his later theoretical discussion, namely, phenomena of electrical conduction from the discretely parametrized circuits of Wheatstone's bridge, to the distributed parameters of the infinite transmission line. Even his comments concerning the elusive essence of electric currents can now be taken for a clever pedagogical "trick" serving to prepare the reader for the general theory soon to be developed.

Of course, this is not to say that already in 1872 Heaviside had carefully laid out a grand plan, involving a nine-year phenomenological introduction followed by another nine years of theory. Nor does the above suggest that the *Electrical Papers* can be read like any normal textbook on electricity and magnetism. One can be almost certain that Heaviside started innocently enough by publishing loosely related studies of detecting circuits and of telegraph lines without knowing that the endeavor would lead to an exposition of Maxwell's theory. The element of continuity in his work is revealed by the way he grew to master his own knowledge. As time went on he read and expanded his theoretical field of view. His comments in the later telegraphic papers discussed in the previous chapter make it clear that by the time he wrote "On Induction Between Parallel Wires" he mastered at least part of Maxwell's theory and understood the astonishing scope of its achievement.

With important elements of Maxwell's theory in hand, Heaviside must have sensed that a complete rewriting of telegraph theory would be in order. He could now discuss the electric current and its relation to the magnetic field in a way that encompasses all his previous work and grounds it in a wider theoretical framework. At the same time he never forgot his original problem, the transmission line. The unbroken continuity of his work as well as the clarity with which he held on to the original questions he discussed in the 1870's will be fully revealed between the years 1885-1887 during which he published his single most influential dissertation, "Electromagnetic Induction and its Propagation." At this point we may only conclude that his introductory comments regarding the theoretical continuity of his work are not without foundation.

Even more importantly, one can already begin to glimpse the unique beauty of Heaviside's *Electrical Papers*: their thematic continuity is deeply rooted in an historical one. It is a very special kind of history, for it does not correspond, even roughly, to the reception of electromagnetic knowledge by one or another specialized community. It is a history of personal discovery by one man who dedicated his professional lifetime to conquering Maxwell's theory and making it his own. From this point of view, Heaviside's papers present a living, absorbing story of discovery, rather than a logical treatise modelled after a formal text on Euclid's *Elements*.

4. Introduction to Field Thinking for the Intelligent Non-Mathematical Electrician

4.1 "Curling": Learning to See Vector Fields.

In all treatises on electricity and magnetism before as well as after Heaviside, the curl furnishes the main analytical tool with which to investigate the nature of magnetic fields around electrical currents. Most treatises require a sound ability to work the machinery of the differential and integral calculus whenever they apply the curl. Heaviside, in contrast, began by showing how much can be done with a basically qualitative sense of this concept—a sense acquired through an appeal to physical common sense rather than to mathematical rigor and symbolical manipulation. The following examples will serve to demonstrate the point.

As his point of departure, Heaviside took for granted the field around a steady cylindrical current distribution as discussed in the beginning of this chapter. From there he proceeded to interchange the roles of current and field, and asked what sort of current should be associated with a uniform magnetic field bound within a long cylindrical tube. The relationship $\text{curl}\,\mathbf{B} = 4\pi\mathbf{C}$ must still hold, so 4π times the total current through any closed curve in space is equal to the line integral of $\mathbf{B}$ along the curve. Since the shape and size of the curve are immaterial, one might as well configure it to the given geometry of the magnetic field so as to render computation easy. All the field lines within the cylindrical space are parallel, and their intensity is constant. A small rectangle, oriented so that two of its sides are parallel to the field lines seems simple enough, for it would make line integration a simple matter of multiplying field intensity by the length of line parallel to it.

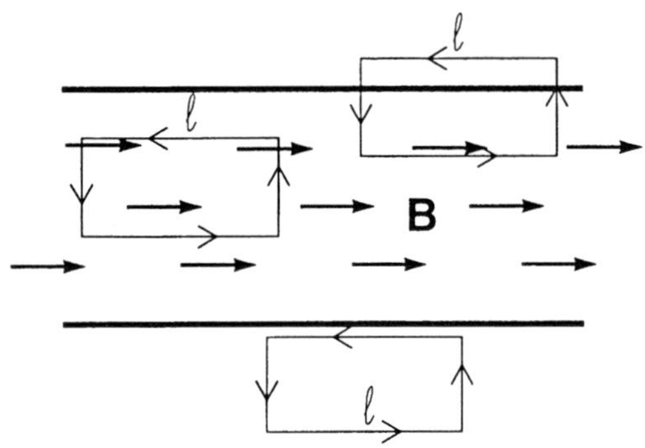

Figure 3.1: The current around a cylindrical distribution of **B** *is confined to the boundary of the distribution.*

It soon becomes apparent (see fig. 3.1) that if the rectangle is placed totally outside the field's cylindrical boundaries, then the line integral is zero, for it involves summing up a constant zero all around the rectangle's circumference. When the rectangle is placed wholly within the cylinder, its two sides that are perpendicular to the field lines contribute nothing (recall that line integration is the sum of all the field components locally parallel to the curve). There remain the two sides that are parallel to the field. Assume that the field points from left to right, and perform the integration counterclockwise around the rectangle. On the bottom part of the rectangle we move with the field lines, hence its contribution would be lB, l being the length of the side. On the top part, we move against the field, hence this contribution would be $-lB$. Obviously, the sum of the two is zero. Thus far, the line integral of the field around the rectangle is zero, which means that no current flows through any loop totally immersed in the field or totally out of the field. This leaves the case of a rectangle whose bottom part is in the field and whose top part is outside it. The top part will contribute $l0 = 0$. The two perpendicular parts will also contribute nothing by virtue of the field being perpendicular to them within the cylinder, and zero outside. Only the bottom part of the cylinder contributes a finite amount, namely lB, which means that there is a total current $lB/4\pi$ flowing through the rectangle. Since integration was carried out counterclockwise, the right-hand rule says that the current must be flowing upwards, out of the plane of the paper. There is no current within the cylinder, and none outside it, so much has already been shown. Therefore, the entire current must be confined to the surface of the cylinder. This can be verified by noting that the total current through the rectangle does not depend on the length of its perpendicular sides. They can be made as short as one likes, as long as they cut the cylinder's bounding surface. Thus, confined to the cylinder's surface there is a current density $B/4\pi$ per unit length measured parallel to the cylinder's axis. It flows in closed circles around the surface, centered on the axis of symmetry, coming out of the paper's plane on top, and going into it on the bottom of the cylinder. In other words, if the right thumb points along the field's direction, then the curved fingers of the right hand now define the current's direction.

This is a simple exercise, which holds special interest because it describes the relationship between current and field in a long solenoid. Of course, it is an idealized solenoid, for the conducting layer is taken to be infinitely thin. But it can be developed a step further, as Heaviside's next alteration of the problem shows. What if the current around the cylinder, still circulating like the surface current of the previous case, were not confined to a surface but evenly distributed in a cylindrical shell of finite thickness? Let the intensity of current density in the shell be C and let t be the shell's thickness (see figure 3.2). From the previous exercise it is clear that each tubular current layer in the shell is associated with a constant cylindrical field $\delta B = 4\pi C \cdot \delta t$, where δt is the infinitesimal thickness of the tube. Therefore, inside the inner boundary of the shell, the contributions of all of the thin tubes within the thickness t add up, to yield:

$$B = \sum \delta B = 4\pi C(b-a) .$$

The first exercise also furnishes the observation that outside a tubular current surface the field is zero. This implies that the field at any position within the shell will be due only to the contributions of the tubular layers above this position. Or, if r is the distance from the axis to any point within the conducting shell, and b is the shell's outer radius, then the field strength at r is simply

$4\pi C\,(b-r)$. To further clarify this, fix the top side of a curling rectangle outside the shell, and its bottom anywhere within the conducting layer (see figure 3.2).

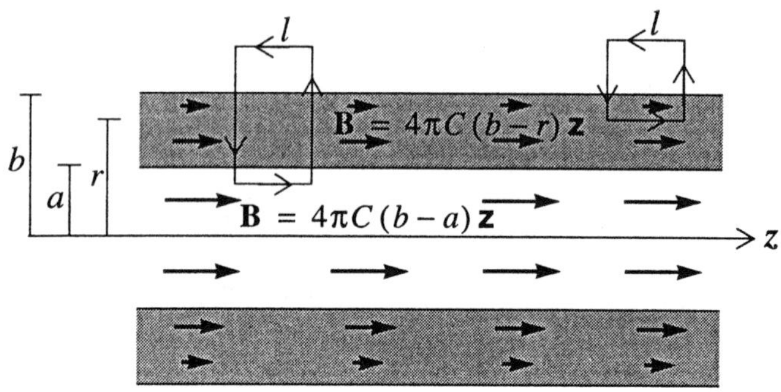

Figure 3.2: Cross section of a coil with inner radius a, outer radius b. The current points out of the page on top, into the page on bottom.

If r measures the distance from the coil's axis to the bottom of the curling rectangle, then the total current flowing through the rectangle is $Cl\,(b-r)$. Now 4π times that amount must be equal to the line integral of the field around the rectangle. Consequently $B(r)$, the magnetic field's intensity inside the conducting layer, falls off linearly from $4\pi C\,(b-a)$ to zero as r goes from a to b. Thus, given a cylindrical shell of inner radius a, outer radius b, carrying a current density C in concentric tubes around the cylinder's axis, the associated magnetic field $\mathbf{B}(\mathbf{r})$ is:

$$\mathbf{B}\,(\mathbf{r}) = \begin{cases} 4\pi C\,(b-a)\,\mathbf{z} & \text{for } r \le a \\ 4\pi C\,(b-r)\,\mathbf{z} & \text{for } a < r < b \\ 0 \quad \text{for } r \ge b \end{cases}$$

In the above, $\mathbf{r}$ is the vector from the axis of symmetry to the point where $\mathbf{B}$ is determined and $\mathbf{z}$ is a unit vector perpendicular to both $\mathbf{r}$ and $\mathbf{C}$ according to the right-hand rule, namely, pointing along the cylinder's axis. In this way Heaviside's text guides the reader from the idealized cylindrical current sheet to the more realistic cylindrical current shell of finite thickness.

Heaviside proceeded to elaborate more examples like the above, always taking advantage of symmetrical configurations in order to simplify computation.[10] Note in particular that while the definition of "curl" as given on page 69 above involves explicit reference to integration, the evaluation of the integrals in question involves no more than grocery-store arithmetic. For the present purposes, "integration" stands for an informal idea of summation, which common sense translates into a specific operation according to what the situation seems to require. This is best exemplified by the manner in which Heaviside explained the possibility of superposing electrostatic fields just a few pages later:

> This is most conveniently expressed by saying that we find the vector sum of the separate forces. This we may denote by $\Sigma(e/r^2)\mathbf{r}_1$, where e/r^2, the intensity of force due to the charge e, is multiplied by $\mathbf{r}_1$, which signifies a *unit vector* drawn along the line from the charge to the point under consideration, thus making $(e/r^2)\mathbf{r}_1$ be the *vector* force due to e. The sign Σ signifies summation. Now, in finding the potential of a vector quantity, such as current, we add together the potentials of the elements into which the current system may be divided, i.e., we find $\Sigma C/r$;...[11]

In the second case, the text refers to a continuous current distribution, and therefore the Σ will translate into integration. Thus, the Σ's used above do not represent a single, well defined algorithm for computation. They are there to signify that the quantity following them must be summed in a manner relevant to its distribution and the particular question at hand. Only later did Heaviside begin to introduce explicit integration procedures. Remarkably, his discussion actually succeeds in conveying the sense that the physical concepts involved are connected by rigorous relationships without nailing the rigor down to the syntactic rules of calculation. As we shall soon see, by proceeding along these lines, Heaviside made the syntax of vector calculus grow as a natural consequence of the physics of magnetic fields and current distributions, not as a pre-existing mathematical language into which one strives to fit these physical concepts. Of course, just as Heaviside promised, his informal introduction of integration does not make the problem easy. It requires a clear distinction between current and current density, and the envisioning of the magnetic field and the current density as two related systems of "flow" in space.[12] Once that

10. *Electrical Papers*, Vol. I, pp. 200-201.

11. *Ibid.*, p. 203.

12. What actually flows there is completely irrelevant. Only the relationship between the flows counts. Heaviside would later generalize this observation into a central pillar of the Maxwellian edifice he began to construct here.

has been done correctly, the geometry of the situation renders computation trivial.

One only has to play a few "curling" games to feel a growing sense of familiarity with the theory of the relationship between electrical current distributions and magnetic fields. It soon becomes evident that given an initial magnetic field, the curl operation generates a well-defined current distribution that, if taken as a new magnetic field, can yield yet another current distribution by "curling," and so on to the formation of an endless series. However, the reverse problem of finding the field given the current presents a difficulty:

> Considering the fundamental relation only, that by finding the line-integral of **B** once *round* any axis we get $4\pi \times$ current-component *along* that axis, if we reverse the operation we discover at once that it fails to work in a suitable manner. We do indeed know from the given value and direction of the current at a given place what the line-integral of magnetic force round it is, but that does not tell us the magnetic force at different points along the line of integration. Some other method is, therefore, wanted.
>
> There are different ways of obtaining the magnetic force from the current. We shall commence with that one of them which has the advantage of telling us immediately in a great many cases the general nature of the magnetic force. This method is expressed in the following statement:- The magnetic force is the vector-potential of the curl of the current. Here we introduce another concept, that of the potential of a vector quantity, and in order to render it intelligible, some explanation becomes necessary.[13]

4.2 Vector and Scalar Potentials: Using Electrostatics as an Analogy.

As its name suggests, the term "vector potential" relates to the idea of potential, be it the potential energy of mechanics, or the electrostatic potential. In the electrostatic case, Heaviside stated without proof or introductory discus-

13. *Electrical Papers*, Vol. I, p. 202. Heaviside rarely used the term magnetic *field.* The word "force" stands sometimes for what we now call the field, sometimes for the mechanical force experienced by a charge immersed in a "force" (meaning a field). There is, however, no ambiguity or vagueness associated with this dual use, and as we shall see, Heaviside was clearly aware of the difference. It is an interesting question why Heaviside, whose sensitivity to rational nomenclature was great, perhaps excessive, never felt the need to distinguish the two by introducing a special word for "force" in the sense of field. Perhaps it was his way of paying his respects to Faraday. However, it must be clearly noted that he adhered to the word "force" in this context quite consciously, despite his full awareness of the different meanings it possesses in different contexts (see page 135 ahead).

sion that the potential at distance r from a charge q is q/r. Without much ado he proceeded to apply this notion to the construction of a vector potential. One must simply keep in mind that now the potential of a vector is sought, not that of a scalar like the charge q. By analogy then, the vector-potential of a single element in a general vector-distribution should be $\mathbf{c}/r$, where $\mathbf{c}$ is the vector element. So a single vector $\mathbf{c}$ defines a full volume distribution, or a vector field $\mathbf{c}/r$. Obviously, the vector potential at some point due to the entire distribution $\mathbf{C}$ would consist of a sum of all of the individual contributions $\mathbf{c}/r$, which Heaviside denoted by $\Sigma\mathbf{c}/r$, and explained that "the sign Σ signifies summation." As before, this is a completely informal idea, which does not specify any way of performing the actual computation. Heaviside's goal was not the computation of sums and integrals, but rather the relationships between fields, currents and charges. It actually seems that explicit dependence on a rigorous mathematical formalism was precisely what Heaviside wished to avoid at this stage:

> In all cases we may find the vector-potential by means of three scalar sums instead of the vector sum; this is most conveniently done by forming the scalar sums of the components in three rectangular directions, and then compounding them. But, though convenient for calculations, this method often very much obscures the matter under consideration.[14]

So far, the vector potential seems simple and problem-free; that is to say, until one recalls that this potential, defined on purely analogical grounds, is supposed to have a concrete connection to the curl of the previous section. "The magnetic force is the vector-potential of the curl of the current."[15] Thus, if $\mathbf{D}$ is the curl of the current density $\mathbf{C}$, then $\mathbf{B}$, the magnetic field associated with $\mathbf{C}$ is:

$$\mathbf{B} = 4\pi\sum\frac{\mathbf{D}}{r}.$$

There is no proof added; no demonstration that the equality indeed holds.[16] The text seems to be asking for the reader's indulgence, as if Heaviside was saying, "take it on faith, play with it a little, learn to use it. Proof may come later. The important thing right now is to see how the idea can actually be put to work." To illustrate the vector potential's place in the scheme, Heaviside

14. *Electrical Papers*, Vol. I, p. 204.
15. *Ibid.*, p. 202.

reverted to the straight current with which he opened the discussion. The curl of a straight, cylindrical current distribution pointing from left to right circulates counterclockwise (when viewed directly against the current) on the surface bounding the current (see pp. 77-79 above). With a little thinking it will become evident that the vector potential of such a distribution consists of a system of concentric circular lines, circulating in the same direction as the curl of the current, increasing from zero at the axis to a maximum at the boundary, and dropping off in intensity as one recedes outward from the boundary. This is purely qualitative, but as far as it goes it furnishes a correct description of the magnetic field around a straight steady current. While no quantitative measurement of any significance can be based on it, this mode of thinking involves working directly with the vector field as a whole, not with the isolated Cartesian components of each field element at a time.

Following Heaviside, we may now think of a curl-related series of vector fields **A**, **B**, **C**, **D**, **E**, . . . as a sort of ladder whose top step is **A**, so that curling is the operation of moving down the ladder a step at a time. "Potting," as Heaviside called the operation of finding the vector potential, moves up the ladder two steps at a time. It is easily seen that curl(pot(**X**)) = pot(curl(**X**)).[17]

4.3 *Introducing the Algebra of Vectors*

At this point, one may raise the objection that actually the entire discussion up to now is far from being independent of a formal mathematical language. After all, is it not heavily dependent on the algebra of vectors? The answer is not simple. Vectors are without a doubt crucial to the discussion. Furthermore, vectors were well known by the time Heaviside did his work. In particular, they were used by Maxwell in his treatise on electricity and magnetism.[18]

---

16. The r in $\Sigma\mathbf{D}/r$ is the distance to a specific, stationary point P in space from each element of the distribution **D**. So r is not the unvarying distance of P from some coordinate origin. It is the constantly varying difference between the vector pointing to P, which is constant, and the vector pointing to the particular element of **D** currently undergoing summation. The latter keeps varying throughout the integration. More formally, one could write:

$$\mathbf{B} = 4\pi\sum\frac{\mathbf{D}}{|\mathbf{R}-\mathbf{r}|},$$

where **R** points to the position of P, while **r** points to the particular element of **D** whose contribution to the vector potential at P is being calculated.

17. *Electrical Papers*, Vol. I, pp. 202-206.

However, there was no such thing as a formal algebra of vectors. Actually, the algebra of vectors as we know it today is precisely what Heaviside was slowly forging out of the physics of currents and magnetic fields. Therefore, the objection above may be turned on itself if one wishes to observe that historically it was the mathematical language that depended on, and grew out of the physical ideas, not vice versa. This is corroborated by the fact that only in June of 1885 did Heaviside publish a first formal summary of the basic elements of vector algebra.[19] In 1882, when he originally published the study now under consideration, he was not even certain of the symbolical convention to use, as he himself stated in a note that he appended in 1892 to the definition of curl quoted above:

> As this is the first use of vectors in this Reprint, it may be appropriately mentioned here that the algebra and analysis of vectors is introduced very gradually. At first the same type was used both for vectors and scalars, but I found later that it was a matter of some practical importance to facilitate the reading and ease the stress on the memory by employing a special type for vectors. So, the German type used by Maxwell being utterly unpractical, I introduced Clarendon type for the purpose in the *phil. mag.*, August, 1886, and later papers, and now do the same in these earlier papers to harmonize. It will be found to be a particularly suitable type, being very neat, easily read, and well adapted for use in formulae along with ordinary type, roman and italic. When only the tensor (or size) of a vector is concerned, the ordinary type is used. Thus C is the tensor of **C**.

Heaviside began his summary of the "Connected General Theorems In Electricity and Magnetism" by pointing out the adverse effect of using Cartesian components for the purpose of working out the ideas involved:

> ... the real meaning of the relations expressed by them becomes hidden away, as it were, beneath a tangled mass of x, y, z's, and can only be recognised by groping about from one equation to another, comparing them, selecting certain equations as important, rejecting others as needless, and, finally, from the few selected

18. In a sense one could date the modern use of vectors to Galileo's parallelogram of forces, which, according to some, was already used explicitly by Leonardo da Vinci (Kenneth D. Keele, q.v., "Leonardo da Vinci, Life, Scientific Methods and Anatomical Works," *Dictionary of Scientific Biography*). For all practical purposes, it represents vector addition. The development of modern vector algebra and Heaviside's role in shaping and popularizing it are discussed in Michael J. Crowe, *A History of Vector Analysis: The Evolution of the Idea of a Vectorial system*, (New York: Dover Publications, Inc., 1967).

19. *Electrical Papers*, Vol. II, pp. 3-7.

> main equations,... determining the essential nature of the relations under investigation.[20]

Enter Hamilton's quaternions. One can become totally engrossed in the explanation of quaternions. However, for the present purposes, Heaviside's amusing account is perfectly sufficient:

> 'Quaternion' was, I think, defined by an American schoolgirl to be 'an ancient religious ceremony.' This was, however, a complete mistake. The ancients ... knew not, and did not worship Quaternions. The quaternion and its laws were discovered by that extraordinary genius Sir W. Hamilton. A quaternion is neither a scalar, nor a vector, but a sort of combination of both. It has no physical representatives, but is a highly abstract mathematical concept. It is the 'operator' which turns one vector into another. It has a stretching faculty first, to make the one vector become as long as the other; and a rotating faculty, to bring the one into parallelism with the other.[21]

Within this powerful mathematical language, vectors are treated as single entities so that one need not constantly break them into their components whenever a manipulation is required. This would suggest that quaternions should have furnished the language of choice for physicists; but they did not. While conservatism may be partially responsible for this, it is by no means the sole reason, said Heaviside, because:

> Against the above stated great advantages of Quaternions has to be set the fact that the operations met with are much more difficult than the corresponding ones in the ordinary system, so that the saving of labour is, in a great measure, imaginary. There is much more thinking to be done, for the mind has to do what in scalar algebra is done almost mechanically. At the same time, when working with vectors by the scalar system, there is great advantage to be found in continually bearing in mind the fundamental ideas of the vector system. Make a compromise; look behind the easily-managed but complex scalar equations, and see the single vector one behind them, expressing the real thing.[22]

The last, innocent-looking sentence is actually a momentous declaration. It asserts that the physicist's vector must be freed of its quaternionic dependence; that one must learn to think of vectors as the fundamental concept underlying the scalar calculations involved. It essentially calls for the invention of a way of thinking about vectors tailored specifically for that purpose. Thus,

---

20. *Electrical Papers*, Vol. I, pp. 206-213.
21. *Electromagnetic Theory*, Vol. I, p. 136.
22. *Electrical Papers*, Vol. I, p. 207.

under the physically sounding title "Connected Theorems in Electricity and Magnetism," Heaviside actually introduced the first elements of a new mathematical language.

A few words are in order concerning the nature of this mathematical invention of Heaviside's. Physicists had been working with vectors for generations before Heaviside. Until Hamilton's quaternions the analysis of vector relations was inseparably linked to a particular choice of coordinate system—Cartesian, cylindrical, and spherical coordinates being the most commonly used. In Hamilton's formulation vectors were treated as single entities, but Hamilton's algebra was an algebra of quaternions, not of vectors, so that the vector could be reached only through the quaternion. This gave rise to results that were often baffling when one wished to interpret the quaternionic vector as representing a directed length in physical space. One such case that Heaviside often pointed to is that in quaternionic algebra the square of a vector always turns out negative.[23] Regardless of all these inconveniences, it should be clear by now that Heaviside did not "discover" or invent the vector, nor did he develop the first mathematical formalism capable of handling vectors. His contribution was of a different nature. He developed what might be called a natural language for physical vectors: a formal algebra of vectors whose rules mimic the physical laws that describe the behavior of the physical entities that Heaviside wished to represent as vectors.

4.4 *Stokes's Theorem: From the Physics of Currents and Fields to the Mathematics of Vectors*

An attentive reader may have noted that while Heaviside's definition of the curl (see page 69 above) is crystal clear, it may turn out to be plain fantasy. After all, who is to say that for every vector field **B**, even if it is "well-behaved," there actually exists a "flow" field **C** such that the line integral of **B** around a closed curve is equal to the integral of **C** over the surface bounded by the curve? Clearly, a proof is wanting that the line-integral of **B** around a closed

23. "A vector is not a quaternion; it never was, and never will be, and its square is not negative; the supposed proofs are perfectly rotten at the core. Vector analysis should have a purely vectorial basis, and the quaternion will then, if wanted at all, merely come in as an occasional auxiliary, as a special kind of operator." (*Electromagnetic Theory*, Vol. III, p. 508).

curve can indeed be replaced by a surface integral of some **C** over any surface bounded by the curve.

There can be little doubt that Heaviside's style of introducing the physical and mathematical concepts discussed thus far is quite novel compared with Maxwell's, or for that matter, most physicists of the time. In many ways, it is also quite different and often refreshing when compared with modern texts on electricity and magnetism. The reader may therefore find Heaviside's proofs of Gauss's and Stokes's theorems almost anticlimactic, as they follow the lines used in the majority of modern textbooks. However, the proofs of Gauss's and Stokes's theorems in Thomson and Tait's classic text as well as in Maxwell's *Treatise on Electricity and Magnetism* are quite different from Heaviside's.[24] The apparent conventionality of Heaviside's proofs is an illusion created because modern texts usually choose a path similar to his. At the time of publication his discussion was by no means standard.

Heaviside's proof of Stokes's theorem deserves a close look. It crisply manifests the difference between a general vector property and its expression in terms of a particular coordinate system. This has the effect of dividing the discussion into two distinct parts, the first being the proof, while the second is actually an argument designed to arrive at an explicit, coordinate-dependent expression for the curl. This is equally manifested in Heaviside's proof of Gauss's theorem. But we shall look at his proof of Stokes's theorem, if only

24. For a decidedly different, and rather beautiful proof of Gauss's theorem, see Maxwell, *A Treatise on Electricity and Magnetism*, (Oxford: At the Clarendon Press, 1873), pp. 19-22. Note that Maxwell does not use the term "convergence" (the negative of divergence) once during the proof. He defers discussing the physical significance of the theorem to p. 30, and only there does he call $dX/dx + dY/dy + dZ/dz$ the "convergence" of the vector (X,Y,Z) for the first time. At this stage, however, the discussion is already firmly grounded in the quaternionic operation of the vector operator $(d/dx, d/dy, d/dz)$ on a vector function. This operation has a scaler component and a vector component, the former being the convergence, and the latter being the curl. Thomson and Tait's proof of Stokes's theorem (*Treatise on Natural Philosophy*, Vol. I, Part I, [Cambridge: At the University Press, 1879], pp. 143-144) as well as Maxwell's (J.C. Maxwell, *A Treatise on Electricity and Magnetism*, Vol. I, pp. 25-27) proceeds from the given expression of the curl in Cartesian coordinates, although the stages of the proof in Thomson and Tait differ from Maxwell's. A slightly less formal approach may be found in Kelvin's 1869 paper, "On Vortex Motion" (*Mathematical and Physical Papers*, Vol. IV, [Cambridge: At the University Press, 1910], pp. 51-55). This proof actually uses the "trick" of subdividing the original curve into a connected grid and noting that the sum of circulations around all the grid elements is equal to the circulation round the original curve. However, here too, the entire discussion is couched in the Cartesian formulation of the ideas involved and requires an explicit expression of the curl before completion of the proof.

because it is the one relevant to Heaviside's main topic, namely, the connection between currents and magnetic fields.

Naturally, before embarking upon the business of proving the theorems, Heaviside could not resist making one more remark about the useless complexity of "analytical proofs":

> Analytical proofs of [these theorems] may be, and are given, which are only to be followed with some difficulty, especially as regards Theorem (B) Stokes's theorem] below; one I have seen in a German work of Theorem (A) [Gauss's], which is comparatively simple, was about six pages long. This is by no means necessary, for without losing the character of exactness, the demonstrations may be given in words; although in the final expression of the theorems symbols are desirable.[25]

Despite the sarcasm, the last sentence is significant. It underlines Heaviside's opinion that the meaning of the theorems can and should be distinguished from the formal rules of manipulating the particular language in which they are expressed. The comment suggests that if this meaning is clearly grasped, it provides the surest guide to the manipulation of the symbols if and when necessary. Heaviside never tired of repeating this point. Just three weeks after he published these words, he wrote again in part IV of this series:

> ...these theorems are of such a nature that they may be reasoned out mentally, without symbolical aid save for final expression; and, in fact, in thinking them out in this way we may obtain a far clearer conception of what the theorems really mean than by working them out analytically, blindfolding the mind, so to speak, and working mechanically.[26]

In Heaviside's proofs, only a most elementary understanding of the derivative as the instantaneous rate of change of a function is required. Integration, as before, is still a vague notion of summation, without any explicitly stated formal properties. As Heaviside himself says:

> A sound knowledge of the fundamental *principles* of the Calculus is desirable, but ... very little of the *practice* thereof, for differentiations and integrations are usually merely indicated, not performed.[27]

To follow Heaviside's discussion, imagine an arbitrary closed curve immersed in the field **B**. Divide the curve into two parts by drawing a line across

25. *Electrical Papers*, Vol. I, p. 208.

26. *Ibid.*, p. 214.

27. *Ibid.*

it (not necessarily straight) which connects any two points on the curve. This produces two closed curves. Now perform the line-integral of **B** along each of the two, taking care to move in a consistent direction along both, say, counter-clockwise. Note that in so doing, we traverse the dividing line twice, but in opposite directions. This means that upon adding the two line integrals, the two contributions of the dividing line will exactly annul each other, leaving only the integral along the original bounding curve. In other words, the sum of the line integrals around the two subdivisions is equal to the line integral around the original curve. Each of the subdivisions may be subdivided in turn into many smaller sections, and the above reasoning ensures that the sum of all the line integrals around all of the little closed paths is equal to the line integral around the original curve. In this manner the curve may be divided indefinitely, such that every infinitesimal surface inside it is associated with the well-defined value of the line integral of **B** around its tiny boundary. We now have a dense grid which may be envisioned as lying on any continuous surface bounded by the original curve. With each infinitesimal element of the surface Heaviside associated a specific number, say, N, such that its product with the element's area is equal to the line integral of **B** around the element's infinitesimal boundary. The sum of all these area products is identical to the sum of all the line integrals around the corresponding infinitesimal loops. By the argument above, the second sum is equal to the integral of **B** around the original closed curve. Therefore, the line integral of **B** around the original closed curve is also equal to the integral of a variable quantity N over the entire surface bounded by the curve. It has now been shown that if a vector function **B** is well defined throughout a given bounded surface, then the line integral of **B** around the boundary may be replaced by a surface integral over the bounded surface.[28] One can see, therefore, why Heaviside opened the discussion saying: "The proof that the surface-integral is always possible is easy." To fully appreciate the point, compare Heaviside's argument to Maxwell's. Maxwell's proof depends on the rigorous parametrization of a general surface, and proceeds in the reverse direction, namely, given an explicit expression for curl(**B**), it is proved that the surface integral of this expression is equal to the line integral of **B** around the boundary of the surface. The proof involves integration by parts, and one shudders at the thought of producing it in spherical coordinates rather than in Cartesian coordinates as Maxwell had done. Naturally,

28. *Electrical Papers*, Vol. I, p. 211.

such an exercise is quite unnecessary since, as Heaviside's proof clearly shows, the property proved is independent of the coordinate system. Furthermore, by requiring an explicit expression of the curl in some coordinate system, Maxwell's proof actually obscures the independence of Stokes's theorem from the choice of coordinates.[29]

We can easily extend Heaviside's argument a step further. Note that many different kinds of subdivisions are possible, making the individual *N*'s, namely, the contributions of the individual surface elements, differ from subdivision to subdivision. Different subdivisions may create completely different grid surfaces that share only the requirement that they all must have the same bounding curve. However, the sum of all of the *N*'s over any particular surface is a constant because the preceding argument ensures that the sum is always equal to the line integral of **B** around the common boundary. Now, this is precisely what one would expect of a non-divergent flow field running through the curve. Because of the non-divergence, the same amount of flow would have to go through any surface bounded by the curve. To see this, construct a closed volume by drawing two different surfaces both bounded by the curve. If the flow is non-divergent, as assumed, then whatever quantity of it that enters the volume through one surface without exiting through it, must exit through the other surface. By the same token, any quantity of flow leaving the second surface without entering through it, must have entered the enclosed volume through the first surface. Therefore, the net amount of flow going through any surface bounded by the curve must be a constant. This suggests that it is always possible to interpret *N* as the amount of flow per unit area due to a non-divergent flow field **C**. This flow field fulfills all the requirements of curl(**B**) as defined on page 69 above, which demonstrates that the definition indeed makes sense.

Heaviside's proof that the passage from a line to a surface integral is always possible also points the way to the actual calculation of curl(**B**). According to the reasoning, curl(**B**) should define a vector field such that the total "flow" it produces through a given infinitesimal area is identical to the line integral of **B** around the infinitesimal area's boundary as explained above. Thus, to fully specify curl(**B**) Heaviside orients three infinitesimal areas around a specific point in space, and "measures" the amount of "flow" that curl(**B**) pro-

29. J.C. Maxwell, *A Treatise on Electricity and Magnetism*, Vol. I, [Oxford: At the Clarendon Press, 1873], pp. 25-27.

duces through each of them. This defines the three space components necessary for a complete specification of curl(**B**). Of course, no real measurement takes place. One actually calculates the flow through each loop by integrating **B** once around its circumference, because that is in fact the flow strength's definition. It is not difficult to show that in Cartesian coordinates, curl(**B**) is:

$$\mathbf{F} = \operatorname{curl}\mathbf{B} = \left(\frac{\partial \mathbf{B}_z}{\partial y} - \frac{\partial \mathbf{B}_y}{\partial z}, \frac{\partial \mathbf{B}_x}{\partial z} - \frac{\partial \mathbf{B}_z}{\partial x}, \frac{\partial \mathbf{B}_y}{\partial x} - \frac{\partial \mathbf{B}_x}{\partial y}\right).$$

A straightforward calculation using the Cartesian expression of the divergence, namely:

$$\operatorname{div}\mathbf{F} = \frac{\partial \mathbf{F}_x}{\partial x} + \frac{\partial \mathbf{F}_y}{\partial y} + \frac{\partial \mathbf{F}_z}{\partial z},$$

will show that as expected, div[curl(**B**)]=0, or that curl(**B**) is non-divergent.

This, then, is the manner in which Heaviside introduced Stokes's theorem (sometimes referred to as the theorem of version).[30] In a similar way he proved Gauss's theorem and obtained the Cartesian expression of the divergence. Throughout the discussion, a subtle shift has occurred. The analysis began as a study of concrete physical concepts, namely, the magnetic field and the electric current envisioned as flow fields. Accordingly, the proof of Gauss's and Stokes's theorems relies heavily on the visualization of flow fields, not merely as a way of rendering the geometry tangible, but for giving meaning to the concepts of curl and divergence. Still, by the time the proofs have been completed, and the expressions for divergence and curl as mathematical operations fully furnished, something that transcends physical flow patterns has been distilled out of the physics and made independent of it. The theorems of Stokes and Gauss describe mathematical properties of the mathematical "creatures" called vectors. One need not always think of vectors as representing anything physical. It suffices to think of them as mathematical entities that satisfy Stokes's and Gauss's theorems in addition to some rules of combination.

---

30. Maxwell noted that, "This theorem was given by Professor Stokes, 'Smith's Prize Examination,' 1854, question 8. It is proved in Thomson and Tait's Natural Philosophy, § 190(j)." (J.C. Maxwell, *A Treatise on Electricity and Magnetism*, Vol. I, [Oxford: At the Clarendon Press, 1873], p. 27 (note).

Thus, Heaviside began with observations concerning the field associated with a straight current. In his next step he generalized these observations into Ampère's law, assumed to be the natural law that describes the local relationship between steady magnetic fields and electrical currents. He then proceeded to create a mathematical language whose rules are a direct extension of this law and of Gauss's law, which associates static charge with electrostatic field. Nothing in all of this ensures that the mathematical relations thus formulated indeed correspond to the physical situation under investigation. Nevertheless, something very important has been achieved: one can now proceed to investigate this hypothetical system in terms of its own internal logic, and see whether it ends up making solid contact with established electromagnetic knowledge. Furthermore, the investigation does not have to be carried out by abstruse symbolical manipulations. Regardless of whether the system will coincide with other known phenomena of electric currents and magnetic fields, it is capable of being understood in terms of visual flow patterns. It would be gratifying to discover that these patterns faithfully represent currents and fields, but even if they do not, one may continue to think in terms of them, and manipulate them visually in one's imagination, rather than work out their algebraic representation. This is explicitly stated by Heaviside in the form of a concrete example:

> If we have any vector or directed quantity, **R**, components X, Y, Z, and we say that $\frac{dX}{dx} + \frac{dY}{dy} + \frac{dZ}{dz} = 0$ or that it has no convergence, we simply imply that the distribution of **R** is similar to that of velocity in a possible case of motion of an incompressible fluid, so that we may, without going wrong, guide ourselves by the material analogy.[31]

This is the deep significance of Heaviside's insistence that the meaning of the theorems is much better grasped when stripped of their symbolical garb.

It seems one could argue back and forth and never arrive at a final decision as to whether Heaviside produced physics or mathematics in these pages. In view of his mode of reasoning, the distinction loses much of its significance. In many ways, his work here is worthy of any abstract mathematician. If abstract mathematics is the study of the relationships that define mathematical entities, be they groups, rings or abstract vector spaces, then what Heaviside did here is abstract mathematics. He gave meaning to the term "vector" by

31. *Electrical Papers*, Vol. I, p. 215. (Note how the word "vector" is used to imply a field).

elucidating some of the formal rules vectors satisfy. However, he extracted these rules from observations on the relationships between electric currents and magnetic fields. Throughout his work he emphasized physical meanings that stand beyond the formalities, grounded in a visual understanding of vectors as flow fields in space. In Heaviside's presentation, the visual understanding precedes the formalism and determines its particulars. This reduces the problem to a hopeless case of the chicken and the egg, for the debate might proceed thus: visualization of flows and their interrelationship preceded the formalism; no, the formalism must have already been there, albeit implicitly, or else Heaviside could not have conceived of the so-called relationships in the first place; no, it would have been impossible to conceive of relationships before noting, in the first place, something that can behave according to relationships; etc., etc. Indeed, the very subtitles of the six sections under the heading, "The Relations Between Magnetic Force and Electric Current," indicate the difficulty of deciding just where the physics ends and the mathematics begins. They are as follows:

Section 1: The Universal Relations between a Vector and its Curl.
Section 2: The Potential of Scalars and Vectors.
Section 3: Connected General Theorems in Electricity and Magnetism.
Section 4: The Characteristic Equation of Potential and its Solution.
Section 5: Relations of Curl and Potential, direct and inverse. Scalar Potential of a Vector.
Section 6: Magnetic Force of Return Current through the Earth, and Allied Matter.

Section 1 begins with Ampère's law and leads to the definition of the curl. Physics or mathematics? Section 3 promises a discussion of theorems in electricity and magnetism, but delivers the proofs of Gauss's and Stokes's theorems. Physics or mathematics? In section 2, the scalar potential and the vector potential are denoted by pot(P) or pot($\mathbf{A}$) and treated as mathematical operations no different than div and curl. Section 5, under the heading, "Scalar Potential of a Vector," a mathematical title for all intents and purposes considering the contents of Section 2, introduces the climactic point of the whole discussion, namely, the reducibility of magnetism to electrical currents. Mathematics or physics?

The issue need not be settled one way or the other to examine how Heaviside continued the analysis now that he had established its rules of reasoning as the rules that describe the properties of vector fields. It will soon become evident, however, that as far as Heaviside himself was concerned even the purest of mathematics was "all experience, after all."

4.5 The Importance of Keeping the Vector in Mind: The Case of the Earth's Return Current and the Essence of Mathematical Manipulation

With the theorems of Stokes and Gauss and the Cartesian expression for the curl and the divergence, Heaviside could have continued his study of currents and fields using the differential and integral calculus. He chose not to. He must have felt that one more example was necessary; one that would clarify the importance of understanding the meaning of curl and divergence. This example contains one of the most fascinating arguments he constructed in this part of the *Electrical Papers*, and well worth following in detail not only for its own beauty, but because it provides a clear insight into Heaviside's mode of mathematical reasoning. The striking approach Heaviside took to the problem is obscured, to some degree, by the manner in which he elected to present the solution. To bring it out, it seems better to outline it differently, beginning in the middle of his analysis, and tying up its ends later on.

The problem appears simple enough to begin with: Let there be a uniform, straight conducting wire of finite length, carrying a constant current C and lying parallel to the ground. Further, let the circuit be closed through the ground, that is to say, ground the two ends of the wire and allow the current to complete its circulation through the earth. What will the magnetic field of this arrangement be like?

The relationship between the current, its magnetic field and vector potential immediately suggests two ways of solving the problem, namely, using either the relationship $\mathbf{B} = 4\pi \text{pot}(\text{curl}(\mathbf{C}))$, or $\mathbf{B} = 4\pi \text{curl}(\text{pot}(\mathbf{C}))$. However, once one tries to translate these relationships into explicit Cartesian differentiations and integrations, some nasty technical problems emerge. The first case involves taking the curl of a current along a line. This gives impossible results if taken literally. The conductor must have a finite diameter, then we take the curl, calculate its vector potential, and find the desired answer by allowing the diameter to approach zero. The second relationship offers no great improvement. It leads to an integral of the form:

$$\int \sqrt{x^2 + y^2}\, dx,$$

which may compel some readers to resort to a table of integrals; and once that obstacle is surmounted, the curl of the result must be taken, and that involves a fair amount of algebra as well.

There exists, however, a third choice. The first thing to do (beginning in the middle of Heaviside's argument), is to think about the return current. The earth is assumed to be a homogenous conductor, which occupies half of space. At one end of the wire, say on the right, current flows into the ground, and the assumed homogeneity of the earth dictates that it must spread evenly from the grounding point in all directions through the earth. At the other grounding point, an equal amount of current flows from the earth into the wire, and once again, it must flow equally from all directions.

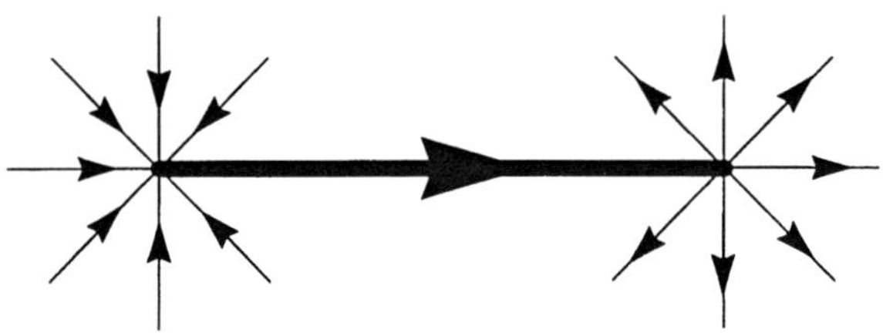

Figure 3.3: View from above of the flow into and out of a conductor grounded at both ends.

For all intents and purposes, the first grounding point is a source that "pours" current into the earth, while the second is a sink into which current is being "drained" from the earth (see figure 3.3). Concentrate on the source first. Since the current in the conductor is C, the source pours that amount constantly into the earth. This means that a total current C must flow through the semi-spherical surface of any inverted dome whose base is a circle on the earth's surface, centered on the source. Let the dome's radius be R. The surface of a half sphere being $2\pi R^2$, the density of the earth current due to the source at distance R from it must be:

$$\mathbf{C}_1 = \frac{C}{2\pi R^2}\mathbf{r},$$

where R is the distance from the source to the point of measurement and $\mathbf{r}$ is a unit vector in the direction of R. Similarly, the density of the earth current due to the sink is:

$$\mathbf{C}_2 = -\frac{C}{2\pi D^2}\mathbf{d},$$

where D is the distance from the sink to the point of measurement, and $\mathbf{d}$ is a unit vector in the direction of D. The total flow at any point is obviously the sum of the two.

Note that save for the factor of 2π these relationships are exactly analogous to the electrostatic fields around a positive and a negative charge. If instead of a normal earth, we allowed the return conductor to completely surround the wire (still making contact with it only at the wire's ends), then the flow pattern everywhere except inside the wire would look exactly like the field lines created by a positive charge and a negative charge separated by a finite distance. Since now the return flow goes through all of space instead of half of it, its density would be $C/(4\pi R^2)$. Clearly, the line along which the conducting wire lies forms the axis of symmetry for this flow system because it coincides with the line connecting the sink and source. This implies that the magnetic field, whatever its specific form turns out to be, will also have the conducting wire as its axis of symmetry.

These observations together with Ampère's law suffice for a complete determination of the field. Ampère's law states that the line integral of the magnetic field around any closed curve is 4π times the total current that passes through the curve. Now, the magnetic field must have no divergence, because by hypothesis it is the curl of the vector potential, and we already know that the divergence of a curl is always zero (see page 92). This implies that the magnetic field's lines must all be closed, and the axial symmetry of the particular problem at hand further suggests that these lines must be circles. So we may as well perform the integration of $\mathbf{B}$ along some circle centered on the axis of symmetry. If the radius of such a circle is h, then the line integral of $\mathbf{B}$ once around is simply $2\pi hB$. This quantity must be equal to 4π times the total current that passes through the circle. As we have already seen, this current may be broken down into two components, one due to a source of strength C, situated R_1 away from the perimeter of the circle, the other a sink of strength $-C$ situated R_2 away from the circle. One may, of course, try to calculate the current due to each of these by integrating the respective densities through a plane circle of radius h situated at the proper distance from the source and the sink. However, a little reflection about the geometry of the situation will again simplify things considerably.

Ampère's law does not specify the particular area over which the current density is to be integrated. In fact, any area bounded by the circle will do, because the law ensures that for all such areas the surface integral of the current will be the same. We already know that the current density of the source is the same all over a sphere of radius R_1 centered on the source. Consider then, the spherical section cut out by the circle of radius h (see figure 3.4).

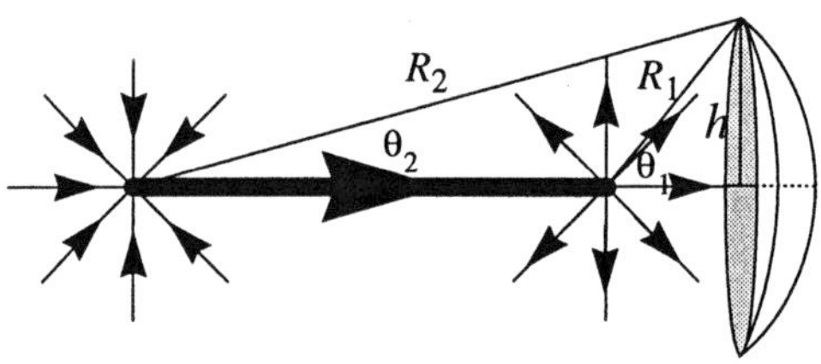

Figure 3.4: Current through a circle of radius h from the sourcing and sinking ends of a grounded wire.

The same amount of current necessarily passes through the spherical section as through the flat circular surface bounded by the same circle, but the current density throughout the spherical surface is a constant. This transforms the surface integration into a simple product of a constant current density and the area S of a dome centered on the axis of symmetry and bounded by an angle θ_1 such that $\sin\theta_1 = h/R_1$. We can calculate the area of such a dome cut out of a sphere of radius R as follows: Consider a radius from the sphere's center, cutting the dome somewhere on its surface and making an angle θ with the axis of symmetry at the source (the axis of symmetry is the line connecting the source and sink points). Move the radius once around the symmetry axis without changing the angle θ. This will describe a circle of radius $h = R\sin\theta$ on the surface of the sphere. Now increase θ by an infinitesimal $d\theta$, and draw another circle in the same manner. Between them the two circles define a band on the sphere. This band has an infinitesimal width $Rd\theta$ and infinitesimal area:

$$da = 2\pi R^2 \sin\theta \cdot d\theta .$$

To find the area of a dome bounded by a central angle θ_1 we only need to integrate the infinitesimal area for θ going from zero to θ_1. The result is:

$$S = 2\pi R^2 (1 - \cos\theta_1) .$$

The total current due to the source through this dome is therefore:

$$C_1 = \frac{C}{4\pi R_1^2} 2\pi R_1^2 (1 - \cos\theta_1) = \frac{C}{2}(1 - \cos\theta_1).$$

Using the same argument, the current through the circle flowing toward the sink turns out to be:

$$C_2 = -\frac{C}{2}(1 - \cos\theta_2),$$

where θ_2 is measured from the sink. Consequently, the total current through the surface bounded by the circle is:

$$C_s = \frac{C}{2} (\cos\theta_2 - \cos\theta_1) .$$

Finally, using Ampère's law, the magnetic field comes out:

$$B = \frac{C}{h} (\cos\theta_2 - \cos\theta_1) . \tag{3-2}$$

It is directed in right handed circles around the axis of symmetry (the extension of the wire). Note that this is the field of the total system, namely, the wire current together with the return current.[32] One may become convinced of this by noting what happens if the wire is extended to infinity at both ends. In this

32. This is actually an important point. The only reason this discussion is possible is that the entire current system is constantly taken into consideration. Indeed, one should be able to convince oneself that a true radial current, such as the one resulting from an isolated source or isolated sink, cannot give rise to a magnetic field, because its curl is everywhere zero, and the field—being the vector potential of the current's curl—must therefore be zero as well.

case, the angles θ_1 and θ_2 from the wire ends to the point where the field is to be determined become 180° and 0° respectively, and B becomes $2C/h$, the familiar magnetic field around an infinite, straight conductor.

The argument is undoubtedly elegant and simple. Instead of the integrations and algebraic manipulations of the curl-pot method, the answer is derived using an argument in which the most complicated calculation involves the integration of $\sin\theta$. But one might ask why bother with the exercise in the first place. After all, the problem asks for the field of a return current that goes only through the earth, not through a return conductor that completely surrounds the wire. That is true enough, but then the required current can be "manufactured" by adding to the one just analyzed two more current systems, one centered on the source point of the wire, the other on its sink point.

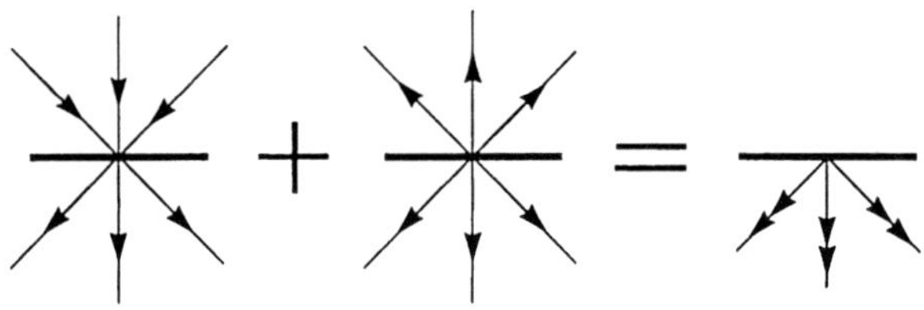

Figure 3.5: Composition of two vector fields

Consider the source once again (figure 3.5). We need an earth current originating from it, such that its density is $C/(2\pi R^2)$. This can be created by adding the current distribution of the source to a second current distribution whose density is $-C/(4\pi R^2)$ above the ground, and $+C/(4\pi R^2)$ under the ground. In other words, the new imaginary current flows radially into the source point above the ground, and radially away from it under the ground. Obviously, above the ground the two distributions (disregarding the wire current) will exactly cancel one another, while underneath they will add arithmetically to yield $C/(2\pi R^2)$, exactly as required. We already know the field due to the first distribution, so it remains to find the field of the second, but then the preceding discussion clearly points the way to the solution. The new current distribution has an axis of symmetry perpendicular to the earth through the source. Let φ be the angle from the new axis to some point underneath the surface. Then by the same argument that we used above the total current out of the dome subtended by φ is:

$$C' = \frac{C}{2}(1 - \cos\varphi)$$

$$\Rightarrow B' = \frac{C}{r}(1 - \cos\varphi),$$

where r is the distance from the perpendicular axis of symmetry to the point at which B' is measured. Above the earth, the situation is slightly different, since here the current's direction no longer corresponds to that of a source but to that of a sink. This means that we must apply to that part the argument that we applied to the sinking part of the previous case, and measure φ from the upper part of the axis of symmetry. The result may be expressed by a single formula once we realize that instead of measuring φ in two different ways we may measure it from the downward direction as we did in the first part, but make certain to take its absolute value (because $\cos(180 - \varphi) = -\cos\varphi$). Hence the magnetic field of the second current distribution lies in circles parallel to the ground, and its intensity is:

$$B' = \frac{C}{r}(1 - |\cos\varphi|). \tag{3-3}$$

To complete the picture we would have to do the same at the wire's sink end, but if we are interested only in the magnetic field near the source,[33] the influence of the sink end can be ignored. There still remains the contribution given by eq. (3-2), which, under the assumption that the sink's influence is negligible, becomes:

$$B = \frac{C}{h}(1 - \cos\theta_1). \tag{3-4}$$

It lies in circles perpendicular to the ground. The magnetic field in the vicinity of the source is the vector sum of **B**' and **B** whose intensities are given by eqs. (3-3) and (3-4) respectively.

There is much to be learned from this example about Heaviside's style of reasoning. J.L.B. Cooper observed of Heaviside that:

33. "Near" means that the distance from the source to the place where the field is reckoned must be much smaller than the total length of the wire.

> As a mathematician he was gifted with great manipulative skills and with a genius for finding convenient methods of calculation.[34]

So far as it goes, this observation is quite accurate. However, it obscures the nature of Heaviside's skill by emphasizing his ability to devise methods of calculation. It creates the impression of one blessed with an extraordinary knack for manipulating mathematical symbols, like the rhyme-maker who has a special gift of playing formal games with English, creating striking combinations by juggling the syntax mechanically. This, however, is not at all how Heaviside worked, and his skill was of a very different kind. He managed to produce extraordinary manipulations because he constantly strove to keep meanings clearly in mind, as he explained in the following remark concerning Stokes's theorem:

> Whilst the expressions for the components vary according to the system of coordinates chosen as most suitable for a special problem, the theorem, on the other hand, is universal, and gives us the inner meaning of the operation. It is far the best in general investigations not to employ any system of coordinates, but to emancipate one's self from their complexity by employing symbols which only relate to the intrinsic meaning of the operations; besides which, there is a great gain in the ease of manipulation.[35]

This attitude is evident already in his early work, where he remarked on a particularly involved manipulation of expansion series pertaining to a transmission line problem:

> ... in working out practical problems requiring arbitrary functions to satisfy certain conditions when expanded in a harmonic series, the physical nature of a particular problem will usually suggest, step by step, the necessary procedure to render the solution complete....[36]

Whenever possible, Heaviside tried to manipulate the physical ideas directly, and the example above is only one striking case in point. In this manner, by manipulating the objects of analysis directly, rather than the symbols of the differential calculus, he was able to find the often ingenious shortcuts that made calculation easy.

---

34. J.L.B. Cooper, "Heaviside and the Operational Calculus," *The Mathematical Gazette*, **36** (1952): 12.

35. *Electrical Papers*, Vol. I, p. 443.

36. *Ibid.*, p. 94.

The primary importance of mathematics for Heaviside did not reside in the ability to do numerical calculations. At the very beginning of his introduction to field thinking for the non-mathematician, Heaviside suggested that "mathematical reasoning does not necessarily involve any calculating in the usual sense" (see page 66). He seems to have perceived mathematics as a form of reasoning that involves, at least in principle, something quite different from the formal rules of symbolical manipulation—of which calculation is a special case. Vectors are not numbers. That with some clever definitions they may be represented by numbers and the rules of numerical combinations, is really an added "bonus," so to speak. This suggests that quantification and the precision that stems therefrom did not summarize the advantages of mathematical thinking for Heaviside. Among the books in the portion of his library preserved in the Institution of Electrical Engineers in London is an English translation of Felix Klein's *Lectures On Mathematics*. In one of his essays, Klein wrote:

> I believe that the more or less close relation of any applied science to mathematics might be characterized by the degree of exactness attained, or attainable, in its numerical results. Indeed, a rough classification of these sciences could be based simply on the number of significant figures averaged in each.[37]

On the margin of his copy of the book, just opposite Klein's observation, Heaviside wrote:

> A wrong idea. It is not just for numerical results that mathematics is applied to physics, but to theorize, and discover new relations.

Heaviside's point receives further clarification with Klein's next observation:

> It follows that a large portion of abstract mathematics remains without finding any practical application, the amount of mathematics that can be usefully employed in any science being in proportion to the degree of accuracy attained in the science.[38]

To which Heaviside replied in the margin:

> This is most certainly erroneous. Nearly all mathematics of physics goes far beyond the limit of exactness.

37. Felix Klein, "On the Mathematical Character of Space Intuition and the Relation of Pure Mathematics to the Applied Sciences," in *Lectures On Mathematics*, (New York: McMillan and Co., 1894), p. 46.

38. *Ibid.*, p. 47.

Heaviside did not argue with Klein's observation that the physicist finds no use for many branches of mathematies; but he rejected the reasoning behind it. Physical measurement never attains the exactness of a numerical calculation, Heaviside observed, yet this never prevented the application of mathematical reasoning to problems of physics. This, Heaviside would probably have said, is because it is the reasoning itself that the physicist wants, not the numerical calculation. If, as Klein noted, there are branches of mathematics without application to physics, the reason is not necessarily the lack of numerical precision inherent to physics; it is rather the failure of these mathematical branches to provide a useful reasoning tool for the particular problem in question. By the same token, one may simplify the route to numerical calculation, whenever possible, by keeping in mind the relationships between the physical concepts involved in a particular problem. To a very large extent, this is the big secret behind Heaviside's extraordinary manipulative skills. His key to mathematical reasoning was formula interpretation rather than formula manipulation.

If the analysis of the magnetic field of a return current through the earth highlights Heaviside's mode of mathematical analysis, then it also reveals the difficulty of following his arguments. As already mentioned, Heaviside did not present his solution in the manner described above. Instead, he began by asserting without proof that eq. (3-2) above describes the magnetic field of an insulated straight wire carrying a current *C* and fixed within an infinite conductor with which it makes contact only at the ends. He then used the Cartesian expression of the curl to demonstrate that the return current by itself, if taken to be identical to the electrostatic field created by a positive and negative charge, has no curl. Since the magnetic field is the vector potential of the curl of the current, this means that the return current by itself contributes nothing to the magnetic field. This observation is important for many later applications, where it must be kept in mind that a purely radial vector field has no curl. Returning now to the original problem and realizing that a radial current has no curl, Heaviside concluded that the earth current's curl must be confined to the earth's surface. Using considerations similar to the ones he used to evaluate the curl of a cylindrical current distribution (see pp. 77-79), Heaviside showed that the curl of the earth current lies in concentric circles around the grounding point. Being the vector potential of the curl of the current, the magnetic field of the earth current turns out to be the sum of the vector potentials of two surface circulations, each centered on an axis perpendicular to the earth's surface at one of the grounding points.

Only after this stage did Heaviside observe that the required earth current can be produced by the addition of a second current system as described above, and use the preceding analysis to calculate its magnetic field. At this point the light of comprehension would probably strike the reader, who may now proceed to use this argument to derive eq. (3-2), which was taken on faith up to this moment. Heaviside must have been aware of the unusual nature of his visual argument:

> As the mode of obtaining equation [(3-3) above] may be considered obscure, it is desirable to check it by direct integration.[39]

He proceeded to derive the field of the second current distribution by calculating the vector potential of the circular curl obtained from the earth-current, noting that this curl is the same as that of the second current system. In the process of calculating this vector potential Heaviside actually made a small error, using a sine where a cosine should have been used. However, this error does not show up in the final answer, because the integral involved happens not to distinguish between $\sin\theta$ and $\cos\theta$. The significance of this error is not that Heaviside made it in the first place, but that he missed it later on. He certainly gave this section of his discussion much thought. This is evident because he originally published it as the fourth section of the study of currents and magnetic fields, but made it sixth and last in the *Electrical Papers*, following the two parts that establish the assumed connection between the curl and the vector potential. Despite this special attention he missed the error, but then, why second guess a tedious calculation once it has produced a result identical to the one arrived at by physical reasoning? All of this suggests that whenever he could, Heaviside reasoned out his problems physically and used rigorous methods only to verify the correctness of an already achieved result.

There are other examples of similar formal "sloppiness" in Heaviside's work. At the same time, his conclusions were rarely wrong. There seems to exist here an uncanny combination of formal carelessness with final results that are nearly always correct. This appears to defy explanation until one realizes that while Heaviside's formal mathematics was sometimes sloppy, his real mathematics, grounded in a keen understanding of physical ideas, was nearly always extraordinarily solid. Herein lies a major key to following his

39. *Electrical Papers*, Vol. I, p. 228.

often cryptic arguments. The reader that attempts to formally manipulate Heaviside's mathematical formulae will invariably encounter great difficulties in understanding his line of thought. Most such difficulties will quickly disappear once the ideas behind the formulae are clearly understood. Heaviside often manipulated ideas directly, rather than through the mathematical symbolism that he used merely to summarize final results and key points (see appendix 3.2 for example).

4.6 "To fit current and magnetic force into the system": From the Mathematics of Vectors Back to the Physics of Currents and Fields

Having established the theorems of Gauss and Stokes, Heaviside resumed his discussion of electric currents and magnetic fields:

> In sections I. and II. we have explained the nature of two operations of frequent occurrence in electromagnetism, viz, from a given vector-function to derive its "curl," and also to find its vector-potential; and some of the relations of these functions were mentioned. But that was for descriptive purposes merely; the only proof that curling the magnetic force gave rise to the current was that derived from the known magnetic force of a straight current, together with the assumption of a certain distribution of force within the current itself.
>
> We have now, to make the matter a little more complete, to prove that the properties ascribed to the vector-potential of current and its curl are really true, and to further develop them.[40]

Heaviside proceeded to do that in two stages that only need to be roughly sketched here. First, he established the relationship between the divergence of a vector field and the field's "scalar potential" along the following lines. It is a straightforward corollary of Stokes's theorem that if a vector field **R** has a curl of zero everywhere, then there exists a scalar function P such that:

$$\mathbf{R} = -\frac{\partial P}{\partial x}\mathbf{i} - \frac{\partial P}{\partial y}\mathbf{j} - \frac{\partial P}{\partial z}\mathbf{k},$$

where **i**, **j**, **k**, are unit vectors in the x, y, z directions respectively. Since this is the relationship between the electrostatic field and its potential, and since P is a scalar function, P is called a scalar potential. However, P exists for any vec-

40. *Electrical Papers*, Vol. I, pp. 213-214.

tor field of zero curl, and it need not always be associated with static electricity. Let there be then some unknown vector field **R** with known divergence σ. The question is how to find P directly from σ, or, if the relationship between **R** and P is taken into consideration, how to solve the equation:

$$\frac{\partial^2 P}{\partial x^2} + \frac{\partial^2 P}{\partial y^2} + \frac{\partial^2 P}{\partial z^2} = -\sigma.$$

Underlying this equation is the assumption that the unknown field **R** must satisfy the condition curl(**R**) = 0. With this in mind, Heaviside proceeded to solve the equation, once again without resorting to any sophisticated algebraic manipulations save for the elementary rule of differentiation:

$$\frac{d}{dx}x^n = nx^{n-1}.$$

Using Heaviside's informal summation symbol, the solution turns out to be:

$$P = A + \sum \frac{\sigma}{4\pi r}, \tag{3-5}$$

where r is the distance from the point at which σ is measured to the point where the value of P is desired, and A is an arbitrary constant. The arbitrariness of A indicates that the potential at any point may be expressed as the potential difference between it and some other point whose potential may be set equal to zero by proper choice of A.[41]

Having established that, Heaviside proceeded in the second stage to the task that he set out to accomplish at the beginning, namely, to demonstrate that if **C** = curl(curl(**A**)), then the operation of going from **C** back to **A**, denoted by pot(**C**), is $\Sigma\, \mathbf{C}/(4\pi r)$. Using the Cartesian expression for the curl of a vector that he derived in the process of proving Stokes's theorem, and adding the condition that the divergence of **A** is zero everywhere, Heaviside went on to study the relationship between respectively paired Cartesian components of **A** and **C**. It turns out that each pair must satisfy the equation of scalar potential, with the component of **A** playing the scalar potential and the component of **C** playing the divergence σ. But the solution to that equation has just been found

41. *Electrical Papers*, Vol. I, pp. 214-218.

in the form of eq. (3-5) above, and thus, compounding the three component solutions, we get $\mathbf{A} = \Sigma\, \mathbf{C}/(4\pi r)$, namely, the desired result.[42]

From this two-stage argument a very important observation emerges. Implicit in the last case is the assumption that the vector function **B** = curl(**A**) has a nonzero curl at least somewhere in space, so that **C** = curl(**B**) is not uniformly zero. This does not preclude the possibility that in certain regions of space curl(**B**) does equal zero. In all such regions, Stokes's theorem ensures that **B** can be derived from a scalar potential. With this in mind, Heaviside went on to calculate the scalar potential of a vector function **C** assumed to be confined to a single closed curve throughout which the magnitude (not the direction) of **C** is a constant *C*. Obviously, Heaviside had a current loop of strength *C* in mind, but for the moment we need not make this specific physical association. The result about to be derived holds for any vector functions **A**, **B**, **C**, satisfying **B** = curl(**A**), **C** = curl(**B**), and **A** = pot(**C**), with **C** confined to a closed curve as stated above. To fix ideas, let **C** circulate right handedly around a straight line pointing upwards along the z-axis. Now, imagine two surfaces bounded by **C**, infinitesimally distant from one another—in other words, a thin shell bounded by **C**. Coat the shell's top surface with divergent "material" of surface density $m = C/dz$, where dz is the infinitesimal distance between the surfaces, and *C* is the constant magnitude of **C**. In the same way coat the bottom surface with convergent "material" of surface density $m = -C/dz$. Note that both *m* and *dz* need not be constant over the surfaces, but their product, *mdz*, must have the constant value of *C*. From the equation of scalar potential and its solution it follows that the top surface must give rise to a vector field $\mathbf{E} = \Sigma\, m/(4\pi R^2)\, \mathbf{r}$. A field of similar form is associated with the bottom surface except that it must be given a negative polarity in accordance with the convergent nature of the surface's coating. Significantly enough, if we now add the fields due to both sides of the thin shell, the compound field will be identical to **B** above except within the space bounded by the two surfaces (For a detailed discussion, see appendix 3.2).

Now, Ampère again discovered that the field of a thin magnetic shell can be reproduced by wrapping the shell's boundary with a conducting wire and running a current $C = mdz$ through it, if *m* is the surface density of magnetic matter on the shell, and *dz* its thickness. At distances much larger than *dz* the two fields become practically identical. The vector property just proved there-

42. *Electrical Papers*, Vol. I, pp. 218-219.

fore suggests that if we associate a vector function **C** with the current, and another vector function **B** with the magnetic field, then the relationship between the field and the current is $4\pi\mathbf{C} = \text{curl}(\mathbf{B})$. The same reasoning implies that if a function **A** be defined such that $\mathbf{A} = \Sigma\,\mathbf{C}/r$, then $\mathbf{B} = \text{curl}(\mathbf{A})$.

Thus, Heaviside has gone full circle. From a single observation concerning the magnetic field around a straight wire he suggested inductively the generalization $4\pi\mathbf{C} = \text{curl}(\mathbf{B})$, where **B** and **C** are vector functions. He then proceeded to show that the scalar potential of any closed loop **C** gives rise to a **B** derivable from the scalar potential of a thin shell, coated with divergent "material" on one side, and convergent "material" on the other. If this material happens to be called "magnetism," then **B** describes a legitimate magnetic field, as verified by experiment. Therefore, adding now the further requirement that all current distributions must consist of closed flow paths, it has just been corroborated that the physical relationship between any current distribution and its associated magnetic field is accurately represented by the vector relationship $4\pi\mathbf{C} = \text{curl}(\mathbf{B})$. To put it in Heaviside's own words:

> Up to the present the vectors **A**, **B**, **C**, have been treated quite apart from whether any of them represented physical quantities (although they may all be possible cases of velocity distribution in an incompressible fluid). The properties are true whether electric current or magnetic force exist or not. But the last step made, the nature of the scalar potential whose space-variation gives **b**, brings us into contact with magnetism, and will enable us to fit current and magnetic force into the system. One experimental truth suffices. A small plane electric current and a small magnet produce exactly similar fields of magnetic force, not too near them, when we reckon the + direction of the axis of the magnet to be from its S. to its N. pole, and that of the current to be such that when we look along the axis in the + direction the current circulates right-handedly about it. ... Let the magnet and the current area be infinitely small, then we may say there is identity of magnetic force at any finite distance. ... After having thus made use of magnetism, we might dismiss it for good. In fact, we might replace every magnetized particle by its equivalent electric current, according to Ampère's theory of magnetization. This theory generally strikes one on first acquaintance as very fanciful, the idea of electric current flowing round molecules being so difficult to swallow. But it explains magnetism—one mystery takes the place of two; and since we neither know what an electric current is nor what magnetism is, it is well to abolish one of them, and there can be no question as to which it should be.[43]

Looking back at the manner in which Heaviside completed his circle of reasoning, one might feel that he was somehow privy to A.N. Whitehead's

Lowell Lectures of 1925 more than forty years before they were delivered. In part of a discussion aimed at demonstrating that no certainty about the physical world can be surmised from the certainty of mathematical reasoning, Whitehead said:

> In the pure mathematics of geometrical relationships we say that, if *any* group entities enjoy *any* relationships among its members satisfying *this* set of abstract conditions, then such-and-such additional abstract conditions must also hold for such relationships. But when we come to physical space, we say that some definitely observed group of physical entities enjoys some definitely observed relationships among its members which do satisfy this above-mentioned set of abstract geometrical conditions. We thence conclude that the additional relationships which we concluded to hold in *any* such case, must therefore hold in *this particular* case.[44]

Clearly, Heaviside constructed his 1882 development of the relationship between magnetic fields and electrical currents in conformity with this general structure of reasoning; and yet, there is a fundamental difference between his conception of mathematics and Whitehead's. Whitehead's discussion seems to assume the preexistence of the mathematical system. This was not the case in Heaviside's experience. He constructed the mathematics along with the investigation of the physics. The chicken-and-egg problem outlined above (page 94) did not disturb Heaviside. He regarded mathematics as an experimental science, and believed that it does not reside *a priori* in the human mind but that it must be induced from experience. In 1900 he wrote a short letter to *Nature* about the teaching of mathematics. He severely criticized the formal manner in which Euclid's *Elements* were being taught. Having expressed his opinion that geometry "is essentially an experimental science, like any other, and should be taught observationally, descriptively and experimentally in the first place", Heaviside concluded:

---

43. *Electrical Papers*, Vol. I, pp. 222-223. Heaviside's confident decision to give the current priority is probably traceable to Maxwell's statement concerning the identity of magnetic fields associated with currents and magnets: "The action of magnets at a distance is perfectly identical with that of electric currents. We therefore endeavor to trace both to the same cause, and since we cannot explain electric currents by means of magnets, we must adopt the other alternative, and explain magnets by means of molecular electric currents." (J.C. Maxwell, *A Treatise on Electricity and Magnetism*, [Oxford: At the Clarendon Press, 1873], Vol. II, § 637, p. 250).

44. Alfred North Whitehead, *Science and the Modern World*, (1963), p. 27.

> If I understand them rightly, it is generally believed by mathematicians that geometry is pre-existent in the human mind, and that all we do is to look at nature and observe an approximate resemblance to the properties of the ideal space. You might assert the same pre-existence of dynamics or chemistry. I think it is a complete reversal of the natural order of ideas. ... it is all experience, after all; although learned philosophers, by long, long thinking over the theory of groups and other abstruse high developments, may perhaps come to what I think is a sort of self-deception, and think that their geometry is pre-existent in themselves, whilst nature's is only a bad copy. Like the old Indian pundit, whose name was something like Bhatravistra, who, after fifty years inward contemplation, discovered God—where—it would not be polite to mention.[45]

The climactic identification of magnetism and electrical current ends the first part of Heaviside's introduction to field thinking for the intelligent and highly motivated non-mathematical electrician. Perhaps the most curious aspect of this often enchanting analysis is that none of the particular results derived therein were new at the time of publication. Still, Heaviside's treatment does contain something that was quite novel at the time. The very manner of reasoning about the physics of currents and fields in this short treatise was without precedent. For the first time in the subject's history, the physics of currents and fields was expressed in terms of the algebra of vectors. Indeed, Heaviside invented this algebra specifically for that purpose. This highlights a rather remarkable general characteristic of Heaviside's work: he often attained the height of his creative efforts not in the derivation of novel results, but in his manner of reasoning out a problem. He added nothing fundamentally new to electricity and magnetism, and by the time he finished his work, the theory he presented was still very much Maxwell's. But along with this seemingly uneventful presentation he had also introduced the algebra of vectors as well as many daring and novel uses and extensions of the operational calculus.

There is a feeling of completeness about this part of Heaviside's study that one would regret to see shattered. His ultimate goal, however, was a full analysis of electricity and magnetism, and this involves time-varying currents and fields, which the preceding study does not address at all. One merely needs to think of the phenomenon of induction to realize how little was actually accomplished in that direction. Somehow, a way had to be devised to incorporate the established basic relations into a theoretical framework that describes change

45. *Electromagnetic Theory*, Vol. III, pp. 513-514.

in electromagnetism; a way that would transform this essentially static study into a dynamic one. All time-varying electrical current systems seem to involve energy transfer. It stands to reason, therefore, to search for the missing link in the energy of the electric current. Indeed, having completed his introduction to vector fields, Heaviside turned his attention to this subject.[46]

4.7 The Energy of Two Current Loops and the Priority of Physics.

Thus far, Heaviside identified magnetism with electric currents, abolishing the concept of magnetism as an irreducible natural phenomenon. In so doing, he showed how this identity, first expressed by Ampère, follows as a natural consequence of a field-oriented view of electricity and magnetism. This does not mean, however, that in reasoning about magnetic fields one should feel compelled to discuss things exclusively in terms of currents. The idea of magnetic matter may turn out to be quite useful, as Heaviside's next argument indicates.

The problem he set himself now was to find the energy of the electric current. The association of a certain amount of energy with the electric current seems quite unavoidable, considering that it takes energy to produce it in the first place. As long as all the energy is expended on current production, the manner of production should have no effect on the current's final energy. Therefore, to calculate the energy one ought to study the current itself. All this is rather colorfully illustrated by Maxwell:

> In the case of the electric current, we find that, when the electromotive force begins to act, it does not at once produce the full current, but that the current rises gradually. What is the electromotive force doing during the time that the opposing resistance is not able to balance it? It is increasing the electric current.
>
> Now an ordinary force, acting on a body in the direction of its motion, increases its momentum, and communicates to it kinetic energy, or the power of doing work on account of its motion. ... Has the electric current, when thus produced, either momentum or kinetic energy?
>
> We have already shewn that it has something very like momentum, that it

46. This motivation for the expectation that energy considerations will provide the missing link is traceable at least in part to Maxwell's influence over Heaviside (see, in particular, J.C. Maxwell, *A Treatise on Electricity and Magnetism*, [Oxford: At the Clarendon Press, 1873], Vol. II, pp. 182-183, and 196-198).

> resists being suddenly stopped, and that it can exert, for a short time, a great electromotive force.
>
> But a conducting circuit in which a current has been set up has the power of doing work in virtue of this current, and this power cannot be said to be something very like energy, for it is really and truly energy.
>
> Thus, if the current be left to itself, it will continue to circulate till it is stopped by the resistance of the circuit. Before it is stopped, however, it will have generated a certain quantity of heat, and the amount of this heat in dynamical measure is equal to the energy originally existing in the current.[47]

Naturally, within the framework of the field view Heaviside was exposing, currents were always current distributions, characterized by a current density given throughout space. Furthermore, a cardinal requirement in Maxwell's scheme is that electrical currents never diverge, and consequently always flow in closed circuits. The current distribution can therefore be regarded as a distribution of current-carrying tubes, each tube bounded by lines of flow, each carrying a constant amount of current (which may, of course, differ from tube to tube) and each closed upon itself. The energy of such a distribution would then be the total amount of work it takes to bring together all of the current tubes that make up the complete distribution. This suggests that one should first determine the energy of two current loops in space. Once this has been done, the energy of any current distribution should be available, at least in principle, by summation. This seems to have been the likely motivation behind Heaviside's decision to begin by finding the mutual energy of two current loops.

The mutual energy could be determined directly in terms of the two current distributions, but the calculations would turn out to be quite involved. Heaviside, however, had just shown how any infinitesimal magnetic shell may be replaced by an infinitesimal current loop flowing around the shell's boundary and *vise versa*. The "trick" of subdividing a closed curve to prove Stokes's theorem may be used here once again. One may regard any current loop, irrespective of size, as the sum of its subdivisions. Each subdivision takes the form of an infinitesimal current loop carrying a current equal in strength to that of the original circuit, while the latter forms the boundary of the network of subdivisions. By the preceding discussion, every infinitesimal subdivision is equivalent to a doubly-coated magnetic shell. Let m be the surface density of

47. J.C. Maxwell, *A Treatise on Electricity and Magnetism*, (Oxford: At the Clarendon Press, 1873), Vol. II, § 551, p. 182.

magnetic "matter" over the top surface, and $-m$ the density of magnetism on the bottom surface. Requiring that $mdz = C$, with dz representing the shell's thickness and C the current in the circuit, ensures the identity of the shell's magnetic field with that of the current loop. Since the original circuit is simply the sum of its subdivisions, its magnetic field is also the sum of all of the infinitesimal magnetic shells. Hence, the entire current carrying wire may be replaced by a thin magnetic shell bounded by the outline of the original current loop, and having a magnetic dipole moment of $mdz = C$ per unit surface area. In this manner, determination of the mutual energy of two magnetic shells replaces the problem of calculating the mutual energy of two current loops.

With the scalar potential of a magnetic shell that he derived as shown in the previous discussion, Heaviside arrived at two equivalent expressions for the mutual energy M of the two loops:[48]

$M = -C \times$ total field through the loop of C due to C',

or:

$M = -C' \times$ total field through the loop of C' due to C,

where C and C' denote the strength of the currents in the two loops.[49]

Consider the first of the two. Written in the more formal language of integrals, it states:

$$M = -C\int_S \mathbf{B}' \cdot \mathbf{dS}.$$

In the above S is any surface bounded by C, and $\mathbf{dS}$ is a vector oriented in the direction of the normal to $\mathbf{dS}$ with magnitude equal to the infinitesimal area dS. If $\mathbf{A}'$ is the vector-potential of the current-loop C', then by the basic electromagnetic relations derived so far, $\mathbf{B}' = \text{curl}(\mathbf{A}')$. Substituting this in the expression for the mutual energy and using Stokes's theorem, the mutual energy turns out to be:

$$M = -C\int_S \text{curl}\,\mathbf{A}' \cdot \mathbf{dS} = -C\oint_c \mathbf{A}' \cdot \mathbf{dl}.$$

The second integral is a line integral around the boundary of S and **dl** is an infinitesimal line element of this path. The bounding curve, however, traces

48. *Electrical Papers*, Vol. I, pp. 231-235.

49. See appendix 3.3 for a more detailed discussion.

the very same path as the current-loop C so that **dl** is everywhere parallel to **C**. Therefore, we may always write:

$$\mathbf{dl} = \frac{\mathbf{C}}{C} \cdot dl.$$

Substitution of this into the last expression for the mutual energy yields:

$$M = -\oint_c \mathbf{C} \cdot \mathbf{A}' dl.$$

Now $\mathbf{A}' = \Sigma(\mathbf{C}'/r')$, where r' is the distance from each summed element of **C'** to where **A'** is reckoned. In general, the summation must be taken throughout space. However, in this particular case the current is confined to a closed curve and the summation reduces to a line integral around C'. Therefore, using the explicit expression for the vector potential, one arrives at a third expression for the mutual energy, namely:

$$M = \oint_c \oint_{c'} \frac{C'C}{r'} \mathbf{dl}' \cdot \mathbf{dl}.$$

Heaviside did not toil to arrive at this last expression only to furnish yet another mathematical statement of the mutual energy. The expression was well known as Neumann's formula (after Franz Ernst Neumann) for the mutual energy of two current loops. Therefore, with its reproduction Heaviside demonstrated the equivalence of results between his field-oriented analysis and the more traditional, action-at-a-distance approach. At the same time he used the argument to highlight an important point that he would later emphasize again and again. Neumann's formula suggests that the mutual energy is actually the sum of the mutual energies of linear current elements, and that any two such elements (of lengths dl and dl') have a mutual energy of:

$$dM = -\frac{C'C}{r'} \mathbf{dl}' \cdot \mathbf{dl}.$$

Hcavisidc, howcvcr, would havc nonc of this. To him, it madc no sense to speak of an unclosed electrical current, and he regarded the above expression as physically meaningless. Therefore:

> ... all we can say at present is that the potential energy of two closed currents is *as if* a pair of linear elements, one taken from the first circuit, the other from the

> second, possessed the mutual energy expressed in [the above formula]; for such elements could not exist alone.[50]

Throughout the discussion Heaviside was forever the physicist. He would readily reshape mathematics and, if necessary, invent new mathematical conventions to fit the physical ideas he wanted to convey. He would not allow the reverse, namely, let mathematical forms dictate new physical relations. Physics was always the master, mathematics always the slave, and mathematical expressions had to be interpreted according to the preexisting physical principles that guided the whole analysis. Even here, in one of his mathematically most innovative dissertations, he did not betray this principle of his.

4.8 The Mutual Energy of Any Two Current Distributions

From the energy of two current loops it is a short step to the energy of any two current distributions. Since all such distributions consist of closed current tubes as described above, one only needs to calculate the energy of each current tube from the first distribution due to the second, and then add them all up. If the vector potential due to the entire second distribution is $\mathbf{A}'$, then the energy of the ith tube from the first distribution is:

$$M_i = -\oint_{c_i} \mathbf{A}' \cdot \mathbf{C_i} dl.$$

However, C_i is total current through a tube of infinitesimal cross section da_i. Hence, if $\mathbf{C}$ is the current density of the first distribution, it is always perpendicular to da_i and parallel to dl_i all along the tube, for these are the properties required by definition from the current tube. Therefore, $\mathbf{C_i} = \mathbf{C}da_i$, and:

$$M_i = -\oint_{c_i} \mathbf{A}' \cdot \mathbf{C} da_i dl = -\oint_{c_i} \mathbf{A}' \cdot \mathbf{C} dv_i,$$

where dv_i is the volume of a small section of the tube C_i. The sum of the M_i's is the mutual energy of the two distributions. Such a summation actually amounts to summing $\mathbf{A}' \cdot \mathbf{C}$ over the space permeated by tubes of $\mathbf{C}$, but since in all other points $\mathbf{C} = 0$, the summation may be extended to all of space with-

50. *Electrical Papers*, Vol. I, p. 237.

out changing the final result. In fact, the elements dv_i need not be confined to sections of current tubes. They may be any arbitrary, infinitesimal volume elements, for once the entire summation has been completed, all current tubes have of necessity been closed, the sum yielding the same result as when confined to sections of current tubes. Thus, using Heaviside's informal Σ to denote the operation appropriate for the case at hand, we have his conclusion that the mutual energy of two arbitrary current distributions is:[51]

$$M = -\sum \mathbf{A}' \cdot \mathbf{C} = -\sum \mathbf{A} \cdot \mathbf{C}'. \tag{3-6}$$

4.9 The Third Expression for the Energy.

Not counting Neumann's formula, Heaviside now had two equivalent expressions for the energy of the electric current: the first in terms of a surface integral (see page 114) and the second involving a volume integral of the scalar product of the current and the vector-potential (eq. (3-6) above). For reasons that will shortly become clear (see section *4.10*), Heaviside produced yet another expression, introduced by the following comment:

> There is another very important and significant expression for the energy, which involves a transformation of an extraordinary character. Expressed generally, it amounts to this:-Let **A** be any vector function of no convergence, **B** its curl, and **C** that of **B**, then $\Sigma\mathbf{AC} = \Sigma\mathbf{B}^2$ the summations being taken throughout all space; or, which comes to the same thing, the first summation extended to all places where **C** exists, and the second over all space where **B** exists.[52]

51. The self-energy of a current distribution is no more than a particular case of the above. Let there be a current distribution **C** with vector potential **A** everywhere. Divide each and every tube of **C** into two parallel tubes, each carrying half the current. This creates two distributions, each having current density **C**/2 and vector potential **A**/2. Their mutual energy is $-1/4\Sigma\mathbf{AC}$. Now the self-energies of the two subdivisions must be added to the above. Each of them may be subdivided as before, giving each secondary subdivision an energy of $-1/16\Sigma\mathbf{AC}$ and yielding a total energy of $-(1/4 + 2/16)\Sigma\mathbf{AC}$. Further subdivision creates eight current systems, each with current **C**/8 and vector potential **A**/8, which must be reckoned in pairs. Each pair has a mutual energy of $1/64\Sigma\mathbf{AC}$, and there are four such pairs. This yields $-(1/4 + 2/16 + 4/64)\Sigma\mathbf{AC}$ for the total energy less the self-energy of each of the eight subdivisions. The pattern should be clear by now. The self-energy of a current distribution **C** is $-\Sigma\mathbf{AC}$ multiplied by the sum of an infinite geometrical series whose first member is 1/4 and whose multiplier is 1/2. The sum of the series is 1/2, and the self-energy of a current distribution is $-1/2\Sigma\mathbf{AC}$ (*Electrical Papers*, Vol. I, p. 243).

52. *Electrical Papers*, Vol. I, p. 244.

Heaviside rarely emphasized any of his results for no reason. In this case, he had a twofold motivation. The first concerns some far-reaching theoretical consequences of the identity. These will be examined in the next section. The second involves the manner in which Heaviside obtained the identity and requires a careful examination of his argument.

Heaviside derived the required identity step by step, beginning with the self-energy of a single current loop, generalizing it to the self-energy of a current distribution, and finally extending it to the mutual energy of any two current distributions.[53] His basic approach to the problem, however, is immediately applicable to the last, most general case, and it proceeds as follows. Let there be two current systems, $\mathbf{C}_1$ and $\mathbf{C}_2$, giving rise to magnetic fields $\mathbf{B}_1$ and $\mathbf{B}_2$, the fields being derivable from vector potentials $\mathbf{A}_1$ and $\mathbf{A}_2$ respectively. For clarity's sake, ignore for the moment the factor of 4π in the basic relationship between current and field, and let the two triplets above satisfy the fundamental requirements $\mathbf{B} = \text{curl}(\mathbf{A})$, and $\mathbf{C} = \text{curl}(\mathbf{B})$. According to eq. (3-6) the two systems have a mutual energy M such that:[54]

$$M = \int_{\text{all space}} \mathbf{A}_1 \cdot \mathbf{C}_2 dv.$$

Therefore, the energy of a single current tube $\mathbf{C}_{2,1}$ of $\mathbf{C}_2$ due to all of $\mathbf{C}_1$ is:

$$M_1 = \int_{\text{all space}} \mathbf{A}_1 \cdot \mathbf{C}_{2,1} dv.$$

Since $\mathbf{C}_{2,1}$ is zero everywhere except within the space enclosed by the tube, the integration reduces to a summation over the tube's volume. Considering

53. *Electrical Papers*, Vol. I, pp. 244-246.

54. Note the omission of the (-) sign. The conventions used in energy expressions require this sign, and whenever he dealt with actual physical systems, Heaviside made certain it appeared. The property he was about to prove here, however, is quite independent of the negative sign, and requires only that **A**, **B**, and **C** satisfy the vector relations $\mathbf{B} = \text{curl}(\mathbf{A})$ and $\mathbf{C} = \text{curl}(\mathbf{B})$. Thus, for the immediate purpose at hand Heaviside apparently elected to drop the (-) sign and concentrate only on the essential requirements. Therefore, the above is really a proof of a mathematical identity rather than a discussion of the properties of electromagnetic systems; however, as Heaviside's argument will show, this did not prevent him from using a very physical line of reasoning relying directly on the properties of magnetic fields and electric currents.

that the tube's orientation is determined by the direction of current within it, any length element **dl** of the tube is always parallel to $\mathbf{C}_{2,1}$. Therefore, M_1 may be re-expressed in terms that reflect the tube's geometry:

$$M_1 = \int_v \mathbf{A}_1 \cdot \mathbf{C}_{2,1} (dlds)_{2,1},$$

where v represents the volume enclosed by the tube, dl an element of length measured parallel to the current in the tube, and ds a cross section of the tube taken at right angles to the current. The current through any cross section of the tube must be constant throughout the tube, or the requirement of the current's non-divergence will be violated. Therefore, writing:

$$\mathbf{C}_{2,1} = \frac{\mathbf{dl}_{2,1}}{dl_{2,1}} C_{2,1},$$

and recalling that $C_{2,1}ds$ is the total current through the tube, the volume integral above reduces to a line integral as follows:

$$M_1 = I_{2,1} \oint_{C_{2,1}} \mathbf{A}_1 \cdot \mathbf{dl}.$$

$I_{2,1}$ is the total current in the tube, and the line integral is taken along the closed path outlined by the tube. Using Stokes's theorem and the requirement that curl(**A**) = **B**, M_1 becomes:

$$\begin{aligned} M_1 &= I_{2,1} \oint_{C_{2,1}} \mathbf{A}_1 \cdot \mathbf{dl} \\ &= I_{2,1} \int_{S_{2,1}} \operatorname{curl} \mathbf{A}_1 \cdot \mathbf{dS} \\ &= I_{2,1} \int_{S_{2,1}} \mathbf{B}_1 \cdot \mathbf{dS}, \end{aligned}$$

where **dS** is an element of any surface bounded by the closed path of $\mathbf{C}_{2,1}$. In fact, the last expression can be arrived at directly by a simple generalization of the expression for the mutual energy of two current loops (page 114) thus: first, generalize the expression for the mutual energy of two current loops to obtain the energy of a single loop carrying a current $I_{2,1}$ immersed in the magnetic field of an arbitrary current system $\mathbf{C_1}$. Any particular tube of $\mathbf{C_1}$ contributes an energy equal to the current $I_{2,1}$ multiplied by the total amount of magnetic field due to the tube (of $\mathbf{C_1}$) passing through the loop of $I_{2,1}$. Obviously, the total energy of the single loop carrying $I_{2,1}$ due to all of $\mathbf{C_1}$ is the sum of all the individual energies contributed by $\mathbf{C_1}$'s tubes, in other words, it is equal to $I_{2,1}$ multiplied by the total magnetic field through the loop of $I_{2,1}$ due to all of $\mathbf{C_1}$. This is precisely what the integral above represents.[55]

Ampère's law may now be used to express $I_{2,1}$ in terms of its associated magnetic field $\mathbf{B_{2,1}}$. This law states that $I_{2,1}$ is the line integral of $\mathbf{B_{2,1}}$ around any path that completely encloses $I_{2,1}$. Since $I_{2,1}$ is the only current associated with $\mathbf{B_{2,1}}$, the integral of the latter around any path not enclosing $I_{2,1}$ is zero. Now, all $\mathbf{B_{2,1}}$ lines are closed because the magnetic field, being the curl of a vector-potential can have no divergence. Therefore, all lines of $\mathbf{B_{2,1}}$ must necessarily enclose the loop of $I_{2,1}$; or, which amounts to the same thing, all lines of $\mathbf{B_{2,1}}$ must necessarily pass through the loop of $\mathbf{C_{2,1}}$, cutting any surface bounded by the latter. Thus, using Ampère's law the energy of the current tube $\mathbf{C_{2,1}}$ that carries a total current $I_{2,1}$ may be written as:

$$M_1 = \oint_{B_{2,1}} \mathbf{B}_{2,1} \cdot d\mathbf{l} \int_{S_{2,1}} \mathbf{B}_1 \cdot d\mathbf{S}, \tag{3-7}$$

where the line integral follows any line of $\mathbf{B_{2,1}}$.

As already stated, the surface integral may be taken over any surface bounded by $\mathbf{C_{2,1}}$. Heaviside called attention to one particular family of such surfaces, namely, the one consisting of the equipotential surfaces of $\mathbf{B_{2,1}}$. He had already shown (see pp. 106-108) that outside the current tube $\mathbf{C_{2,1}}$ (whose cross-section may be made arbitrarily small) the field $\mathbf{B_{2,1}}$ is the gradient of a scalar potential $\Omega_{2,1}$. It is not difficult to see that $\mathbf{B_{2,1}}$ cuts perpendicularly through any surface over which $\Omega_{2,1}$ is constant. Since the surface is equipotential by construction, there cannot be any component of $\mathbf{B_{2,1}}$ parallel to it, for throughout the surface $\Omega_{2,1}$ never varies and therefore its gradient is identically zero in any direction along the surface. From this it follows that in a coordinate system whose x-y plane coincides with the equipotential surface at

55. One wonders why Heaviside elected not to use this simple argument. Perhaps he wanted to show his readers how one can argue from the expression involving the vector potential, instead of constantly having to refer back to the expression on p. 87.

the point where $\mathbf{B}_{2,1}$ is measured, $\mathbf{B}_{2,1}$ has no x-component and no y-component. Therefore all of $\mathbf{B}_{2,1}$ lies along the z-axis, or perpendicular to the equipotential surface.

Reconsidering all of this, the following conclusions emerge: all lines of $\mathbf{B}_{2,1}$ go through the loop of $\mathbf{C}_{2,1}$; $\mathbf{B}_{2,1}$ is everywhere derivable from the scalar potential $\Omega_{2,1}$ (except within the tube, but then its cross section may be made indefinitely small); $\mathbf{B}_{2,1}$ cuts each and every equipotential surface of $\Omega_{2,1}$ perpendicularly. To satisfy all three requirements, each equipotential surface must be bounded by the loop of $\mathbf{C}_{2,1}$, or else it would give rise to lines of $\mathbf{B}_{2,1}$ that pass outside the loop, violating Ampère's law. In eq. (3-7) above, then, **dl** is a line element parallel to $\mathbf{B}_{2,1}$ and **dS** may be taken for an element of an equipotential surface cut at right angles by each and every line of $\mathbf{B}_{2,1}$. Take now, the particular case Heaviside analyzed first, namely, the one involving the self energy of a single current loop. In this case, the fields $\mathbf{B}_1$ and $\mathbf{B}_{2,1}$ are one and the same, bringing eq. (3-7) to the following form:

$$M_1 = \oint_{B_1} \mathbf{B}_1 \cdot \mathbf{dl} \int_{S_1} \mathbf{B}_1 \cdot \mathbf{dS}, \tag{3-8}$$

with **dl** parallel to $\mathbf{B}_1$ and **dS** perpendicular to both. At this point, Heaviside concludes:

> It is important to notice that *any* tube of $\mathbf{B}_1$ will do for the line-integral, and *any* surface bounded by $\mathbf{C}_1$ for the surface-integral. But all the tubes of $\mathbf{B}_1$ are cut by the surface, and when it is an equipotential one they are cut perpendicularly. The same being true for the whole series of equipotential surfaces, it follows that the product of the line and surface-integral in [eq. (3-8)] includes every place where $\mathbf{B}_1$ exists; hence we may ignore the tubes completely, and write
>
> $$\sum \mathbf{A}_1 \cdot \mathbf{C}_1 = \sum \mathbf{B}_1^2$$
>
> extending summations through all space; whence the proposition is proved in the case of an infinitely fine tube of **C**.[56]

56. *Electrical Papers*, Vol. I, p. 245.

In the more general case illustrated by eq. (3-7), the fields $\mathbf{B}_1$ and $\mathbf{B}_{2,1}$ are not parallel, which suggests substituting their scalar product instead of the square in Heaviside's special case. The identity thus becomes:

$$\int_V \mathbf{A}_1 \cdot \mathbf{C}_{2,1} dV = \oint_{B_{2,1}} \mathbf{B}_{2,1} \cdot \mathbf{dl} \int_{S_{2,1}} \mathbf{B}_1 \cdot \mathbf{dS}$$
$$= \int_V \mathbf{B}_1 \cdot \mathbf{B}_{2,1} dV, \tag{3-9}$$

with the volume integrals reckoned through all of space. The last integral reflects only the energy of a single current tube $\mathbf{C}_{2,1}$ out of the system $\mathbf{C}_2$ due to the entire current system $\mathbf{C}_1$. To obtain the total mutual energy, add the contributions from all current tubes in $\mathbf{C}_2$, thus:

$$M = \int_V \mathbf{A}_1 \cdot \mathbf{C}_2 dV = \int_V \mathbf{A}_1 \cdot (\mathbf{C}_{2,1} + \mathbf{C}_{2,2} + \mathbf{C}_{2,3} + \ldots) dV$$
$$= \int_V \mathbf{B}_1 \cdot (\mathbf{B}_{2,1} + \mathbf{B}_{2,2} + \mathbf{B}_{2,3} + \ldots) dV$$

$$\therefore M = \int_V \mathbf{B}_1 \cdot \mathbf{B}_2 dV. \tag{3-10}$$

This concludes the demonstration of the proposition stated in the quotation on page 121 for the most general case of any two vector systems $\mathbf{A}_1$, $\mathbf{B}_1$, $\mathbf{C}_1$ and $\mathbf{A}_2$, $\mathbf{B}_2$, $\mathbf{C}_2$ satisfying $\mathbf{B} = \text{curl}(\mathbf{A})$ and $\mathbf{C} = \text{curl}(\mathbf{B})$. In the particular case of electrical current systems, the relationship between **A**—the vector-potential, and **B**—the magnetic field, remains as above. The relationship between the current and the field, however, involves an additional factor of 4π; thus: $4\pi\mathbf{C} = \text{curl}(\mathbf{B})$. The identity in eq. (3-10) that represents the mutual energy of two electrical current systems if **C** represents the current, becomes (see appendix 3.3 for further clarifications):

$$M = \int_V \mathbf{A}_1 \cdot \mathbf{C}_2 dV = \frac{1}{4\pi} \int_V \mathbf{B}_1 \cdot \mathbf{B}_2 dV. \tag{3-11}$$

As already stated, this last expression has some far-reaching consequences for the rest of Heaviside's analysis. However, he elected to postpone their discussion by two weeks, and instead offered the following observation as a concluding thought:

> I think it was a philosopher who propounded the theory that men always thought in some language; an Englishman in English, for example; always, if he knew no other language; otherwise he might think in any one he knew. Not to raise the obvious objection that persons dumb from birth should, according to this, have no thoughts at all, the theory is certainly proved to be false by an examination of such a transformation as the above. As regards the case of a single tube of **C**, if only the geometrical conditions are pictured in the mind, the division of space into small cubes by the tubes of $\mathbf{B}_1$ cut across by the equipotential surfaces of $\mathbf{B}_1$, the transformation becomes as self-evident as an axiom, and no form of words or sentences is necessary. The less one is cumbered with them the better. And although the extension to an arbitrary system is less easy, it is still easier to be pictured than logically demonstrated. The transformation might have been seen quite intuitively; it is only when one has to prove it to some one else that clothing the thoughts in words becomes necessary; and, even then, the clothes do not correspond to the original thoughts, but to those arising in the act of description, and both words and thoughts require to be readjusted, perhaps two or three times, before they will mutually fit with any decency. The Cartesian transformation, breaking up each of the vectors **A**, **B**, **C** into three rectangular components, is short enough, but is gifted with a total absence of visible reason and significance.[57]

One can hardly ask for a more explicit statement from Heaviside concerning his view of mathematical reasoning. Throughout the development of ideas so far, he remained wonderfully faithful to this view. One would also do well to reconsider his comments on Faraday (see page 66) in light of this conclusion. According to Heaviside's idea of mathematical reasoning, there is a sense in which Faraday displayed a close affinity to mathematical thinking: he reasoned out his problems in terms of the very concepts he was studying, not in terms of their expression in the language of the differential and integral calculus. It is not surprising that Heaviside held Faraday's work as a shining example to all non-mathematicians who were interested in a predominantly mathematical subject.

57. *Electrical Papers*, Vol. I, p. 246.

4.10 Where is the Energy?

In his quest for the energy of the electric current Heaviside obtained four expressions reflecting this energy. He had already dismissed one of them, Neumann's formula, as having no meaningful physical significance past providing a correct account of the total energy. The remaining three, taken for the self-energy of a single current loop **C** are as follows:

1) The energy is ½C times the total field due to **C** through its own loop. This actually amounts to a surface integral of **B** across any surface bounded by **C**.
2) The energy is ½ the volume integral of the scalar product $\mathbf{A} \cdot \mathbf{C}$ throughout all of space, which effectively reduces to a volume integral through the space permeated by the current in the loop.
3) The energy is ½the volume integral of $\mathbf{B}^2$ throughout space, which effectively reduces to the volume integral through the space permeated by **B**.

All three expressions yield the same quantity of total energy. Indeed, Heaviside already demonstrated their identity. Each of the expressions consists of a sum that suggests a well-determined volume distribution of energy. It is fascinating, however, that while the expressions agree as to the total amount of energy, they differ with regard to the volume density of the energy distribution. Where, then, is the energy?

Heaviside noted that the first expression locates the energy on a surface bounded by the current loop. However, there exists an infinite number of such surfaces, any of which will do for the calculation. The theory so far propounded offers absolutely no criterion according to which one such surface is preferable to another. The decision to restrict the energy to one out of infinitely many equally likely candidates would be quite arbitrary. Therefore, Heaviside concluded that the first expression could not provide a useful guide as to the location of the energy.

The second expression, considered to be fundamental by Maxwell and most of his British followers,[58] seems to be a better candidate. Here the energy appears as distributed throughout the space permeated by electric currents, for the integration yields nonzero results only where **C** is nonzero. Further-

58. J.Z. Buchwald, *From Maxwell to Microphysics*, (1985), pp. 58-59. As Buchwald noted, this expression measures the amount of energy *controlled* by a well-defined volume element, but it does not localize the energy itself.

more, the electric current appears to be the carrier of electric energy; indeed, this very observation led to the examination of its energy in the first place. It would appear likely, then, to localize the energy along the current's flow lines. Heaviside, however, argued otherwise. The energy in the second expression depends on **A**, the current's vector potential, as well as on the current itself. The former depends not only on the current within the volume element for which the energy is sought, but on **C**'s distribution throughout space (recall that $\mathbf{A} = \Sigma \mathbf{C}/r$). In a mechanical system of stresses and strains, Heaviside pointed out, the energy stored at a particular point depends only on the stress and strain at that point, not on their intensity and manner of distribution elsewhere. Therefore, he refused to allow localization of the energy according to an expression that renders it dependent upon conditions at points other than the ones at which the energy is sought. This argument is particularly interesting. It obviously reflects a general objection to anything that might imply action at a distance; but it actually says much more than that: it provides the first clear indication of how far Heaviside was willing to push the analogy between electromagnetism and mechanical systems of stress and strain. Heaviside rarely pointed to the abstract idea of action at a distance as a sufficient reason for rejecting anything. His objections were usually more specific, as in the case above. In fact, if criticism of action at a distance were high in his mind, he could easily have picked Neumann's formula as a case to argue against. In this formula, individual current segments are depicted as interacting across space. Heaviside, however, elected to discard the expression as physically meaningless because it contradicts the notion that all currents must be closed, not because of its action-at-a-distance character (see page 115). Moreover, were action-at-a-distance such a strong argument, Heaviside could have employed it against the first case. There too, the energy depends on the current strength at the surface's boundary as well as on the strength of the magnetic field's component locally perpendicular to the surface. Again, Heaviside preferred to criticize the expression on the grounds of its indeterminacy rather than accuse it of suggesting action at a distance.

This leaves only the third expression, which displays none of the above shortcomings. It distributes the energy throughout the space permeated by the magnetic field, obliterating the need to locate it on one of infinitely many equally qualified surfaces. It makes the energy explicitly dependent only on the field at the point where the energy is assumed stored, and the field there is quite independent of the field elsewhere. Thus the energy *density*, or energy per unit volume due to the presence of a current distribution **C** is $1/(8\pi)\mathbf{B}^2$.

Essential to note here is that the question of energy localization signifies far more than an aesthetic issue raised by an action-at-a-distance abhorring field theorist. Localized transformations in the energy of the electric current will provide Heaviside with a way of tying his field-oriented view to the known phenomena of time-varying currents. Therefore it is important to note that Heaviside concluded this examination of the steady electric current by associating its energy with the magnetic field rather than directly with the current. Whenever a localized energy transformation occurs, the electromagnetic field there, not the current, furnishes the transforming mechanism. Thus, for all dynamical purposes, the electromagnetic field becomes the fundamental physical entity to consider. In this manner Heaviside provided the first step toward establishing the preeminence of the electromagnetic field in all electrodynamic investigations.

4.11 Energy Conservation, Ohm's Law, and the Nature of the Electric Current.

Oliver Heaviside worked in a world that had already known that when the charged plates of a condenser (capacitor, to the modern reader) are conductively connected, the charge on the plates disappears. It had already been observed that with the disappearance of charge, other phenomena, previously completely missing, suddenly appear. The most prominent of these involved magnetic effects, clearly dependent upon the conductor's shape and position in space. It had been demonstrated that the strength of the magnetic effects can be correlated with the rate at which charge disappears from the condenser's plates. It had been further suggested that all of this may be accounted for by the assumption that the disappearing charge actually flows along the conductor, creating what was known as the electric current. This last assumption made Heaviside raise an eyebrow:

> There is an enormous difference between a state of electrification and the electric current which may produce it or be derived from it, and we cannot from electrostatic properties say what properties will be developed when a static charge disappears, and energy is transferred from a dielectric to a conductor. In the resulting phenomenon all trace of the electrostatic properties is lost, there is no external sign of any electricity in the conductor, whilst entirely new properties

> come into existence. As the static charge disappears, and the action upon static electricity likewise, or rather as the latter occurs and we infer the former, action upon a magnet at once appears in its place. The relation symbolised by Q = Ct does not, without special hypothesis, imply the motion of electricity from place to place, meaning by electricity the quantity we make acquaintance with in electrostatics, in spite of the law that the current is the same in all parts of the circuit, and that it is virtually equivalent to convection of electricity.[59]

This statement creates an unpleasant problem. In transferring a charge Q across a potential difference E an amount of energy equal to EQ is expended. Therefore, if electric current is static charge in motion, then a current of C charges per unit time flowing under a potential difference E volts must transfer EC units of energy per unit time. Heaviside's refusal to identify the current with static charge in motion completely undermines this line of thinking. To remove this difficulty, Heaviside put together the following argument. First, energy never disappears, it is only transferred and transformed. This is the principle of energy conservation in its nineteenth century generality. If EQ of energy disappears from the condenser, then EQ of energy must be found somewhere else. One may wish to associate it with the electric current. If so, one ought to remember that current is *potentially* static electricity. That is to say, a constant current C flowing for a time t into a condenser will charge it with an amount $Q = Ct$ of electricity. As before, this does not mean that the current consists of moving static charges. It does mean, however, that if the current flows under an E.M.F. of strength E, then it must transfer ECt energy in time t or the principle of energy conservation will be violated. Therefore:

> ...although with the galvanic cell we have no disappearance of static electricity, or, indeed, any sign of a large store of it to be drawn upon, yet the phenomena, being the same as regards the conductor, must be virtually equivalent to the discharge of static electricity, and we may apply the same formulae exactly as in the case of the condenser,
>
> $Q = Ct$ and $W = EQ$,
>
> which are now truths, though not truisms. The Q is now not electricity *in esse*, but *in posse* (i.e. without hypothesis to account for the absence of electrostatic force from the electricity we may assume to be moving).[60]

Up to this point Heaviside merely produced definitions. He knew with certainty that in the truly static case, $W = EQ$. Defining the strength of the cur-

59. *Electrical Papers*, Vol. I, p. 296.
60. *Ibid.*, p. 300.

rent as the rate of potential static charge transfer, he arrived at the inescapable conclusion that EC measures the rate at which the current transfers energy. The question is whether this definition makes sense. In Ohm's law and Joule's law Heaviside found the confirmation he was seeking. Joule's law states that the amount of heat generated by a current C during a time t is R_jC^2t. R_j is not merely a numerical constant of proportionality. The quantity Ct denotes electric charge according to the definition of current as charge per unit time. Charge by itself is not energy; it must be multiplied by a potential difference to yield energy. This means that R_jC must denote potential difference. Now, in an electrical circuit of resistance R, Ohm's law says that $E = RC$. Therefore, if all that the current does in this circuit is produce heat, then the conservation of energy and the definition of current as potential charge per unit time imply that a quantity of heat $W = EQ = (RC)(Ct)$ must be produced in time t. Comparing this with Joule's law, we see that Joule's constant is always identifiable with the circuit's ohmic resistance. In this way Heaviside used Ohm's law and Joule's law to legitimize the interpretation of $EC = RC^2$ as the rate at which the current transfers energy without ever having to regard the current as consisting of moving static charges.

4.12 The General Role of Energy Considerations In Heaviside's Work

Heaviside's use of energy considerations to connect and legitimize the relationships of current and electromotive force is by no means restricted to his analysis of electric circuits. It provides an important clue to his understanding of Maxwell's theory in its most general form. However, a word of caution is in order here. While the principle of energy conservation undoubtedly plays a crucial role in Heaviside's scheme of things, it is the concept of force around which he built his formulation of Maxwell's theory. Unfortunately, Heaviside's inability to resist a caustic remark readily enhances the impression that he considered energy to be the only meaningful physical idea:

> When the sage sits down to write an elementary work he naturally devotes Chapter I. to his views concerning the very foundation of things, as they present themselves to his matured intellect. It may be questioned whether this is to the advantage of the learner, who may be well advised to 'skip the Latin', as the old dame used to say to her pupils when they came to a polysyllable, and begin at Chapter II. If this be done, Prof. Tait's 'Properties of Matter' is such an excellent

scientific work as might be expected from its author. But Chapter I. is metaphysics. There are only two Things going, Matter and Energy. Nothing else is a thing at all; all the rest are Moonshine, considered as Things.[61]

As a rule, however, further reading usually reveals qualifications to such sweeping statements. Perhaps the best counter to the above may be found in a discussion of energy transfer Heaviside wrote in 1891. If the density of energy stored in space is T, and if it is conveyed at a velocity $\mathbf{q}$, then the divergence of $T\mathbf{q}$ in a given volume element measures the rate at which the volume element is being depleted of energy. Thus, the equation

$$\text{div}\,(T\mathbf{q}) = \frac{dT}{dt}$$

is exactly analogous to the equation of continuity of matter in fluid flow. Heaviside proceeded from this simple idea to the following observation:

This brings us to Prof. Lodge's theory of the identity of energy. (Phil. Mag. 1885.) Has energy personal identity, like matter? I cannot see it, for one; and think it is pushing the principle of continuity of energy, which Prof. Lodge was writing about, too far. *It is difficult to endow energy with objectivity, or thinginess, or personal identity, like matter* [my italics]. The relativity of motion seems to be entirely against the idea. Nor are we able to write the equation [above] in all cases. Energy may be transferred in other ways than by convection of associated matter. If we atomise the energy we can then imagine the $\mathbf{q}$ above to be the velocity of the energy, not of the matter. But as the science of dynamics is at present understood, we cannot make use of this idea profitably, I think.[62]

Here, as on so many other occasions, Heaviside was faithful to Maxwell's views:

We cannot identify a particular portion of energy, or trace it through its transformations. It has no individual existence, such as that which we attribute to particular portions of matter.

The transactions of the material universe appear to be conducted, as it were, on a system of credit. Each transaction consists of the transfer of so much credit or energy from one body to another. This act of transfer or payment is called work. The energy so transferred does not retain any character by which it can be identified when it passes from one form to another.[63]

61. *Electrical Papers*, Vol. II, pp. 91-92.

62. *Electromagnetic Theory*, Vol. I, p. 75.

63. J.C. Maxwell, *Matter and Motion*, (1991), § 109, p. 90.

Heaviside regarded the conservation of energy as a fundamental and universal principle that all physical theories must satisfy. At the same time, he explicitly stated that in and of itself, the principle is insufficient for the description of physical processes, even when they directly involve energy transformations. All one learns from it is that when a certain amount of energy disappears, it must be found elsewhere in a different form. It leaves no clue whatsoever as to the nature and whereabouts of the different form, nor can it fully describe the process, or mechanism by which the transformation takes place:

> The statement sometimes made, that the laws of induction follow of necessity from Ampère's forces and conservation, is of too broad a nature. If we modify the statement, and say that the laws of induction are consistent with Ampère's actions, and with conservation, there will be nothing to be objected to. That this is not a mere difference of tweedledum and tweedledee may be easily seen from the history of the subject, if it be not sufficiently evident by itself. We may, indeed, from the existence of Ampère's forces and a conviction of the truth of the conservation principle, conclude certainly that some other actions occur, but the principle merely asserts that energy is never lost, that energy put into a system from outside must necessarily be either stored up or make its appearance somewhere in some form or other; but what the form may be depends upon the mechanism—on the dynamical connections—and conservation does not tell us what they are, nor what will happen. There must be other information given.[64]

Thus, it takes additional information, such as Ohm's law, to specify the new forms energy assumes after a transformation.[65] As for the mechanism, Heaviside always sought it in the idea of force. Indeed, he crisply stated his opinion that energy transformations could not occur without the action of a force, and added to it the unwavering requirement that the transformation

64. *Electrical Papers*, Vol. I, p. 282.

65. It seems that Heaviside had in mind Helmholtz's derivation of Faraday's law, independently arrived at by Kelvin a little later. Maxwell discusses the derivation, stating that Helmholtz and Kelvin "... shewed that the induction of electric currents discovered by Faraday could be mathematically deduced from the electromagnetic actions discovered by Oersted and Ampère by the application of the principle of the Conservation of Energy." (J.C. Maxwell, *A Treatise on Electricity and Magnetism*, [1873], Vol, II, § 543, pp. 176-177. Maxwell proceeded to outline the argument using Ampère's law and the principle of energy conservation as represented by Joule's law. Implicitly, however, the discussion requires Ohm's relationship between current, voltage, and resistance in addition to Joule's law, and this probably prompted Heaviside's comment.

would occur only where the force acts and nowhere else. Thus, in the case of a body moved upward against gravity:

> Force of some kind is equally involved in kinetic and potential energy, when there is change from one to the other. Force produces relative motion and its kinetic energy, and the latter cannot be utilised without force. In the above simple case the conservation of energy only means that when kinetic energy is lost by the body moving against the force, such loss is perfectly recoverable in the return motion, since the force remains the same in the same places. This last remark, indeed, contains the reason why the energy is conserved, or returnable to the kinetic form.[66]

Heaviside emphasized energy considerations because he was aware that the actual "forces" he dealt with, viz., the electric and magnetic force-fields, are not forces like the mechanical force of a spring or the weight of a mass. Electric "force," for example, must be multiplied by a charge to yield mechanical force. Therefore the force field itself is not a force in the mechanical sense of the word, its dimensions being force per unit charge. It does, however, bear a remarkable analogy to mechanical force through the idea of activity, which is fundamental to the understanding of Heaviside's work in general.

The word "activity" is closely related to Newton's "action," which he explained in an alternative formulation of his third law that is all too often ignored:

> For if we estimate the action of the agent from the product of its force and velocity, and likewise the reaction of the impediment from the product of the velocities of its several parts, and the forces of resistance arising from the friction, cohesion, weights and acceleration of those parts, the action and reaction in the use of all sorts of machines will be found always equal to one another. And so far as the action is propagated by the intervening instruments, and at last impressed upon the resisting body, the ultimate action will be always contrary to the reaction.[67]

For the force-minded Heaviside, the identification of Newton's "action" with power, or energy per unit time, offered a way to formulate a sound electrodynamic field theory in keeping with the physical primacy of forces. Newton's action became for Heaviside the activity of a mechanical force—the rate at which it performs work. It is given by the product $\mathbf{F} \cdot \mathbf{V}$, where $\mathbf{V}$ is the ve-

66. *Electrical Papers*, Vol. I, p. 292.

67. Isaac Newton, *Principia*, Motte's translation, revised by Cajory, (Berkeley: University of California Press, 1962), p. 28.

locity of a body under the action of the force. Hence, if a body moves upward at a velocity **V** under the action of a force **F**, then $\mathbf{F} \cdot \mathbf{V}$ measures the rate at which the force converts energy of unspecified nature into potential and kinetic energy associated with the moving body. If **V** is the horizontal velocity of a body in a medium that resists the motion with a force **F**, then $\mathbf{F} \cdot \mathbf{V}$ is the rate at which **F** converts the body's kinetic energy into heat.

In the electrical case, the electric current density **C** acts as a generalized velocity with respect to the electric field **E** considered as a generalized force. **E** is not a force, **C** is not a velocity, and Heaviside went to great lengths to instill in his readers the understanding that nothing was known about the real essence of electric "force," charge, and current. Remarkably, however, the product $\mathbf{E} \cdot \mathbf{C}$ yields energy density per unit time. It measures activity, or the rate at which an electrical current system delivers energy through the action of an electric "force." Thus, the idea of activity, identified in the electrical case with the product $\mathbf{E} \cdot \mathbf{C}$, secures the analogy between the electric field **E** and any legitimate mechanical force. It does so by establishing a concrete connection between any energy transforming mechanical force and the particular system described by **E**, **C**, and their various interconnections. From this point of view the electric force and the electric current become part of a generalized mechanism for the balanced transfer of energy as Newton's third law requires.

In the general case of electromagnetism as in the limited case of electric force and current density, the principle of energy conservation did not serve Heaviside as an analytical tool with which to deduce the particular laws of the electromagnetic system. It served him as a theoretical test of validity and as an aid to tracing the operating forces in a dynamical system. As he later said with regard to "the grand Principle [of energy conservation], as contained in Newton's Third Law":

> The Principle cannot be violated. It conserves itself automatically in any self-consistent dynamical connection. The question nowadays is, How are we to account for the energy lost or gained? What force is concerned? It must be implicitly contained in the connections already given, though it is not yet displayed.[68]

The particular connections that define the electromagnetic system must be such that they can always describe the transfer of energy into and out of a given space permeated by the fields in accordance with energy conservation.

68. *Electromagnetic Theory*, Vol. III, p. 406.

As we shall see (pp. 159-161), Heaviside reached a high point in his reformulation of Maxwell's electrodynamics when he demonstrated the ability of electromagnetic fields to effect such a transfer of energy. Energy conservation, then, is a necessary condition that the system must fulfill. It is by no means sufficient for the specification of its particular relationships, for which "there must be other information given."[69]

Through its function in Heaviside's rendition of Maxwell's theory, the concept of activity further illuminates the intricate relationship of mathematics and physics in Heaviside's style of reasoning. Heaviside had already shown himself to be the consummate physical thinker, abhorring the notion of giving

---

69. Insufficient emphasis on the primacy of forces in Heaviside's electrodynamics combined with his unconventional notation can lead to the identification of nonexistent energy functions in his work. An instructive instance is provided by Buchwald and Lervig's analysis of Heaviside's seminal paper, "On the Forces, Stresses, and Fluxes of Energy in the Electromagnetic Field" (*Electrical Papers*, Vol. II, pp. 521-574. See J.Z. Buchwald, "Maxwell's Apostle and Maxwellian Apostate," Centaurus, 28 (1985): 315-316, 328). In his paper, Heaviside studied the general case of fields in anisotropic media in relative motion. Following his usual approach, he attempted to locate the operating forces by tracing elements of the energy transfer in the system to the activities of the operating forces. This required of him to consider changes in the overall energy in a volume element of the system, i.e., the time derivative of $U + T$, where U is the energy of the electric force, and T is the energy of the magnetic force. In an anisotropic medium, the electric energy is $(1/2)\mathbf{ED} = (1/2)\mathbf{E}c\mathbf{E}$, where c is a linear operator. Heaviside's mode of investigation calls for a distinction between energy changes due to variations in the force, and energy variations due to changes in the properties of the medium. Hence, he wrote: $\dot{U} = \mathbf{E}\dot{\mathbf{D}} - (1/2)\mathbf{E}\dot{c}\mathbf{E} = \mathbf{E}\dot{\mathbf{D}} - \dot{U}_c$. Heaviside explained the meaning of $\dot{U}_c$ as follows: "$\dot{U}_c = \frac{1}{2}\mathbf{E}\dot{c}\mathbf{E} = \frac{1}{2}\frac{d}{dt}(\mathbf{ED})_c$ ... representing the time-variation of U due to variation in the c's only" (p. 544). It should be clear from this that the meaning of the strange subscripted notation is: take the derivative of U with respect to t, allowing only c to vary. Heaviside used this notation consistently throughout this paper (see, e.g., p. 562) and on later occasions when he returned to the subject (*Electromagnetic Theory*, Vol. III, p. 427). However, between the unorthodox notation and an undue emphasis on energy considerations, Lervig took Heaviside's $\dot{U}_c$ to mean: take the time derivative of the energy function U_c. Lervig then proceeded to show with an elegant argument that such an energy function makes no physical sense, and contains an inherent contradiction ("Apostle and Apostate", p. 328, note 9). The problem can be illustrated directly by force considerations. Positing the existence of an energy function U_c suggests that the energy of a compressed spring may be decomposed so as to have a component that assigns identifiable energy to the spring's elasticity. This idea is obviously meaningless dynamically, and indeed, Heaviside never used it. However, it is perfectly sound, dynamically speaking, to consider *variations* in the energy of a compressed spring when the compressing force is kept constant while the elasticity alone varies. This is precisely what Heaviside considered in the course of his force-oriented analysis.

priority to abstract mathematical formalism. Here, however, he revealed the other side of the coin and consciously gave up, at least temporarily, the attempt to turn the electromagnetic field into a concrete physical entity in its own right. The mechanism that transfers electromagnetic energy is no more than a formal connection between entities whose real physical nature is a complete unknown, even though they can be grasped in terms of analogies to fluid flow and mechanical stress systems. In a sense, Heaviside's idea of electromagnetic theory may be described as the art of talking about things electromagnetic without knowing what they really are. Naturally, Heaviside paid tribute to the hope that all of electromagnetism would be reduced to "matter and motion" someday, and said on one occasion that, "We are set down in space, to march with time, and have matter in motion everywhere around us."[70] But like his statement about everything besides energy being moonshine, this one too should be taken with a grain of salt. Concretely, Heaviside did little to bring about the reduction of electromagnetism to matter and motion, and often expressed his skepticism concerning the feasibility of such a project in the foreseeable future:

> Ewing's recent improvement of Weber's theory of magnetism seems important. But as in static explanations of dynamical phenomena the very vigorous molecular agitations are ignored, it is clear that we have not got to the root of the matter. We want another Newton, the Newton of molecular physics. Facts there are in plenty to work upon, and perhaps another heaven-born genius may come to make their meaning plain. Properties of matter are all very well, but what is matter, and why their properties? This is not a metaphysical inquiry, but concerns the construction of a physical theory.[71]

For Heaviside, the ability to express the theory of electromagnetism in sound dynamical terms provided the best indication that an eventual reduction to matter in motion may be possible. Beyond that he consistently refused to speculate.

Related to Heaviside's special brand of electrodynamics is his seemingly ambiguous use of "force." He employed it sometimes to denote the *actual* force experienced by one body in electromagnetic interaction with another, and sometimes to denote the electromagnetic force *field*. Terminological ambiguities like these often pinpoint important theoretical difficulties, stemming

70. *Electrical Papers*, Vol. I, p. 334.
71. *Electromagnetic Theory*, Vol. I, p. 42.

from a physicist's attempt to deal with an ill-understood situation. There is an element of that here as well, associated with the unknown nature of electricity; but it must be understood that there is no theoretical vagueness involved, and that the ambiguity itself is more illusory than real. Heaviside would have been the first to admit that the force exerted on a charge through its electrostatic interaction with another charge, and the electrostatic field (or "force") associated with a charge, are two distinct concepts. The first is a real mechanical force, acting at a specific point and capable of compressing a spring or lifting a weight. The second is an idea of force per unit charge, representable as a vector distribution throughout space. By itself it will never compress a spring or budge a speck of dust. At the same time, Heaviside surely would have rejected the notion that using the word "force" to denote both concepts involves any ambiguity whatsoever. Both fall squarely within the category of generalized force as part of a dynamical system that constitutes, as described above, a generalized mechanism for the transference of energy. Indeed, from this point of view, the word "force" is more illuminating and conveys more information about the electrostatic field then the general term "field." Many vector fields may be envisioned; very few are capable of inclusion into a generalized energy-transferring mechanism. Heaviside would have observed that beyond its role in the transfer of energy as part of a system involving electric currents, charges and magnetic fields, nothing is known of the electrostatic field. All attempts to reduce it to matter and motion had failed, and fanciful speculations were left. The latter might prove extremely useful as guides to further work, but they should not be admitted into the scientific body of knowledge on an equal footing with the known laws of electromagnetism as expressed by Maxwell's equations. There exists no ambiguity here, Heaviside probably would have concluded, but quite the opposite: by using the word "force" in the above context, one makes a razor-sharp distinction between what one knows about the electromagnetic interaction and whatever additional speculative ideas one might entertain about it. All this is beautifully illustrated in a comment Heaviside made in 1900, which ties together both the question of using the word "force" for describing the field and the importance of keeping the dynamical mechanism in mind:

> Electric and magnetic force. May they live for ever, and never be forgot, if only to remind us that the science of electromagnetics, in spite of the abstract nature of the theory, involving quantities whose nature is entirely unknown at present, is really and truly founded upon the observation of real Newtonian forces, elec-

> tric and magnetic respectively. I cannot appreciate much the objection that they are *not* forces; because they *are* the forces per unit electric and magnetic pole. All the same, however, I think Dr. Fleming's recent proposal that electric force and magnetic force shall be called the voltivity and the gaussivity a very good one; not as substitutes for with abolition of the old terms, but as alternatives; and beg to recommend their use if found useful, even though I see no reason for giving up my own use of electric and magnetic force until they become too antiquated.[72]

In conclusion, Heaviside's physical thinking is really thinking in terms of dynamical analogies made sound by their ability to yield a mathematical expression for the transfer of energy. If his mathematics was physical through and through, then his physics was every bit as mathematical. There is simply no way to disentangle the two in Heaviside's thought, and the very idea of doing one without the other was quite unacceptable to him. All of this adds up to a rather uncommon and marvelously coherent style of analysis. In many ways it characterizes Heaviside better than any of his individual contributions to physics, mathematics, and electrical communications. Unfortunately, this style emerges piecemeal from nearly ten years of publication, beginning with his early telegraphic papers, and culminating in the four essays that were described here under the collective theme of an introduction to field thinking for the non-mathematical electrician. Anyone who reads only part of Heaviside's work runs the risk of observing only one side of his multifaceted style and forming a seriously misguided idea about his view of physical and mathematical knowledge.

72. *Electromagnetic Theory*, Vol. III, pp. 1-2.

4.13 "On Explanation and Speculation in Physical Questions"

Heaviside viewed Maxwell's theory as a system of mathematical laws that define a generalized mechanism for energy transfer without in any way explaining the nature of electricity and magnetism. This view seems to have served him as a model for legitimate scientific knowledge in general. He gave sharp expression to this general idea in a short essay, entitled "On Explanation and Speculation in Physical Questions," which he published in January of 1884. This essay relates specifically to his discussion of the Peltier and Thomson thermoelectric effects and their analysis by German physicists, particularly Clausius. His remarks, however, are quite independent of the particular aspects of the two effects, and refer to the more fundamental question involving the essence of electricity. Indeed, following a short introduction, Heaviside's essay quickly turns into a discussion of what makes legitimate scientific knowledge and what marks the legitimate goals of scientific research.

Clausius complained, as Heaviside pointed out, that William Thomson (Kelvin) did not explain in any way the cause of the thermoelectric forces. That, Heaviside continued, should not be taken as a weakness in Thomson's study, but rather as its strength:

> It is true that Sir W. Thomson abstained from vain speculation and went straight to the point at once. There are reversible heat effects at the junctions of different metal when currents pass across them. There may be (although then unknown) similar effects in the metals themselves. There is no resultant E.M.F. in a circuit of one homogeneous metal, however it may be heated, and however its section may vary. The effects must be subject to the law of conservation of energy, that is, the First Law of Thermodynamics. They are very probably subject to the Second Law as well. Now, with these data, develop the laws governing the E.M.F.'s, without unnecessary hypotheses. Such is the method followed in Sir W. Thomson's papers (what ever may have been his private speculations), the truly scientific method in the strictest sense of the word, bearing in mind its derivation, and what science ought to mean—viz., knowledge, and discarding the vague extended meaning it has gradually acquired in the mouths of the unscientific.[73]

Contrast this, Heaviside wrote, with Clausius, who:

> ...preludes his investigation, which, it may be remarked, has the same object and result,... by speculations on the causes of contact force in general, and of the ther-

73. *Electrical Papers*, Vol. I, p. 331.

> moelectric force in particular, using hypotheses, which appear based entirely on the materiality of electricity supposed to act directly at a distance on other electricity, and to be attracted differently by different kinds of matter. It is not easy to express these hypotheses in terms of less gross ideas of electricity without at the same time making them become mere ghosts, of no tangibility and of little utility. Such speculations should, in my opinion, be kept entirely apart from, and in particular should not precede, and so apparently form the groundwork of, a mere development of laws not in any way dependent on the hypotheses; so long as the object of inquiry is the laws, not the causes thereof.[74]

Perhaps the most significant statement in the last quotation is Heaviside's acknowledgment that the two studies have the same result. The issue is then clearly not which of the two investigations produces the surer guide to the phenomena. There seems to exist something else about Clausius's work that disturbed Heaviside. It may appear at first that he objected to Clausius's tendency to suggest hypotheses concerning the nature of electricity. It will soon become evident that Heaviside actually regarded speculation as a central ingredient of good scientific work. What he objected to was the elevation of speculation to the status of explanation. However, before any further elaboration of the point, it is essential to note that Heaviside's objections to Clausius's remarks stem from two independent roots. The first, clearly stated in the last paragraph, originates in Heaviside's observation that Clausius's hypotheses are unnecessary for the formulation of the laws governing the Thomson and Peltier effects. Heaviside expanded on this side of the matter in a separate section on chemical contact forces.[75]

The second objection involves the relationship between speculation and legitimate scientific knowledge. Heaviside dedicated his essay, "On Speculation and Explanation in Physical Questions," to the elaboration of this point. It boils down to a rather clear statement of what may and may not be expected from scientific research. The basic idea he attempted to convey in the essay is quite simple: some speculations are scientific; others are unscientific. The unscientific speculation is submitted as a causal explanation of some puzzle. It is unscientific because science is not about finding causes, but about formulating the laws of the phenomena:

> It is human nature to speculate, and there will be always plenty of scope for speculation until everything is found out, which will not be for some few million

---

74. *Electrical Papers*, Vol. I, p. 331

75. *Ibid.*, pp. 337-342.

> years. We want to know the causes of things, why such and such things happen. Well, the first preliminary would be to find the laws of the phenomena. That is work for the scientific man, and usually difficult work, requiring scientific training and reasoning. When laws are established ... we may speculate on their causes. Or, since it would often be very tedious to wait until sufficient facts are known, we may speculate on the causes of phenomena without knowing anything about the laws governing them. Now this may be done by any one. Not that any one can find out a probable explanation of something strange, but any one can speculate. The more imaginative a man is the better for his speculative powers. Also, if he be unscientific, it is not desirable for him to know too much of the facts of the case, because facts are very unaccommodating, and form a great drawback to the free exercise of the speculative faculties.[76]

That is not to say, however, that there is no room for speculation in scientific work. On the contrary:

> ... in the hands of the philosopher (not meaning metaphysicians who appropriate the title), with a proper attention to facts arranged in correct perspective, and in especial with a due attention to geometrical and quantitative relations in regard to space, time, motion, energy, etc., speculation becomes a very different thing from the above, and may be most usefully employed in forming hypotheses, which, though they may be themselves very improbable, may be provisionally of great utility, not merely to hang the facts together, but, on account of the inquiries they suggest, to serve as stepping-stones to a truer theory. Imagination is required no less than before, but it must be guided by strong sense and understanding.[77]

As an example of such a speculation, Heaviside mentioned Maxwell's hypothesis of molecular vortices to explain electricity and magnetism. He actually lamented that Maxwell chose not to give an account of this hypothesis in his *Treatise on Electricity and Magnetism.* Not that Heaviside had much faith in the truth of Maxwell's molecular vortices. He regarded them to be "exceedingly unlikely to correctly represent the reality."[78] Heaviside felt, however, that by electing not to describe his speculations Maxwell robbed future students of a hypothesis that could have proved very useful in its suggestiveness. Speculation is more than acceptable, then, as long as it is always conscientiously distinguished from sound formulation of natural laws. As Heaviside put it in a later essay:

---

76. *Electrical Papers*, Vol. I, p. 332.
77. *Ibid.*,p. 333.
78. *Ibid.*, p. 334.

> I am not objecting to the use of the imagination. That would be absurd; for most scientific progress is accomplished by the free use of the imagination (though not after the manner of professional poets and artists when they touch upon scientific questions). But when one, by the use of the imagination, has got to a definite result, and then sees a stricter way of getting it, it is perhaps as well to shift the ladder, if not to kick it down. For I find that practically, in reading scientific papers, in which fanciful arguments are much used, it gives one great trouble to eliminate the fancy and get at the real argument. Nothing is more useful than to be able to distinctly separate what one knows from what one supposes.[79]

Speculations and hypotheses then, have an essential role in the development of sound scientific knowledge, but one must not forget that they are the means to an end, and never an end in themselves.[80] The end, as mentioned before, is to formulate the laws of the phenomena, and a science reaches its mature state only once these laws have been expressed mathematically. Therefore:

> Political Economy can never be a science in the same sense as Electricity, even if what takes place in electrical phenomena remain for ever unknown. But just in proportion as a branch of knowledge rises from being a heterogeneous collection of facts and imperfect laws to being a system, consistent in all its parts, so does it become scientific, and under the rule of exact relations. So long as there is uncertainty as to exactly how much a certain effect amounts to under given circumstances, it cannot be a finished science. ... Consider the present science of Electricity, with its various units, measuring instruments, and methods. Who have made this possible? The mathematicians. It would be very little use accumulating piles of facts without having the mathematicians to sift them, discover

---

79. *Electrical Papers*, Vol. I, p. 423.

80. In light of this, Heaviside would have people exercise greater care in their usage of the words "speculation" and "hypothesis," and reserve the word "theory" for special occasions: "Examples of useful scientific speculation are innumerable. They are usually gifted with importance by being termed theories, thus leading the uninitiated to take them for more than they are worth. Hypothesis would be a better name than theory, because its sound and associations suggest something suppositious and to be received with caution; whilst theory, on the other hand, has also the much more important meaning exemplified in Fourier's 'Theory of Heat,' Maxwell's 'Theory of Electricity' (not the vortex hypothesis to be mentioned), or Rayleigh's 'Theory of Sound,' which have very little to do with speculations, but are mainly rigid developments of established laws. But it would certainly lead to a considerable loss of dignity were an investigator to speak of 'my hypothesis' or 'my speculations' on, for instance the cause of magnetism, instead of the usual 'my theory.' For it is very well recognised that dignity, or the appearance thereof, has a very imposing effect on all, save those who take the trouble to look below the surface. Which is why lord mayors are dressed up in robes and chains, and the judges wear horsehair wigs." (*Electrical Papers*, Vol. I, p. 333).

> the numerical relations, test various theories with the mathematical touchstone, and gradually turn chaos into system, as they have done in Electricity.[81]

Once one confuses the end with the means, scientific knowledge disappears. Speculations are then advanced as explanations by "the unscientific who cannot bear to have no explanation, who would rather worship a false god than none at all."[82] True to his love of the caustic remark, Heaviside could not resist a poke in the ribs of his German rivals:

> Then there are poetical explanations of natural phenomena. As might be expected, these are *very* bad. The illustrious Goethe's explanation of colour should be a caution to poets to the end of time to keep to their poetry. ... He, in his complete confidence in his theory, astonishing ignorance of the subject, contempt for Newton's theory, and hatred of the methods of the French mathematicians who had developed the laws of polarisation mathematically, displayed many of the characteristics of the unscientific explainer, whilst the complete ignoration of the great poet's theory by the scientists was no less characteristic of them.[83]

An explanation, Heaviside wrote, can be scientific without being real, that is to say, without truly explaining anything. It is scientific by virtue of being useful for the formulation of mathematical laws and one need not, indeed should not, attach to it a measure of reality it does not possess. As for real explanation, Heaviside concluded, we may sometimes feel that something is explained when we succeed in reducing it to something else that is already known. Such is the case of the kinetic theory of gases, "that remarkable triumph of hard-headed men", which reduces the laws of gases to the laws of matter in motion under the hypothesis that a gas consists of many small particles in motion. But even this explanation is more illusory than real, for:

> After that, there is the nature of molecules, and of matter in general. And even if we resolve all matter into one kind, that one kind will need explaining. And so on, for ever and ever deeper down into the pit at whose bottom truth lies, without ever reaching it. For the pit is bottomless.[84]

Explanation then, or finding causes, is not the goal of scientific work. The goal is scientific knowledge which consists of the laws of phenomena, and

81. *Electrical Papers*, Vol. I, p. 335.

82. *Ibid.*

83. *Ibid.*

84. *Ibid.*, p. 337.

these describe and relate phenomena to one another but do not explain them. Everything else is speculation; and while speculation may prove extremely useful to the development of natural laws, it is only a means to an end that should never be considered legitimate scientific knowledge. Hence, Heaviside's remarks seem to imply that by requiring of Kelvin to furnish an explanation of the thermoelectric current, Clausius in effect required of him to be unscientific.

This image of legitimate scientific knowledge harmonizes remarkably well with Heaviside's scientific work. Time and again he refused to take seriously attempts to explain the nature of electric charges, currents, and magnetic poles. Important for him was the system of laws that relate electricity and magnetism to one another and provide the dynamical basis for energy transfer. Thus, just as Maxwell's theory provided Heaviside with the art of talking about electricity and magnetism without knowing what they really are, so did his idea of scientific knowledge in general provide him with a justification for discussing natural phenomena without ever explaining them. Heaviside remained faithful to this philosophy to the end of his life.

5. Heaviside's Rough Sketch of Maxwell's Theory

Heaviside took well over a year, from November 1882 to March 1884, to complete his extensive introduction to field thinking with the various case studies he used to illustrate his arguments. Then, in seven short weeks from January to February of 1885 he published what he entitled a "Rough Sketch of Maxwell's Theory" in four consecutive papers in *The Electrician*.[85] In these

85. *Electrical Papers*, Vol. I, pp. 429-449. In the interim he published an important study entitled "The Induction of Current in Cores." (*Electrical Papers*, Vol I, pp. 353-416). It is Heaviside's first extensive application of Maxwellian notions to a specific problem. In this series he showed that magnetic induction penetrates the core from the outside, decreasing in intensity as it diffuses inward (pp. 359-361). Several weeks later, he made his first allusion to the analogy between this case and the flow of an incompressible fluid endowed with mass through a pipe (pp. 378-384, esp. p. 384). Just prior to starting his series "On Electromagnetic Induction and Its Propagation," Heaviside published a short paper entitled "Remarks on the Volta Force, Etc." It shows that at least by the time he wrote this paper, his work was receiving some attention from other physicists. Heaviside wrote the article in response to some remarks Oliver Lodge had made with respect to material Heaviside had published in 1884.

papers Heaviside presented for the first time a comprehensive summary of Maxwell's ideas.

The presentation may be broken into three parts. The first discusses the interaction of matter with electric and magnetic forces—a subject that Heaviside ignored during his introduction to field thinking. The second is primarily a pedagogical tool. It consists of a purely nontechnical argument designed to highlight the problems involved in perceiving the conduction current as the carrier of electric energy. With this problem outlined, Heaviside proceeded to suggest the only other alternative, namely, that the energy is conducted not through the conductor, but through the surrounding dielectric. Armed with this notion to motivate the rest of this discussion, Heaviside completed the rough sketch in the third part. Once completed, Heaviside used the ideas developed in his rough sketch to solve the problem of the second, pedagogical part. He showed how Maxwell's theory gives rise to a continuous energy flux throughout space. Finally, he showed how Maxwell's theory depicts the propagation of the electric and magnetic forces in the form of waves in the dielectric.

Rather than following Heaviside's order of presentation, we shall first study his rough sketch in the five parts of this section. We shall then briefly examine the place of the ether in Heaviside's scheme of things, and conclude by reexamining the conducting circuit in view of the Maxwellian outlook. As a point of departure, we may take Heaviside's following comment on the nature of the electric current:

> It so happened that my first acquaintance with electricity was with the dynamic phenomena, and after I had read with absorbed interest that instructive book, Tyndall's "Heat as a Mode of Motion." This may explain why, when it came later to book learning regarding electricity, I had the greatest possible repugnance to all the explanations, and could not accept the electric current to be the motion of electricity (static) through a wire, but thought it something quite different. I simply did not believe, except so far as mere statements of experimental facts were concerned.[86]

From Tyndall, Heaviside learned in the first place that heat need not be considered as a special kind of material essence that permeates bodies in different degrees:

86. *Electrical Papers*, Vol. I, p. 435.

> Why should heat be generated by mechanical action, and what is the real nature of the agent thus generated? Two rival theories have been offered in answer to these questions, which are named respectively the *material theory*, and the *dynamical*, or *mechanical, theory* of heat. ... The material theory supposes heat to be a kind of matter—a subtle fluid stored up in the inter-atomic spaces of bodies. ... The dynamical theory, or, as it is sometimes called, the mechanical theory of heat, discards the idea of materiality as applied to heat. The supporters of this theory do not believe heat to be matter, but an accident or condition of matter; namely, *a motion of its ultimate particles*.[87]

Tyndall aligned himself squarely with the supporters of the second view, and Heaviside, as his own comment testifies, followed suit. Heaviside also learned from Tyndall that whatever may be the molecular interactions that constitute heat, they always involve the conversion of two basic forms of energy, "potential," and "actual," or "dynamic," and that the sum of the two always remains the same:

> ... as potential energy disappears, actual energy comes into play. *Throughout the universe, the sum of these two energies is constant.* To create or annihilate energy is as impossible as to create or annihilate matter; and all the phenomena of the material universe consist in transformations of energy alone. The principle here enunciated is called the law of the *conservation of energy*.[88]

Heaviside internalized this lesson as well, but learned to qualify it. We have already seen (page 132) that while he believed that all physical interactions involve energy transformations, he did not consider the energy transformations themselves sufficient for describing the interactions. Instead, he treated the transformations as guides to the underlying force structures that actually performed the transformations.

The third major lesson Heaviside learned from Tyndall is that there exists a suggestive analogy between heat and electricity. When the end of a cool poker is thrust into the fire, the entire poker gradually heats up, Tyndall explained. However, this heating up consists of the gradual communication of a state of local motion from molecule to contiguous molecule, and should not be thought of as a material flow of some kind inside the poker:

> The motion, in this instance, is communicated from atom to atom of the poker, and finally appears at its most distant end. ... Convection we have already defined to be the transfer of heat, by sensible masses of matter, from place to place; but

87. John Tyndall, *Heat: A Mode of Motion*, fourth edition, (1870), pp. 23, 25
88. *Ibid.*, p. 132.

> this molecular transfer, which consists in each atom taking up the motion of its neighbours, and sending it on to others, is called the *conduction* of heat.[89]

With this in mind, and with the knowledge that the better heat conductor between two metals is almost always also the better conductor of electricity, Tyndall suggested that, "We have every reason to conclude that heat and electricity are both modes of motion."[90] The most obvious inference to be made from this is that the electric current is actually a process by which some form of motional, or "dynamic" energy is conducted through the wire. As we shall see, this may have been what Heaviside thought until he encountered Maxwell's *Treatise*, in which he discovered an entirely different way of thinking.

5.1 Taking the Presence of Matter into Consideration

The world of force fields that Heaviside painted thus far lacks a major ingredient, namely, the interaction of the electric and magnetic force fields with matter. Without the benefit of Heaviside's philosophical and methodological introduction, the construction of this ingredient might seem like too tall an order. It would appear to require a detailed theory of the electric and magnetic structure of matter before its interaction with electric and magnetic fields could be elucidated. Having read the four papers described here under the collective title, "An Introduction to Field Thinking for the Intelligent Non-Mathematical Electrician," Heaviside's student would probably know better than to expect any such theory. Nothing is known about the nature of electricity and magnetism properly speaking, Heaviside would say, so a theory of the electromagnetic structure of matter is obviously out of the question. However, for the purpose of formulating the laws of interaction between matter and the two force fields, the necessary properties of matter may be summed up by three coefficients. They are the specific electric conductance, or conductivity (k), the specific electric capacitance, or capacity (c), and the specific magnetic inductance, or inductivity (μ).

The electric conductivity represented to Heaviside all that was known about matter from Ohm's law. In a conductive material characterized by conductivity k, the presence of an electric field $\mathbf{E}$ is always accompanied by an electric current of density $\mathbf{C}$ such that $\mathbf{C} = k\mathbf{E}$. In the most general case $\mathbf{E}$ and

89. John Tyndall, *Heat: A Mode of Motion*, fourth edition, (1870), p. 191.
90. *Ibid.*, pp. 194-195.

C have different magnitudes and different directions, and k**E** does not stand for the arithmetic product of a single quantity k by the magnitude of **E**. Instead, it becomes an operation that changes both the direction and magnitude of **E**. The conductivity k would then be a 3x3 matrix rather than a single number. Much of Heaviside's more advanced work revolves around the detailed study of such general cases. However, while they often represent Heaviside's highest achievements in the application of Maxwell's theory to difficult problems, these studies do not shed further light on his general reasoning. Therefore, our present purposes do not require going beyond the elementary case of k as a single number representing an isotropic conductor.

5.2 Electric Displacement and the Case against 4π

Not all matter is conductive. The presence of electric force in a dielectric is not accompanied by an electric current. This does not mean, however, that dielectrics do not interact with the electric force. In a dielectric, the electric force is always accompanied by electric displacement. The best way to understand the concept of electric displacement, Heaviside explained, is to think of the literal meaning of the word. One ought to think of a free electric charge placed in a dielectric as displacing electricity in all directions around itself. The total amount of electricity displaced through a surface surrounding the free charge is equal in quantity to the displacing charge:

> According to Maxwell's remarkable theory there is a real displacement of electricity all along a tube of force, proportional in amount to the strength of force at any place, and in the same direction in general (i.e., in an isotropic medium, whose specific capacity does not vary in different directions); the whole displacement across one section being the same as that across any other. If this amount be e for a certain tube, then at the commencement of the tube there is an amount e of positive, and at its end an equal amount of negative electrification, or of *free* electricity, to distinguish it from the electricity displaced in other parts of the tube, which gives no indication of its presence, for a similar reason that a uniformly magnetized magnet shows no signs of "magnetism" save at the ends of the lines of magnetization.[91]

Thus, a charge q in a dielectric is associated with an electric displacement vec-

91. *Electrical Papers*, Vol. I, p. 256.

tor **D** such that:

$$\mathbf{D} = \frac{q}{4\pi R^2}\mathbf{r},$$

where **r** is a unit vector in the radial direction and R is the distance to the point at which **D** is reckoned. Regardless of the dielectric material in which q resides, it will give rise to **D** as stated above. Thinking back to Gauss's theorem, this immediately suggests that any space distribution of charge, $q(x,y,z)$, will be associated with a well-defined electric displacement **D** such that $q(x,y,z) = \mathrm{div}(\mathbf{D})$.

Dynamically, Heaviside observed, a dielectric permeated by electric displacement **D** should be thought of as a compressed electrical spring. In particular, **D** may be considered as dynamically analogous to the mechanical strain in an elastic solid under stress. As such, one would expect the electric displacement to be accompanied by a related stress, namely, the elastic force that develops along with the strain. Of course, in this case the strain is an electrical one so it must not be regarded as an actual, physical displacement of a definite material entity. It becomes instead a generalized displacement associated with the electric force as a generalized force and the current density as a generalized velocity. Indeed, the electric force **E** is precisely the generalized force to associate with the displacement **D**. In the mechanical case we consider $\mathbf{X} = \mathsf{e}\mathbf{F}$ to represent the connection between the strain **X** and stress **F**, where e sums up the elastic properties of the material. In the electrical case, there is an equivalent representation of electrical "elasticity" c, such that $\mathbf{D} = \mathsf{c}\mathbf{E}$. In most practical cases, using the standard units of electricity in Heaviside's time, the relationship turns out to be $\mathbf{D} = (c/4\pi)\mathbf{E}$, where c is the coefficient of specific capacity for the particular dielectric involved.

As Heaviside correctly noted, the additional factor of 4π reflects a peculiarity of the traditional electric units, which had been adopted in accordance with Coulomb's law for the force between two charges in a vacuum. To fully appreciate this point, compare Coulomb's law to Heaviside's field-oriented point of view. Coulomb's law may be taken as a definition of the unit electrostatic charge as a result of the following idealized experiment: charge two conducting balls by using the same procedure, then measure the force between them by measuring the extent to which they compress an attached spring. By placing the balls at different distances it may be found that the force between them is inversely proportional to the distance squared. Using that relationship, and under the assumption that having been charged by the same procedure the two balls carry equal charges, the force law may be written as $F = q^2/r^2$, q being the charge on each ball. This defines the unit charge as the amount of electricity that exerts a unit force at a unit of distance. Heaviside, however, regarded an electric charge as the source of an electric displacement field. The amount of charge then represents the strength of the source. If the charge q is the sole source within a given space of unit capacity, and if displacement cannot disappear by itself, then by $1\mathbf{E} = \mathbf{D}$ the total amount of force through the surface of a sphere centered on q must be equal to q. The strength E of the force field at any point on the sphere's surface would then be $q/4\pi r^2$, and the force between two equal charges would now be $qE = q^2/4\pi r^2$. The unit of charge may once again be defined as the amount of charge that exerts a unit of force at a unit of distance. Since both systems employ the same units for force and distance, the difference between the two force laws makes it clear that a unit charge in the second system is $(4\pi)^{1/2}$ multiplied by a unit charge in the first system. In the first system div($\mathbf{E}$) = $4\pi q$; in the second div($\mathbf{E}$) = q, and all other field equations reflect the geometry of the field view directly without an artificially added factor of 4π. For Heaviside, who believed that the mathematics of physics should reflect the physical ideas as faithfully as possible, there was no question that the second system of units should be the one to use in electromagnetism formulated after Maxwell. From Heaviside's point of view, the traditional choice of units seemed quite unnatural, and in later years he fought it bitterly. At this point, he only commented:

> The excrescence 4π is a mere question of units, and need not be discussed here. The 4π's are particularly obnoxious and misleading in the theory of magnetism. Privately I use units which get rid of them completely, and then, for publication, liberally season with 4π's to suit the taste of B.A. [British Association for the Advancement of Science] unit-fed readers.[92]

5.3 *The Cardinal Feature of Maxwell's Theory: The Displacement Current.*

Heaviside had already remarked that in Maxwell's theory, a free charge is regarded as the source of electric displacement throughout the space surrounding it (see page 146). Now, Heaviside continued, it must take time for a di-

92. *Electrical Papers*, Vol. I, p. 432.

electric medium into which free electricity has been introduced to settle into its new, strained equilibrium. The electric displacement starts out as zero everywhere, and ends up as $q/4\pi r^2$ (using Heaviside's "rational units"). While the displacement is being established, more and more electricity is displaced through each and every surface that bounds the free charge. By definition, this flow must constitute electric current. Hence, the rate at which the electric displacement varies is electric current for all practical and theoretical purposes. Of course, owing to the elastic character of electric displacement this current cannot flow forever. It will cease once the electric stress that resists the displacement counteracts the electrical "pressure" exerted by the free charge. Also, since the dielectric current is purely an elastic displacement, it encounters no "frictional" resistance and produces no heat. All the energy of the electric displacement is therefore recoverable. To sum up, in pure conductors characterized by zero specific capacity, only a conduction current will flow; in pure dielectrics where the conductivity is zero, only a displacement current can exist; but in electric media characterized by finite conductivity *and* specific capacity, the electric current is the sum of the conduction current and the displacement current. That sum, not each of its components separately, is always continuous. The displacement current, Heaviside observed, was the cardinal feature that enabled Maxwell to formulate his electromagnetic field theory as a dynamically complete scheme:

> The most remarkable and distinguishing feature of Maxwell's theory of electromagnetism is his dielectric current, whose introduction into the theory gives us a dynamically complete system, with propagation of disturbances in time through the medium surrounding and between conductors, doing away with the mathematically expressible but practically unimaginable, instantaneous actions of currents upon one another at a distance, and similar, though more simply expressible, instantaneous forces between imaginary accumulations of the "electric fluid." Some might say that the distinguishing feature is his dielectric displacement. But it is scarcely that, strictly. For that a dielectric medium was put into a state of polarisation by electric force was Faraday's idea, and this polarisation is only another name for electric displacement. ... [T]he important step made by Maxwell was the recognition that changes in the displacement constitute a real electric current (though without dissipation of energy in Joule-heating), and that the electric current, whether conductive or not, is always continuous. Accumulations are done away with altogether, and with their abolition the fluid or fluids become meaningless.[93]

Shortly it will become clear why Heaviside considered Maxwell's a dynamically complete theory. First, however, a basic understanding of the displacement current, or as Heaviside refers to it in this particular passage, the dielectric current, is necessary. Consider an open loop of conductive wire of specific conductivity *k* and specific capacity zero. Connect each of its ends to a conductive plate of the same material, so that the two plates are separated only by a very small distance. Let the space between the two plates be filled with a dielectric of capacity *c* and conductivity zero. Start an electric current in the system by connecting a battery in series with it. Since the conducting loop is broken and a dielectric material fills the space between the two plates, the current cannot continue flowing for long. According to the old fluid theories that Heaviside referred to, the electric current, regarded as some sort of substantial flow, accumulates on the conducting plates. These are the "accumulations" Heaviside mentioned; and they create an electromotive force opposed to that of the battery. When a sufficient amount of charge accumulates on the plates, the two forces become equal, and the current ceases to flow. The amount of charge that needs to be deposited on the plates to oppose a given electromotive force depends on the plates' capacity to store electric "substance": the greater the capacity, the greater the amount of accumulation under a given force.

In Maxwell's view, Heaviside would say, what occurs in the above electrical arrangement is quite different. Electric force permeates the conducting wire and an electric current accompanies it. The current is now just the generalized velocity associated with the electric force, not to be confused with material flow of any kind. As current "flows" in the circuit, electric displacement occurs throughout the dielectric material separating the two conducting plates; that is to say, electric strain begins to develop in it. The rate at which this generalized strain develops between the plates is equal to the current in the conductor. Obviously, the current cannot last forever, because as the strain grows, so does its elastic opposition to the change. All current ceases once the stress in the dielectric exactly counterbalances the force generated by the battery. There is no accumulation of any substance on the plates. The distribution of "charge" that we are accustomed to envision on the surface of the plates is replaced by the divergence of the displacement lines that terminate abruptly on the plates (recall that the conducting plates are characterized by a specific

93. *Electrical Papers*, Vol. I, pp. 476-477.

capacity of zero and hence do not support electric displacement). The dielectric between the two plates becomes analogous to a compressed spring. The total "charge" on the plates actually represents the total displacement of this generalized spring from its neutral state. Dynamically speaking then, charge is a generalized displacement. The energy stored in the capacitor is no longer that of a force acting at a distance between two charged plates. Rather, it is stored throughout the dielectric as the energy of a strain, similar to the energy of a compressed spring and perfectly analogous to the energy stored in an elastically strained solid. Not surprisingly, when Heaviside turned his attention to the nomenclature of electromagnetism (see chapter IV), he preferred to speak of the "elastance" of a capacitor rather than its capacitance. "As for the positive and negative charges," he explained in an account of the charged capacitor, "they are numerically equal to the total displacement through the condenser. They are located at the places of, and measure the amount of discontinuity of the elastic displacement, and that is all."[94]

In view of this statement, it may be surprising that Heaviside relied so heavily on the image of electricity displaced through space to introduce the displacement current. He could have justified the perception of variable displacement as a legitimate electric current using purely dynamical arguments. The expression for the energy of the electric force in Heaviside's units is $(1/2)\,\mathbf{E}\cdot\mathbf{D}$. If **E** is a generalized force, as has been assumed throughout, then for the above product to yield energy, **D** must be a generalized displacement. Therefore the time derivative of **D** must be a generalized velocity associated with **E**. Consider, however, that the generalized velocity associated with **E** is the electric current density. It follows that for all practical purposes we may consider the time variation of the electric displacement as an electric current density. Heaviside, however, always gave precedence to the ideas behind the formalism. To most readers of *The Electrician*, a capacitor was a charge-storage device. Heaviside's first order of business was to undermine this view. As he shrewdly observed in connection with the introduction of novel terminology and formalism:

> Ideas are of primary importance, scientifically. Next, suitable language. As for the notation, it is an important enough matter, but still only takes the third place.[95]

94. *Electrical Papers*, Vol. II, p. 80.
95. *Ibid.*, p. 25.

5.4 *Magnetic Induction and Completion of the Rough Sketch*

The conduction current, electric displacement, and the derived displacement current complete the sketch of the interaction between matter and the electric force. In the magnetic case things are quite similar, with one important qualification due to the apparent nonexistence of a magnetic conduction current. Magnetic induction is the magnetic equivalent of electric displacement.[96] Magnetic strain accompanies magnetic force in matter. Thus, $\mathbf{B} = \mu\mathbf{H}$ expresses the relationship between magnetic induction **B** and the accompanying magnetic force **H**. The coefficient μ is the coefficient of magnetic "elasticity," or the inductivity of the magnetic medium. A fundamental difference between electric and magnetic force is the lack of a magnetic equivalent of the electric conduction current. That is to say, there exists no magnetic heat production similar to the heat produced by the electric current in a resistive conductor. However, in perfect analogy with the dielectric or displacement current, there is a magnetic "elastic," non-dissipative current wherever magnetic induction is varying. The rate of variation of **B** is a magnetic current just as the time rate of change of electric displacement is electric current.

For an astonishingly wide range of theoretical and practical applications, k, c and μ are all that one needs to know about matter to study various electromagnetic effects. This includes the propagation of signals along telegraph and telephone lines, as well as electromagnetic waves in dielectrics—without a doubt the most exciting electromagnetic phenomenon of the 1880s. As far as Heaviside was concerned, the three simple connections between electric and magnetic forces and their generalized fluxes—electric current, electric dis-

96. "A comparison is often made between distributions of magnetic induction and of electric current. There is, however, a far more satisfactory analogy between magnetic induction and electric displacement in a dielectric, which may be pushed much further before correspondence ceases." (*Electrical Papers*, Vol. I, p. 471). Correspondence ceases once it is stipulated that $d\mathbf{B}/dt$ is the total magnetic current. Like the total electric current the magnetic current never diverges. In the electric case, however, the displacement current by itself can diverge, and that divergence is the rate at which displacement develops in a particular volume element. If, however, the magnetic conduction current consists only of the time variation of magnetic induction, and if, as we required, the total current cannot diverge, then an accumulation of free magnetic charge is impossible. Hence, the lack of a magnetic conduction current is intimately tied to the absence of magnetic monopoles. Note also that this discussion reveals that the existence of free electricity requires the presence of matter (see page 187 for further discussion; see also J.Z. Buchwald, *From Maxwell to Microphysics*, [Chicago: The University of Chicago Press, 1988], pp. 25-29).

placement, and magnetic induction—made a complete outline of the interaction of fields with matter:

$$\mathbf{D} = c\mathbf{E}, \tag{3-12}$$

$$\mathbf{C} = k\mathbf{E}, \tag{3-13}$$

$$\mathbf{B} = \mu\mathbf{H}. \tag{3-14}$$

In addition to these, the electric displacement **D** is related to the space distribution of free electricity q by:

$$\operatorname{div}\mathbf{D} = q\,(x, y, z)\,. \tag{3-15}$$

In a similar way, one may expect the space distribution of free magnetism, say $m(x,y,z)$, to be related to the magnetic induction by:

$$\operatorname{div}\mathbf{B} = m\,(x, y, z)\,.$$

One could argue that since **B** is always derivable from a vector potential **A** through the relationship **B** = curl(**A**), it follows necessarily that the divergence of **B** is zero, because the divergence of a curl always vanishes. In all likelihood, Heaviside would have objected to this argument on the ground that it represents a complete reversal of the correct order of reasoning. The only reason we may write **B** = curl(**A**) is because div(**B**) is known experimentally to be zero. The vector potential **A** is an extremely useful device, he would add, but dynamically speaking, the force **H** and its related strain—the magnetic induction **B**—take precedence. A dynamically sound theory must be expressed in terms of the dynamically important concepts, namely, the generalized forces, displacements, and velocities. Otherwise, all semblance of a mechanism disappears. Heaviside therefore observed that all known experimental data indicated the absence of free magnetism, and that to satisfy this observation the magnetic equivalent of eq. (3-15) should be:

$$\operatorname{div}\mathbf{B} = 0. \tag{3-16}$$

The time variation of electric displacement and magnetic induction constitutes electric and magnetic non-dissipative currents. Writing **G** for the magnetic current and Γ for the electric current, Heaviside has:

$$\mathbf{G} = \frac{d\mathbf{B}}{dt}, \tag{3-17}$$

$$\Gamma = \mathbf{C} + \frac{d\mathbf{D}}{dt}. \tag{3-18}$$

Up to now, all we really have are two separate theories, one of electricity, the other of magnetism. The name "electromagnetism" derives from the realization that electricity and magnetism are fundamentally related. In Maxwell's theory, Heaviside explained, the relationship between electricity and magnetism is expressed by two laws, namely, Ampère's law and Faraday's law. The former relates electric current to the magnetic field as follows:

$$\mathbf{C} = \operatorname{curl} \mathbf{H}'.$$

However, with Maxwell's displacement current in mind, Ampère's law must be corrected to include the total current Γ, or:

$$\Gamma = \operatorname{curl} \mathbf{H}' = \mathbf{C} + \frac{d\mathbf{D}}{dt}.$$

H' in the above equations is generally not the same as **H** in the previous equations. In Heaviside's discussion, eqs. (3-12) through (3-18) involve the total electric and magnetic forces, that is, the sum of internal and externally impressed forces. Ampère's law, however, relates the current only to the internal magnetic force, denoted here by **H'**. Heaviside defined an impressed force as one that transfers energy into the system under consideration:

> ... we may define impressed electric force thus. If **e** be the impressed electric force at a point, and Γ the electric current there, **e**Γ of energy is taken into the electromagnetic system there per unit volume per second. Similarly, we may define the impressed magnetic force **h** at a point, by saying that if there be a magnetic current **G** there, **hG** of energy is taken in per second per unit volume by the electromagnetic system there.[97]

Symbolizing the impressed magnetic force by **h**, we have **H'** = **H** - **h**, and

97. *Electrical Papers*, Vol. I, p. 449.

Ampère's law becomes:

$$\text{curl}\,(\mathbf{H}-\mathbf{h}) = \Gamma = \mathbf{C}+\frac{d\mathbf{D}}{dt}. \tag{3-19}$$

Expressing the electric current in terms of the magnetic force, we obtain the first explicit cross connection between the electric and magnetic forces (using Heaviside's units):

$$\text{curl}\,(\mathbf{H}-\mathbf{h}) = k\mathbf{E}+\frac{d}{dt}c\mathbf{E}.$$

The second fundamental relationship in Heaviside's scheme is provided by Faraday's law, which relates the magnetic current to the internal electric force as follows:

$$-\text{curl}\,(\mathbf{E}-\mathbf{e}) = \mathbf{G} = \frac{d\mathbf{B}}{dt}. \tag{3-20}$$

Expressing the magnetic induction in terms of the magnetic force immediately transforms the above into another direct relationship between electric and magnetic force, thus:

$$-\text{curl}\,(\mathbf{E}-\mathbf{e}) = \mathbf{G} = \frac{d}{dt}\mu\mathbf{H}.$$

The two circuital equations (3-19) and (3-20) are the core of Heaviside's "duplex" system, which describes Maxwell's theory of electromagnetism in terms that are symmetric for the magnetic and electric aspects. With the circuital equations in hand, Heaviside finally proclaimed: "We have now a dynamically complete system."[98]

As the previous sections demonstrate, Heaviside regarded a region of space permeated by an electromagnetic field as analogous to a mechanically strained medium. He wanted a theory that could furnish a complete description of the forces and associated displacements in such a strained system. However, in the electromagnetic case, the forces are not real forces, and the displacements are not real displacements. Nevertheless, Heaviside had good reasons for dignifying **E**, **D**, **H** and **B** with the titles of forces and displacements. The electric and magnetic force fields **E** and **H** are closely related to

98. *Electrical Papers*, Vol. I, p. 449.

the actual mechanical forces experienced by electrified bodies and magnets immersed in them. For example, an electrified body carrying a charge q immersed in an electric field **E** experiences a mechanical force $\mathbf{F} = q\mathbf{E}$. Hence **E** represents the force per unit charge. In addition, **E** may be paired with $\dot{\mathbf{D}}$ and **H** with $\dot{\mathbf{B}}$ to create expressions for the electric and magnetic activities. This suggests the possibility of regarding **D** and **E** as associated generalized displacement and generalized force, and subjects them to analysis in terms borrowed from generalized classical mechanics. The same could be done with **B** and **H**. Maxwell's equations as reformulated by Heaviside would now denote the equations of motion in the electromagnetic system, establishing the basic connection between forces and displacements. In this sense, then, the circuital equations could indeed be regarded as the core of a dynamical system. Unfortunately, the description of the duplex system as "dynamically complete" is still somewhat premature. Heaviside required that a dynamically complete system provide a full account of the transfer of energy through contiguous space elements, but he had yet to show that his system could indeed fulfill this requirement. Fortunately, Heaviside did not leave this issue unattended for long. He addressed it immediately following the discussion above, in one of the most illuminating sections of his work, entitled "The Equation of Energy and its Transfer."[99]

5.5 Circuits, Forces and the Equation of Energy Transfer: The Origins of Heaviside's Duplex Equations

In his rough sketch of Maxwell's theory, Heaviside completed the presentation of Maxwell's equations in a new format that he maintained throughout the rest of his work. There is nothing particularly difficult about the transformation from the potential formulation that Maxwell used to Heaviside's duplex equations (eqs. (3-19) and (3-20)). Maxwell's first fundamental equation is the definition of the electric current as the curl of the magnetic force. This is also Heaviside's first circuital equation. The second circuital equation is merely a matter of taking the curl of both sides of Maxwell's second fundamental equation,

$$\mathbf{E} = -\dot{\mathbf{A}} - \nabla P,$$

99. *Electrical Papers*, Vol. I, pp. 449-450.

and noting that curl $\dot{\mathbf{A}} = \dot{\mathbf{B}}$ and that the gradient of a scalar potential always has a zero curl. Maxwell himself was aware of the possibility of formulating the equations directly in terms of the fields, but elected not to use this formulation as the basic one. The question we must therefore ask is, what were Heaviside's reasons for preferring the duplex system. Heaviside often complained about the unwieldy nature of Maxwell's potential formulation,[100] but he never explained what precisely drove him to the specific form of the two circuital laws (eqs. (3-19) and (3-20)). His argument concerning the localization of the energy in the field accounts for his emphasis on the electric and magnetic forces and their associated displacements. By itself, however, this cannot account for the separation of the internal and external components of the forces, which was unique among all other formulations of Maxwell's theory. In the absence of an organized account by Heaviside on the origins of his duplex equations, any consideration of these origins must remain somewhat speculative. However, while Heaviside never fully explained the origin of his format, his work yields powerful indications regarding the motivations behind it.

We have already seen that the thematic break between Heaviside's transmission-line work and his introduction to field thinking is more apparent than real. From the beginning, his approach to the presentation of field concepts reflected an ongoing desire to obtain a comprehensive theory of electric circuits. His duplex equations actually represent the successful completion of the basic groundwork for such a theory. To see this, we may begin by applying Heaviside's second circuital equation to an isolated circuit containing voltage sources together with resistive, capacitive, and inductive elements. Integrating the equation over the area bounded by the circuit, we obtain:

$$-\int_S \text{curl}\,(\mathbf{E}-\mathbf{e})\cdot \mathbf{dS} = \frac{d\Phi}{dt},$$

where Φ is the total magnetic induction enclosed within the circuit. Using Stokes's theorem, this becomes:

$$\oint_C \mathbf{e}\cdot \mathbf{dl} = \oint_C \mathbf{E}\cdot \mathbf{dl} + \frac{d\Phi}{dt}.$$

100. *Electrical Papers*, Vol. II, pp. 92-93; 173.

The contour integrals are taken around the circuit. The first one therefore measures the sources of impressed voltage in the circuit (batteries, for instance), while the second one sums up the voltage drop over the various resistive (R), capacitive (S), and inductive (L) elements in the loop. In other words:

$$E_{\text{imp}} = V_R + V_S + V_L + \frac{d\Phi}{dt}.$$

The time variation of the magnetic inductance enclosed by the loop may be expressed in terms of the total current and loop's self-inductance L', which should not to be confused with the self-inductance L of discrete inductive elements such as coils in series with the loop. By redefining the total self-inductance in the circuit as the sum of L and L', the contribution to the voltage drop around the loop owing to variations in the total induction through the loop may be subsumed under the inductive voltage drop in the circuit:

$$E_{\text{imp}} = V_R + V_S + (L + L')\frac{dC}{dt}.$$

This yields the general voltage law:

$$E_{\text{imp}} = \sum V_{\text{int}}. \tag{3-21}$$

This direct and intuitive way of transforming the second circuital equation into the familiar form of Kirchhoff's voltage law must have seemed very promising to Heaviside in his quest for a comprehensive framework for circuit theory. In particular, it explains why Heaviside elected to express the magnetic current in terms of the curl of the difference between internal and impressed forces: in application to practical circuits, this form naturally yields the required separation between voltage sources and voltage drops in a circuit.

By itself, however, the second circuital equation does not provide a complete description of an electric circuit. It is a simple exercise to show that Kirchhoff's current law emerges naturally from the first circuital equation, by taking its divergence, and then integrating it over the volume of any junction between conductors. One has only to distinguish between surfaces through which current enters the junction and surfaces through which it leaves the junction. Moreover, this derivation of the current law automatically generalizes to cases that involve capacitors, wherein the divergence of **D** is not zero. This argument, however, might suggest that the proper companion to the second circuital equation is not the first circuital equation but its divergence. Heaviside preferred eqs. (3-19) and (3-20) as the fundamental equations of motion in the electromagnetic system because they complemented his requirement of dynamical completeness. To see this, note first that eq. (3-21) is analogous to Newton's third law, which states:

$$\mathbf{F}_{\text{imp}} = \sum \mathbf{F}_{\text{int}}.$$

This is a general law, which applies to both isolated and coupled systems. It is useful, however, to distinguish between an isolated system in which all the forces are internally balanced, and coupled systems in which overall balance is achieved through some sort of mutual interaction. Under such circumstances, especially when the coupling forces are not explicitly specified, it is more useful to reformulate the third law in terms of the activity of the forces in the system:

$$\mathbf{F}_{\text{imp}} \cdot \mathbf{U} = \sum (\mathbf{F}_{\text{int}} \cdot \mathbf{U}) - \dot{W},$$

where **U** represents the generalized velocities associated with the internal and impressed forces within the system, while $-\dot{W}$ represents all net energy flow into the system that does not originate in the activities of the impressed and internal forces (such as heat radiation from without).

In the analogous electrical case, imagine that the circuit we have been examining is inductively coupled to some external electrical environment. In such a case, the inductive voltage drop should include a reference to the mutual inductance between the circuit and its environment. There are, however, various ways to couple an electrical system to its environment. Once again, therefore, the activity formulation is advantageous:

$$E_{\text{imp}} C = \sum (VC) - \dot{W},$$

where C is the current in the circuit and $-\dot{W}$ is the rate at which any coupled external system transfers electrical energy to the circuit.

When we come to examine the possibility of expressing the dynamics of the electromagnetic field in an analogous way, it becomes immediately clear that the second circuital equation is insufficient. The electromagnetic field includes two interrelated forces, electric and magnetic, each with its associated generalized displacement and velocity. The second circuital law relates the electric force to the magnetic current. We need the first circuital equation to relate the electric current to the magnetic force. When both equations are used in the context of the activity principle, they yield a remarkable result that demonstrates how the two equations together describe a dynamically complete system. Writing the activity principle for an infinitesimal volume permeated by an electromagnetic field, we have in analogy with the above:

$$\mathbf{e} \cdot \Gamma + \mathbf{h} \cdot \mathbf{G} = \mathbf{E} \cdot \Gamma + \mathbf{H} \cdot \mathbf{G} - \dot{W}.$$

Next, consider that in the field case energy must flow continuously between contiguous space elements. Therefore, to satisfy continuity of energy we must require:

$$\dot{W} + \operatorname{div} \mathbf{w} = 0,$$

where $\mathbf{w}$ represents the density of energy flow through the system. Substituting into the activity principle, we obtain:

$$-\operatorname{div} \mathbf{w} = (\mathbf{E} - \mathbf{e}) \cdot \Gamma + (\mathbf{H} - \mathbf{h}) \cdot \mathbf{G}.$$

Substituting for the electric and magnetic currents from the two circuital equations, this transforms into:

$$\begin{aligned} -\operatorname{div} \mathbf{w} &= (\mathbf{E} - \mathbf{e}) \cdot \operatorname{curl} (\mathbf{H} - \mathbf{h}) - (\mathbf{H} - \mathbf{h}) \cdot \operatorname{curl} (\mathbf{E} - \mathbf{e}) \\ &= -\operatorname{div} [\, (\mathbf{E} - \mathbf{e}) \times (\mathbf{H} - \mathbf{h}) \,] . \end{aligned}$$

In other words:

$$\mathbf{w} = (\mathbf{E} - \mathbf{e}) \times (\mathbf{H} - \mathbf{h}) . \tag{3-22}$$

Thus, in addition to determining all of the energy transformations between the electric and magnetic forces within the volume element, the circuital equations also specify the energy current through the element in terms of the electromagnetic forces inside it. The energy current flows perpendicularly to the plane defined by the internal electric and magnetic forces. It points in the direction of the right thumb when the curled fingers of the right hand indicate the shortest rotation that would bring the direction of electric field into congruence with the direction of the magnetic field. Its intensity is the product of the intensities of the electric and magnetic fields multiplied by the sine of the angle between them.

Newton's third law requires that when mechanical forces propagate action through a mechanical system, they must conserve the total action at all times (see page 131). Heaviside has just demonstrated that as generalized forces, the electric and magnetic fields satisfy the same requirement in an electromagnetic system. His duplex equations therefore provide a complete dynamical description of the electromagnetic field, and, as he concluded: "The appropriateness of [the second circuital equation] as a companion to [the first circuital equation] is very clearly shown."[101]

To recapitulate, with his duplex field equations Heaviside could preserve a close affinity to circuit theory through an analogy with Newton's third law and the activity principle. At the same time the reformulated circuital equations provided him with a smooth, natural transformation to Kirchhoff's circuit laws. This dual nature of the relationship between field and circuit theory through the activity principle accounts for all the special characteristics of Heaviside's duplex equations. In particular, the required affinity to circuit theory accounts for the place Heaviside gave to impressed forces in the scheme.

Heaviside's equation of energy transfer also highlights the importance of the displacement current. It is not necessary to have a displacement current to formulate an electromagnetic theory consistent with the principle of energy conservation. However, without the displacement current the electric force has no activity in a dielectric. Consequently, there is no way of describing a continuous transfer of energy through the dielectric medium that separates conductors. From such a theoretical point of view energy will simply disappear in one place and reappear in another, without showing up in the intermediate space. Thus, only with the inclusion of the displacement current does the theory become dynamically complete in Heaviside's sense of the word.

Finally, Heaviside made an important qualification regarding Maxwellian electrodynamics. To obtain a picture of energy transfer one must have explicit knowledge of the electromagnetic force system. If there exists an indeterminacy in the specification of the forces, then certain energy transformations may indeed be taking place right under one's nose, but they cannot be identified:

> It is like this. If a person is in a room at one moment, and the door is open, and we find that he is gone the next moment, the irresistible conclusion is that he has left the room by the door. But he might have got under the table. If you look

101. *Electrical Papers*, Vol. II, p. 174.

there you can make sure. But if you are prevented from looking there, then there is clearly a doubt whether the person left the room by the door or got under the table hurriedly. There is a similar doubt in the electromagnetic case in question, and in other cases. Thus, we can unhesitatingly conclude from the properties of the magnetic field of magnets that the mechanical force on a complete closed circuit supporting a current is the sum of the electromagnetic forces per unit volume..., but it does not follow strictly that the so-called electromagnetic force is the force really acting per unit volume, for any system of forces might be superadded which cancel when summed up round a closed circuit. So, in the transfer-of-energy case, there may be any amount of circulation of energy in closed paths going on (as pointed out in another manner by Prof. J.J. Thomson), besides the obviously suggested transfer, provided this superposed closed circulation is without dissipation of energy. Or, if **W** be the vector energy-current density, according to the above mentioned rule, we may add to it another vector, say **w**, provided **w** have no convergence anywhere. The existence of **w** is possible, but there does not appear to be any present means of finding whether it is real, and how it is to be expressed.[102]

6. There Must be Ether

In 1927 Albert Abraham Michelson wrote in contemplation of the theory of relativity that:

> The existence of an ether appears to be inconsistent with the theory.... But without a medium how can the propagation of light waves be explained? How explain the constancy of propagation, the fundamental assumption (at least of the restricted theory) if there be no medium?[103]

Abraham Pais observed that, "This is the lament not of a single individual but of an era, though it was an era largely gone when Michelson's book came out."

The force of Michelson's "lament" is most beautifully illustrated in a picture by M.C. Escher. It is a very simple depiction of what appears to be the sun or moon peeking behind some upside down tree branches. Both the tree branches and the sun or moon are rippled in a manner that suggests most forcefully that one observes their reflection in a water surface, perturbed by the ripples of two pebbles just thrown into it. Indeed, the English, German, and

---

102. *Electrical Papers*, Vol. II, pp. 92-94.

103. Reproduced in Abraham Pais, *Subtle is the Lord: The science and life of Albert Einstein*, (Oxford: Oxford University Press, 1982), p. 115.

French translations of the picture's title read respectively: "Rippled Surface," "Gekräuselte Wasserfläche," and "Cercles dans L'eau."

Note, however, how cunningly Escher entitled his picture by the single word "Rimpeling," which denotes the act of being rippled but avoids the specification of what precisely is being rippled. A closer examination of the picture quickly reveals that nowhere is the water surface, or any other surface for that matter, explicitly drawn. There are only correlated ripples in the tree branches and moon. The observer's mind adds in the water.[104] Apparently then, correlated phenomena such as waves suggest almost intuitively the existence of a supporting medium.

On further reflection this elegant and tempting explanation of the nineteenth-century's need for an ether seems quite problematic. The wave propagation of electromagnetic fields emerges mathematically from Maxwell's equations without any further assumptions. They state that given the required initial and boundary conditions, the intensity of the electric and magnetic forces will vary periodically in space and time. Strictly speaking, one does not require an ether, or any other kind of medium for this conclusion to hold. This is perhaps best demonstrated by the fact that the basic relationship between the electric and magnetic fields survived the transition from Newtonian to special-relativistic dynamics without alteration. Could it be then that an entire generation of natural philosophers was fooled by something like Escher's clever picture?

Like Michelson, Heaviside regarded the ether as the medium that propagates electromagnetic waves. Note, however, how he explained the basic necessity of having an ether in his rough sketch of Maxwell's theory:

> Ether is a very wonderful thing. It may exist only in the imaginations of the wise, being invented and endowed with properties to suit their hypotheses; but we cannot do without it. How is energy to be transmitted through space without a medium?[105]

---

104. The picture is reproduced in Douglas R. Hofstadter, *Gödel, Escher, Bach*, (New York: Vintage Books, 1980), p. 257. It would seem that Hofstadter himself was taken by Escher's ruse, as he also refers to the picture in terms of water reflections (see *ibid.*, p. 256). Escher himself, it appears, knew exactly what he was doing: "Two systems of concentrically expanding ripples are generated by ascending air bubbles or by fallen raindrops. They disturb the stillness of the reflection of a tree with the moon behind it. The circles, perspectively seen as ellipses, are the only means to suggest the water surface." M.C. Escher, *Escher on Escher: Exploring the Infinite*, (1989), p. 59.

105. *Electrical Papers*, Vol. I, p. 433.

For Heaviside, the idea of energy not stored in some material body would have been quite absurd. All forms of energy contemplated at the time were invariably associated with the presence of matter as something quite distinct from energy. There is no sense of speaking about the energy of a compressed spring without the spring; or of the energy stored in a strained solid without the solid; or of the potential energy of a stone at a certain height above the ground without the stone; or of the kinetic energy of a moving body without the body. At the time Heaviside worked, the basic framework in which to ground the question of energy storage and transfer was Newtonian dynamics. It would not permit him to conceive of the transfer of energy without a material medium. But in Heaviside's version of Maxwell's theory this general requirement received a far more specific form. Magnetic force, for example, is always associated with magnetic induction. Heaviside made it crystal clear in his rough sketch that magnetic induction is a magnetic polarization of the space permeated by the force. It is the magnetic analogue of electric displacement. Time variations in the magnetic induction constitute a magnetic current analogous to the electric displacement current; it is the generalized velocity associated with the magnetic generalized force. Their product is the activity of the force, or the rate at which energy is being transferred by the varying condition of the generalized strained state defined by the magnetic induction and associated force. But this very mode of speech necessarily presumes something that could be strained, or magnetically polarized. The idea of magnetic induction in empty space simply makes no sense. If magnetic induction is a polarization, then something must be polarized. It cannot be an empty space. Even if it is a generalized strain, or generalized displacement, there still must be some material basis to support the generalized strain, unless one is willing to let the generalized force and associated displacement become purely mathematical conventions. Heaviside expressed this view explicitly at the beginning of his three-volume treatise on electromagnetic theory:

> All space must be conceived to be filled with a medium which can support displacement and induction. In the former aspect only it is a dielectric. It is, however equally necessary to consider the magnetic side of the matter, and we may without coining a new word, generally understand by a dielectric a medium which supports both the fluxes mentioned.[106]

106. *Electromagnetic Theory*, Vol. I, p. 21.

Thus, even in space devoid of matter, where the force and the flux are numerically equal, the physical distinction between them must be maintained. Indeed, this distinction underscores the word "dynamics" in Heaviside's conception of electrodynamics. This notion is of crucial importance if Heaviside's rendition of Maxwell's theory is ever to be properly understood. In the Maxwellian view as expounded by Heaviside, the ether comes into play *before* electromagnetic waves. From this point of view, electromagnetic waves, pure and simple, do not account for Michelson's insistence on the need for ether; rather their electrodynamic establishment demands that there must be ether.[107]

7. Recapitulation: The Straight Conducting Wire Revisited

We have seen that Heaviside began his introduction to field thinking with an examination of a straight conducting wire carrying a steady current. A re-examination of the same wire provides a most striking way of appreciating the new outlook advocated by Heaviside throughout the papers examined in this chapter. The original wire had been envisioned as carrying a constant current, evenly distributed throughout the wire's cross section. The vast majority of Heaviside's prospective readers were accustomed to think of this current as a

107. Considering this, it appears that the theory of relativity did not render the ether unnecessary only from the observational point of view. By undermining the distinction Newtonian dynamics makes between matter and energy, it pulls the theoretical rug from underneath Heaviside's dynamical insistence on the presence of a material medium wherever energy exists. So it comes to be that in a modern textbook on electrodynamics, expressly written from the special relativistic point of view, one may read: "At this point we must interject a small bit of philosophy. It is customary to call **B** the magnetic induction and **H** the magnetic field strength. We reject this custom inasmuch as **B** is the truly fundamental field and **H** is a subsidiary artifact. We shall call **B** the magnetic field and leave the reader to deal with **H** as he pleases." (Melvin Schwartz, *Principles of Electrodynamics*, [New York: McGraw-Hill Book Company, 1972], p. 156). Or, consider the following from another textbook, this time concerning the electric displacement: "In the approach we have taken to electric fields in matter the introduction of **D** is an artifice which is not, on the whole, very helpful. We have mentioned **D** because it is hallowed by tradition, beginning with Maxwell, and the student is sure to encounter it in other books, many of which treat it with more respect than it deserves." (Edward M. Purcell, *Electricity and Magnetism*, [New York: McGraw-Hill Book Company, 1963], pp. 332-333). Needless to say, such interjected philosophy would have been utter gibberish to Heaviside, but then Heaviside did not know relativity, while Schwartz and Purcell were writing for a world in which the Newtonian sense of dynamics had been effectively transcended.

flow inside the conductor that conveys electrical energy from one place to another. Relying precisely on this conception of the electric current, Heaviside invoked Tyndall to express dissatisfaction both with the material view of the electric current and with a Tyndall-like notion that the current signifies conduction of energy through the wire by the communication of local motion:

> ... is there not the fact that we can send a current into a long circuit, and that it plainly travels along the wire, taking some time to arrive at the other end? Does that not show that electricity travels through the wire? To this I should have answered formerly, when filled with "Heat as a Mode of Motion," that it is a fact that there is a transformation of energy into the battery, and that this energy is transmitted through the wire, there suffering another transformation, viz., into heat; that when the current is set up steadily, the heat is generated uniformly; that the electric current in the wire is therefore some kind of stationary motion of the particles of the wire, not exactly like heat, but having some peculiarity of a directional nature making the difference between a positive and a negative current; but that there was no evidence in the closed circuit of any motion of electricity through the wire, but only of a transfer of energy through the wire.
>
> However, leaving personal details of no importance to anyone but myself, let us consider the transmission of energy through a wire.[108]

Now consider the following: in accordance with the constancy of the current throughout the conductor's cross section, there must exist a uniform electric force parallel to the wire's axis, such that $\mathbf{C} = k\mathbf{E}$ all along the wire. According to Ampère's law there must also be a magnetic force inside the wire. It is distributed in concentric circles around the wire's axis. The magnetic force is always the same at a constant distance from the axis all along the wire, but it diminishes in strength as one approaches the axis. Finally, there are no impressed forces inside the wire (except at the battery end), which makes the internal forces identical to the total forces. With the two forces thus specified, consider that the energy current, being the vector product of the electric and magnetic forces, is perpendicular to the planes defined by the electric and magnetic forces. These planes are cylindrical surfaces, distributed concentrically around the wire's axis. Therefore, the energy current in the wire flows radially inward from the wire's outer skin. It diminishes as it progresses toward the axis, since the magnetic force weakens as we approach the axis. Thus, in its flow toward the axis the energy current is characterized by a negative divergence. In other words, energy is somehow released within each vol-

108. *Electrical Papers*, Vol. I, p. 436.

ume element of the conductor as the energy current flows into it. To see where this energy goes, we need only write the equation of energy explicitly, reflecting the absence of impressed forces:

$$0 = \operatorname{div} \mathbf{w} + \mathbf{E} \cdot \Gamma + \mathbf{H} \cdot \mathbf{G}$$

$$\Rightarrow -\operatorname{div} \mathbf{w} = \mathbf{E} \cdot (\mathbf{C} + \dot{\mathbf{D}}) + \mathbf{H} \cdot \dot{\mathbf{B}}.$$

The underlying assumption of the entire exercise is that the situation is a steady one, and therefore the current and both magnetic and electric forces are everywhere constant in time. Hence, there is no displacement current and no magnetic current, which implies:

$$-\operatorname{div} \mathbf{w} = \dot{W} = \mathbf{E} \cdot \mathbf{C}.$$

Ohm's and Joule's laws imply that the product on the right-hand side represents the rate of heat production per unit volume in the conductor. The picture is now complete: as energy flows from the conductor's periphery inward to its axis, it gradually dissipates as heat.

To the nineteenth-century reader of *The Electrician* this should have been rather astonishing. No energy whatsoever flows from the battery *along* the conducting wire to wherever it may lead. The conductor is no more than a "sink" in which energy flows from the outer boundary inward to be wasted in the form of heat. Obviously, the energy must somehow reach the conductor's skin, and the only possible manner for it to do so is through the surrounding dielectric. Indeed, in the dielectric, no current—electric or magnetic—flows at all; conduction is impossible in it, and since the situation is a steady one by assumption, there exists no time variation in the magnetic induction and electric displacement. There is only a steady magnetic force around the conductor in accordance with Ampère's law, and a steady electric force whose lines start off at the battery poles and enter the conductor throughout its length almost perpendicularly. However, while no electric and magnetic currents are flowing in the dielectric, the two force distributions define a non-dissipative *energy* current that starts at the battery and flows outward throughout the space surrounding it. Part of this current enters the conductor to be dissipated as heat. The remainder reaches the electrical machinery attached to the end of the wire, there to do whatever work the machinery is designed to perform.

Elsewhere, Heaviside explained this point in greater detail.[109] In a long, homogeneous conducting wire, one end of which is at potential V_1 and the other at V_2, the electric force is $(V_1 - V_2)/L$, where L is the wire's length. The force within the conductor is parallel to the wire, for no current flows radially in it. This would also be the tangential component of the electric force just outside the wire, where a radial component does exist. The latter is determined by V, the potential difference between the wire and its return. In most situations, Heaviside noted, the proximity of the return compared to L would render the perpendicular component much larger than the tangential one. Hence, the electric-force lines enter the conductor almost perpendicularly. The slight deviation from the perpendicular is highly significant. It diverts the energy current in the dielectric from being exactly parallel to the wire, and makes it enter the wire's surface at a very slight incline. Once inside, the energy flow immediately becomes perpendicular to the conductor's axis. Thus, the force configuration in the dielectric specifies exactly how the energy reaches the conductor, and completes the energy-transfer picture.

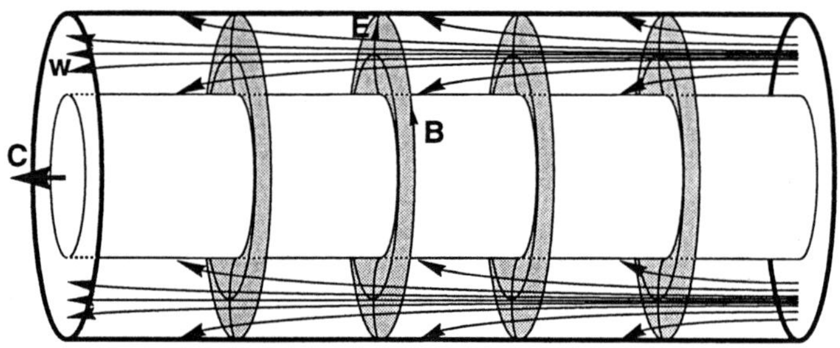

Figure 3.6: Schematic representation of energy flow in a coaxial cable. Current **C** *flows to the left in the central conductor, and to the right in the surrounding sheath; the circular magnetic field* **B** *centered on the axis of symmetry and slightly curved electric field* **E** *define the curved circular planes and the energy flux density* **w** *that flows perpendicularly through them. Note that in general the energy does not flow symmetrically to the two conductors. The flow would be skewed toward the more resistive of the two.*

109. *Electrical Papers*, Vol. II, pp. 94-95.

To further clarify the picture, consider the energy flux in a segment of coaxial cable, qualitatively represented in figure 3.6. The situation in the cable may be thought of in terms of an analogy to water flow in the gap between two concentric pipes. Assume that flat rubber gaskets are firmly pressed against the inside wall of the external pipe and against the outside wall of the inner pipe. Now apply steady pressure from the right to the water between the pipes. The water will press on the rubber gaskets, and the friction between the gaskets and the two pipes will cause them to deform in a shape roughly similar to that of the **ExH** surfaces. Should there be no friction between the gaskets and the walls, the gaskets will slip along with the water without deforming. This is equivalent to a perfect conductor, in which the energy flow is perfectly parallel to the conducting surface.

The salient point is that the conductor of energy in the Maxwellian view is the dielectric, not the electrically conducting wire. Not surprisingly, Heaviside preceded the announcement of this fascinating result with a rather dramatic paragraph:

> Consider the electric current, how it flows. From London to Manchester, Edinburgh, Glasgow, and hundreds of other places, day and night, are sent with great velocity, in rapid succession, backwards and forwards, electric currents, to effect mechanical motions at a distance, and thus serve the material interests of man.
>
> By the way, is there such a thing as an electric current? Not that it is intended to cast any doubt upon the existence of a phenomenon so called; but is it a current—that is, something moving through a wire?[110]

Heaviside apparently felt it very important to drive home the novelty of this view in the strongest possible manner. His first step was not to show how the perception of the dielectric as the energy conductor follows from the connections between the electric and magnetic forces in Maxwell's theory. Instead, he devoted a full article to a discussion of the problems involved with viewing the conduction current as an energy carrier.[111] Then, with these problems in mind, he stated that "It becomes important to find the paths along which the energy is being transmitted.... As the present section is argumentative and descriptive only, we cannot enter into mathematical details...." Without analytical derivation he proceeded to suggest the alternative view. The

110. *Electrical Papers*, Vol. I, p. 434.

111. *Ibid.*, pp. 434-441.

Maxwellian derivation of this result along the lines above appeared almost six weeks later.

Thus, Heaviside used the problematic view of the conduction current as energy carrier to motivate the rest of his rough sketch of Maxwell's theory. The reader who had gone through this argumentative and descriptive part would have been made well aware of a forthcoming fundamental shift in outlook.

8. Conclusions

8.1 Heaviside as a Teacher

Heaviside's publications between 1882 and 1884 suggest that at some point between 1880 and 1882 he set for himself a new goal. His earlier publications, surveyed in chapter II, deal exclusively with various problems associated with the telegraph circuit. He designed several new signal detectors and analyzed others, already in use, with the idea of improving their performance. These papers create an image of Heaviside as a competent engineer, eager to publish his own innovations and positive contributions to his field of expertise.

Heaviside's paper on "The Relations Between Magnetic Force and Electric Current," published in six parts between November 1882 and January 1883, changes this image. The paper is devoted to the study of basic, well known concepts like Ampère's law, and the mathematical theorems of Stokes and Gauss. It is quite safe to say that in this paper as in the three that followed it, Heaviside contributed very little by way of new electromagnetic knowledge. At the same time it would be a grave mistake to suggest that there was nothing innovative or original in these papers. Quite the opposite is the case, for the manner in which Heaviside presented the material was without precedent. Mathematically speaking, these papers were the most original of Heaviside's thus far. In them he began the introduction of a new mathematical language, namely, the algebra of vectors.

There exists a great difference, however, between the spirit in which these innovations were made and Heaviside's contributions in his earlier papers. In his early papers, Heaviside's goal was the innovation, be it a more sensitive Wheatstone arrangement for the reception of a telegraph signal, or a hitherto unexplored transmission-line problem. In the papers described here under the heading "Introduction to Field Thinking for the Intelligent Non-mathematical Electrician," Heaviside did not publish his vector algebra for the purpose of impressing mathematicians with a new contribution to their field.[112] It was purely a means to an end, never an end in itself. With the algebra of vectors Heaviside hoped to make Maxwell's electromagnetic theory more amenable to the uninitiated. Vector algebra was for Heaviside from 1882 to 1884 primarily a pedagogical tool. From an engineering innovator he seems to have turned into a self-appointed teacher of Maxwell for the nonexpert. As we have already noted, this should not be taken to indicate that Heaviside turned his back on the world of practical engineering. Pedagogical goals aside, both his early papers and his work from 1882 to 1884 express his constant desire to form a comprehensive understanding of the electric circuit.

Throughout the work we examined in this chapter, Heaviside kept one idea constantly in mind: make the physical picture painted by Maxwell's theory visible at all times. Granted, he said, the theory is a mathematical one. But it is not about quaternions, nor about the most powerful method of solving mathematical problems. If the physics is mathematically expressed, then the mathematics had better be physical. As we shall see in the next chapter, this attitude constantly prompted him to find fault with attempts to express Maxwell's basic ideas in what may have appeared to others as the mathematically more powerful Lagrangian formulation or Hamilton's principle of least action. Such exercises, he believed, would invariably hinder rather than aid the comprehension of the theory.

For Heaviside, the importance of Maxwell's theory resided in its ability to encompass circuit theory through a continuous system of dynamically interconnected force fields and associated fluxes. It provided him with an electromagnetic equivalent of Newton's dynamics of motion. These considerations led him to reformulate Maxwell's equations. He assigned only a secondary

112. Heaviside himself did not consider vector algebra an outstanding achievement from the purely mathematical point of view. Physically, he considered Hamilton's quaternions awkward and unnatural; mathematically, he regarded them as the more impressive contribution. He expressed both sentiments with his typical sarcasm: "... apart from practical application, and looking at it from the purely quaternionic point of view, I ought to also add that the invention of quaternions must be regarded as a most remarkable feat of human ingenuity. Vector analysis, without quaternions, could have been found by any mathematician by carefully examining the mechanics of the Cartesian mathematics; but to find out quaternions required a genius." (*Electrical Papers*, Vol. II, p. 557).

and purely mathematical role to the vector potential, so prominently featured in Maxwell's original formulation. In a discussion with Kelvin concerning the vector potential, Heaviside depicted it as a metaphysical concept, and added:

> I am myself accustomed to mentally picture the electric and magnetic forces or fluxes, and their propagation, which takes place at the speed of light or thereabouts, because they give the most direct representation of the state of the medium, which, I think, must be agreed is the real physical subject of propagation.[113]

Similar arguments stood behind Heaviside's refusal to describe the electromagnetic system in terms of an energy-based formulation such as the principle of least action. He believed that it effectively hides the dynamical structure. Whenever he addressed energy problems, Heaviside expressed them by the principle of activity, which explicitly ties them to the dynamical structure through Newton's third law. Thus, he derived the energy flux from the electric and magnetic force distributions by examining their activity. The importance Heaviside attached to energy considerations derived from their usefulness in directing attention to operating forces. He regarded these forces as the basic building blocks of a dynamical system and as the desiderata of physical investigation. From this point of view, it is hardly surprising that Heaviside emphasized the principle of activity, since it expresses the conservation of energy in terms of the operating forces.

It would probably be an unwarranted exaggeration to suggest that motivated by a desire to teach the physical essence behind Maxwell's theory, Heaviside was led to insist upon emphasizing the electric and magnetic forces. It does appear, however, that the two sentiments, basically independent, harmonized and supported one another in his work. Imperative to realize in this connection, is that Heaviside did not consider Maxwell's theory as the final word on electromagnetism. He actually regarded Maxwell's theory as a vastly incomplete, skeleton framework:

> The theory of electromagnetism is then a primary theory, a skeleton framework corresponding to a possible state of things simpler than the real in innumerable details, but suitable for the primary effects, and furnishing a guide to special extensions.[114]

Heaviside, however, was not interested in extending the theory to cover those innumerable details, altering and remodelling it whenever necessary. It is

113. *Electrical Papers*, Vol. II, p. 491.

almost as if deep down inside, he remained the telegraph operator who kept struggling to put the electromagnetic phenomena he encountered in order. He found that order in Maxwell's theory, and grew to believe that it must be the theory of the future in the sense of the quotation above. Thereafter, he devoted his efforts to the detailed study of the theory's inner structure. This he must have regarded as the task to fulfill before any further meaningful extensions could be worked out. He was well aware of several effects that taxed the theory and possibly pointed out the need for reformulation and extension, for he said quite frankly at the end of his most sophisticated analysis of this inner structure:

> ... it is desirable to be apologetic in making any application of Maxwell's stresses or similar ones to practice when the actual strains produced are in question, bearing in mind the difficulty of interpreting and harmonizing with Maxwell's theory the results of Kerr, Quincke, and others.[115]

Thus, Heaviside was not motivated by the desire to reformulate Maxwell's theory in terms most suitable for the study of such effects, nor was he eager to be the first to clarify their meaning to the world. He wanted to highlight the fundamental physical essence of the skeleton theory in its connection to the primary phenomena, namely, the phenomena associated with electric circuits. All of this makes him appear more as a teacher of Maxwell's theory than as a theoretical physicist doing original research under its guidance.

If, however, we seriously wish to regard Heaviside as a teacher of Maxwell's theory, we must ask ourselves who were his intended students. An examination of this question will reveal that the image of Heaviside as a teacher must be carefully qualified.

---

114. *Electromagnetic Theory*, Vol. I, Preface. This quotation, and Heaviside's discussion on explanation and speculation in physical questions, should always be born in mind when one encounters pronouncements such as the following in his text: "In view of the extreme relative simplicity of Maxwell's views, and their completeness without any artificial contrivances to save appearances, and in their modernness, referring to modern views regarding action at a distance, one is *almost* [my italics] constrained to believe that the dielectric current, the really essential part of Maxwell's theory, is not merely an invention but a reality, and that Maxwell's theory, or something very like it, is *the* theory of electricity, all others being makeshifts, and that it is the basis upon which all future additions will have to rest, if they are to have any claims to permanency." (*Electrical Papers*, 1, p. 478). With the skeletal nature of Maxwell's theory in mind, it will become quite clear that Heaviside did not insert the word "almost" into this paragraph merely to serve the requirements of polite expression.

115. *Electrical Papers*, Vol. II, p. 574.

8.2 For Whom Was Heaviside Writing?

It appears that Heaviside had quite a clear image of the reader he desired to address, even if that image later proved rather misguided. The words "non-expert" and "uninitiated" must be carefully qualified when used to describe this intended reader. Heaviside was never a popular-science writer, although a careless reading of the paragraph quoted at the beginning of this chapter may suggest that a popular account of Maxwell's theory was precisely his goal. This interpretation of the passage can only be maintained at the cost of ignoring Heaviside's warning that there is no royal road to knowledge. The subject may not require advanced technical skills in mathematics, he wrote, but it is difficult in itself, and requires "hard thinking and rigid fixation of ideas" at all time. A popular-science writer would not be likely to ward off his potential readers with an ominous threat of this kind. This initial remark already casts a doubt on the image of Heaviside as a would-be popularizer of electromagnetic theory. Further reading of the papers following the introductory remarks will set it aside for good.

The aim of an honest popular text is not to impart to the reader a working knowledge of the subject matter. Its aim is to inform the reader about the scientific state of the art in some field of research, with the understanding that the reader will accept the information on faith and will not thereby expect to have acquired a critical understanding of the subject. He will become acquainted with the main terminology used by working scientists. If blessed with a good memory, he will be able to quote certain results in a semi-rationalized manner without really being capable of reproducing the reasoning that ties them together. Meaningful questioning of the scientific knowledge thus communicated is rarely, if ever, possible; so is its application to relevant problems that occupy working scientists.

Heaviside's papers, by contrast with the above, have a much more ambitious goal behind them. The reader who braves through these papers will have acquired enough knowledge of Maxwellian electromagnetism to solve specific problems with it.[116] It would seem, therefore, that Heaviside wished to give his reader far more than a purely descriptive summary of Maxwell's central

116. For example, no further knowledge will be required in order to solve each and every problem in the chapter entitled "Maxwell's equations" in a text like *The Electromagnetic Problem Solver* (Staff of Research and Education Association, Dr. M. Fogiel, Director, [New York: Research and Education Association, 1983], pp. 604-610).

ideas. He intended to supply his reader with the essence of the theory together with a method of applying the knowledge to concrete electromagnetic problems. The algebra of vectors and the long introduction to field thinking are the main building blocks of this method. Heaviside devoted very little space, comparatively, to the actual outline of Maxwell's theory. The rough sketch of Maxwell's theory occupies a mere 22 pages at the beginning of "Electromagnetic Induction and its Propagation," as opposed to the 158 pages that constitute the introduction to field thinking.[117] Whatever else it may be then, Heaviside's is not the work of a scientific popularizer, and consequently, his intended reader could not have been the usual consumer of popular-science texts.

Who, then, was Heaviside aiming at? Surely, it could not have been the Maxwellian experts, although Heaviside naturally would have welcomed their attention. The handful of Maxwellian physicists in the early 1880's were mostly products of the finest British universities. None of them could be described as a non-mathematician. It would appear then, that the reader Heaviside had in mind when he wrote his introductory remarks was somewhere between the Maxwellian expert and the general reader. The manner in which Heaviside took some basic electromagnetic knowledge for granted implies that the imaginary reader should have been somewhat versed in electrical matters. Furthermore, although Heaviside did not require of his reader to have advanced skills in working the differential and integral calculus, he did require of him to understand the meaning of its basic operations. In short, he required an intelligent reader, not merely in the limited sense of the word as bright or talented—a rather slippery notion to begin with—but one possessing some concrete prior knowledge. Above all, the imaginary reader was required to be very highly motivated. Note that Heaviside explicitly referred to "earnest students." Without an earnest desire to acquire a working knowledge of the theory, no novice, regardless of how bright or how well informed in its basics, would ever persist through these papers of Heaviside's. It appears then, that Heaviside was thinking of an electrical or telegraph "practician" such as he had been before he embarked on his solitary exploration of Maxwellian electromagnetism. In other words, it seems that intentionally or not he modeled his imaginary reader primarily upon himself.

The previous section suggested that Heaviside might be regarded more as a teacher than as a research physicist. This statement must now be more pre-

117. *Electrical Papers*, Vol, I, pp. 429-451 and pp. 195-353, respectively.

cisely qualified in view of the above. The theory Heaviside was trying to impart to other electrical engineers was not new. However, he taught it in a highly original way that was specifically intended to provide a basic theoretical framework for electrical engineering. If Heaviside does not appear as a typical research physicist, this should not be taken to imply that he did not do original research; if he appears more as a teacher, the particulars of what he taught cannot possibly be accounted for by the general observation that he was interested in teaching electromagnetism. His original presentation of Maxwell's theory is at one and the same time an original investigation of the underlying principles of the telegraph system he encountered between 1868 and 1874. As we shall see in the next chapter, this dual character of his approach to electricity and magnetism provides both the key to the rest of his work and to the enigmatic image it acquired over the years.

9. Summary

In many ways Heaviside's great work was still ahead of him. The explicit formulation of vector algebra, his daring extensions of the operational calculus, his insightful study of the transmission line—all these were still in the future. However, in one important respect his work was finished by the beginning of 1885. He fully outlined his basic understanding of Maxwell's theory along with a methodological approach to problem solving. In all of the subsequent developments and important contributions that he made, these aspects remained rigidly fixed. They are quite essential to the understanding of his achievements. They also underline the problems he encountered in a world that was coming to terms with the magnitude of his work and his illuminating insights into the fundamentals of Maxwell's theory. These issues will be examined in the next chapter. The main points to be kept in mind thus far are as follows:

Heaviside's initial motivation to study Maxwell's theory was grounded in his desire to comprehend the phenomena he encountered as a telegraph operator. As he grew to master this theory in his own terms, he interrupted his investigation of transmission-line problems and embarked upon a detailed exposition of the theory. Heaviside seems to have found Maxwell's theory especially attractive because it provided him with a comprehensive description of electromagnetic energy transformations in terms of continuous force distributions. His "conquest" of the theory is best symbolized by his reformulation of it as a dual set of relationships between the electromagnetic forces and their fluxes. However, by itself, the force formulation of energy transfer cannot account for the particular form of Heaviside's duplex equations. The outstanding features of his duplex formulation can be accounted for only in conjunction with the affinity between the dynamics of forces encapsulated in the activity principle and the principles of circuit theory.

As he traced the physical essence of Maxwell's theory to the "dynamical connections" furnished by the two circuital laws, Heaviside challenged the prominent role Maxwell gave to the vector potential in his original exposition. For the same reason he found fault with later formulations of the theory in terms of Lagrange's equations or the principle of least action. While he explicitly recognized their mathematical value, Heaviside felt that such reformulations of the basic theory obscured its basic physical nature.

While he considered Maxwell's theory to be the electromagnetic theory of the future, Heaviside did not regard it as the final electromagnetic truth. Indeed, his entire exposition of electromagnetism rested on the observation that nothing, in effect, was known of the real essence of electricity and magnetism. Heaviside's guiding principle was that the forces that constitute the dynamical structure of the theory are not true mechanical forces, but generalized ones. That is to say, each can be paired with a generalized velocity such that their mutual product yields activity, or energy per unit time.

In general, Heaviside regarded Maxwell's theory as a skeleton theory of electromagnetism, simpler in a multitude of details from the reality it purported to portray. He perceived its value in its ability to account for the phenomena associated with electric circuits, and in furnishing a useful guide for further extensions.

Heaviside's discussion of the theory indicates that his primary interest was not in its application to puzzling effects at the forefront of scientific research. Instead, he gave first priority to the theory's inner structure. His attempt to convince the uninitiated of its worthiness relied on pointing out the comprehensive manner in which it assimilated familiar phenomena, relatively well understood by other means. In this respect he seems to have been motivated more as a teacher, and less as a research physicist eager to apply a novel outlook to hitherto unassimilated effects.

Heaviside presented his exposition of Maxwell's theory in conformity with an explicitly stated view of the nature of scientific knowledge. He believed that in the final analysis, scientific knowledge explains nothing; that what some people regard as an explanation is really a speculation. While he considered speculations as invaluable to the development of further scientific knowledge, he required that they be distinguished from the true goal of scientific inquiry that seeks to formulate the laws of the phenomena.

Along with his exposition of the theory, Heaviside began to develop the mathematical tools it requires in the form he believed best suited for the job. The first outcome of this endeavor was the algebra of vectors.

Heaviside believed strongly that useful physics must be mathematical, and he considered virtually all of the important developments in electromagnetism to have been impossible without mathematics. However, he did not consider numerical precision to be the important contribution of mathematics to physics. He suggested that the real value of expressing physical ideas mathematically is that it enables one to conduct further investigations under the guidance of mathematical reasoning. At the same time, he strongly objected to the view that mathematical knowledge is somehow prior to physical knowledge, and he believed that in the final analysis all mathematical knowledge is based on experience. Thus, while his physics was thoroughly mathematical, he strove to make his mathematics as physical as possible. In particular, he saw the essence of mathematical reasoning not in the manipulation of formulae according to a set of formal rules, but in seeing through them to the meaning they convey. Accordingly, he sought to mold the mathematical formalism so as to best reflect the physical meanings behind it. One might say that the essence of Heaviside's mathematical reasoning was formula interpretation rather than formula manipulation.

It would appear from all of the above that by and large Heaviside strove for an ideal moderation. He advocated intense use of mathematical reasoning but cautioned against allowing it to dominate over the ideas one conveys and analyses with its aid. He devoted a great deal of time to studying the inner structure of Maxwell's theory, but explicitly noted that it had not reached its intended goal of providing a comprehensive formulation of the laws of electromagnetic phenomena. He believed that without speculative imagination there is no scientific knowledge, but criticized the attempt to elevate such speculations beyond their status as a means to an end. He argued strongly against constructing physical theories on the basis of action at a distance, but never presumed to suggest that such theories were conclusively refuted. In one respect, however, Heaviside seems to have violated this ideal of moderation. His style of writing is anything but moderate. He seems to have written in fulfillment of the observation made about the ancient Romans to the effect that they could not resist a good joke even if it meant making an enemy. Remarkably enough then, Heaviside's books suggest time and again that moderate views need not be expressed in moderate terms.

Chapter IV **From Obscurity to Enigma**

> The mere theorist is a dreamer—the mere pracititoner is a drudge; but combine the two and you gain a man who carves his name in the history of the world....
>
> *The Electrician*, 1878.

1. Introduction

In 1892 Heaviside wrote the following comment in accompaniment to the publication of his *Electrical Papers*:

> But in the year 1887 I came, for a time, to a dead stop, exactly when I came to making practical applications in detail of my theory, with novel conclusions of considerable practical significance relating to long-distance telephony (previously partly published), in opposition to the views at that time officially advocated.[1]

Behind these words stands the intriguing story of Heaviside's longest and in many ways most remarkable series of articles, "Electromagnetic Induction and its Propagation," or as he used to call it, "E.M.I.& P." The circumstances that accompanied the series' publication are closely bound to Heaviside's emergence from the obscure anonymity under which he had worked until the later part of 1887.

The real significance of Heaviside's "dead stop" is not easily seen. A first examination suggests that the statement refers to certain difficulties he encountered during 1887 in publishing some articles. We shall presently see that these difficulties amounted to the disruption of a grandiose scheme of publication. It was artfully designed to fuse practically all of Heaviside's work since 1872 into a harmonious development of circuit and field theory. The analysis of transmission lines, and in particular Heaviside's newly discovered distortionless transmission line, provided the common thread that tied together the many aspects of this complex publication scheme.

Heaviside's plan of publication materialized slowly. From March of 1885 until April of 1886, Heaviside continued to develop "E.M.I.& P." along the lines he had drawn in his "Rough Sketch of Maxwell's Theory" (see chapter III). In April of 1886 he decided to reorient the series in the wake of the dis-

1. *Electrical Papers*, Vol. I, p. ix.

covery of the skin effect by David E. Hughes two months earlier. In August of 1886 he began to present Maxwell's theory from the beginning, with emphasis on its connection to linear circuits, and to the theory of transmission lines in particular. At the same time, he began to publish "On the Self-Induction of Wires" in *The Philosophical Magazine* as a counterpart to his discussion in *The Electrician*'s "E.M.I.& P."[2] In the beginning of 1887, in the course of collaborating with his brother on the redesign of Newcastle's telephone system, Heaviside discovered the distortionless transmission line. In short order he realized that it provided a remarkably useful focal point for the theme he was developing in "E.M.I.&P." and in "On the Self-Induction of Wires." With this in mind, he continued to develop the two complementary discussions through 1887. Then, toward the end of 1887, the entire project disintegrated when *The Electrician* and *The Philosophical Magazine* refused to continue the publication of their respective series.

The first signs of trouble began to appear already in the beginning of 1887. In February, Oliver Heaviside and his brother, Arthur West Heaviside, wrote a joint paper on aspects of telephone communication.[3] In his contribution to the joint paper, Oliver Heaviside discussed for the first time the possibility of practical distortion-free telephony. The brothers communicated their paper to the *Journal of the Society of Telegraph Engineers and Electricians*, but the appointed referee, Mr. W.H. Preece, judged the paper devoid of novelty and it was not published.[4] In July of 1887, *The Philosophical Magazine* decided to terminate "On the Self-Induction of Wires" without publishing Heaviside's latest contribution to it, which contained his second attempt to publish the condition for distortionless telephony.[5] In September, Heaviside tried without success to publish a short letter, entitled "Mr. W.H. Preece on the Self-Induction of Wires."[6] In the meantime, *The Electrician* continued to publish parts 40-47 of "E.M.I.& P." Here Heaviside finally published the distortionless condition,[7] but then the journal's editor since 1878, Charles Henry Walker Biggs, was replaced by his assistant, William Henry Snell. The latter informed Heaviside that an inquiry among the journal's readers revealed that no one was

2. *Electrical Papers*, Vol. II, pp. 168-323.
3. *Ibid.*, pp. 323-354.
4. *Ibid.*, p. 324 (note).
5. *Ibid.*, pp. 308-309, 322-323.
6. *Ibid.*, pp. 160-165.
7. *Ibid.*, p. 123.

reading his articles. As a result, part 48 of "E.M.I.& P." was never published. Instead, a short note, dated November 30, 1887, was appended to part 47 explaining that: "The author much regrets to be unable to continue these articles in fulfillment of Section XL., having been requested to discontinue them."

The above may give the impression that Heaviside had really struck a solid wall. Consider however, that in December of 1887, just as the last installment of "E.M.I.& P." appeared in *The Electrician*, *The Philosophical Magazine* resumed publication of Heaviside's work with one of his most important papers, "On Resistance and Conductance Operators." In February of 1888 the magazine began to publish a new series by Heaviside on electromagnetic waves. It took a few more months for *The Electrician* to recommence publication of Heaviside's work. In August of 1888 the journal published a letter of his, in which he called attention to his disrupted work with some thinly veiled criticism of the editor who claimed that none of this work had been read.[8] This was followed in October by another letter that Heaviside dedicated primarily to exposing Preece's ignorance in matters concerning the basic principles of signal propagation along wires.[9] In November *The Electrician* resumed regular publication of Heaviside's contributions with a short series on the subject of electromagnetic waves, the physical status of electromagnetic potentials, and the effects associated with moving charges.[10]

As it turned out, Heaviside saw his work in print more often in the year 1887 than in most others. Except for September and November, readers of *The Electrician* could find his contributions at least once every month. Between January and December the journal printed parts 35-47 of "E.M.I.& P." In addition to this monumental series, *The Electrician* published Heaviside's fourth communiqué on nomenclature, as well as a short remark on the theory of the telephone and hysteresis. At the same time, *The Philosophical Magazine* carried parts 5, 6, and 7 of "On the Self-Induction of Wires." Altogether, 1887 accounts for nearly a third of the second volume of *Electrical Papers*, which consists of practically everything Heaviside wrote between 1885 and 1891. Between *The Electrician* and *The Philosophical Magazine*, Heaviside actually published continuously throughout 1887, despite the fact that both journals requested the discontinuation of the series mentioned. Furthermore, while the

8. *Electrical Papers*, Vol. II, p. 487.

9. *Ibid.*, pp. 488-490.

10. "Electromagnetic Waves, The Propagation of Potential, and the Electromagnetic Effects of a Moving Charge," *Ibid.*, pp. 490-499.

rejection of a number of papers must have been annoying, the journals involved gave apparently reasonable justifications for their decisions. "On the Self-Induction of Wires" was initially scheduled to run in four parts. By July 1887 Heaviside saw part 7 in print, mailed part 8 to the editor, had part 9 in preparation and still there was no end in sight (Heaviside's notebooks contain rough drafts of part 9). The editor's decision to stop publication under these circumstances does not seem wholly unwarranted. Mr. Snell of *The Electrician* justified his decision to terminate "E.M.I.& P." on the basis of readers' disinterest. The *Journal of the Society of Telegraph Engineers and Electricians* could rely on an unfavorable review by a respected referee for the rejection of Arthur and Oliver Heaviside's joint paper. All things considered, Heaviside undoubtedly encountered some difficulties in seeing his work published according to his wishes; but his statement of having come to a "dead stop" in 1887 seems rather exaggerated. After all, bearing in mind his total output in 1887, the rejected papers and the discontinued series do not amount to a dead stop; so why speak of one?

Heaviside was an eccentric recluse and there may have been an element of overreaction in his complaint, but it would be an error to leave it at that. In the first place, it appears that *The Electrician* was paying him for his articles (see chapter I). The discontinuation of "E.M.I.& P." therefore meant the loss of his only steady income, meager as it may have been. Furthermore, we shall see later in this chapter that Heaviside had reasons to suspect that the official justifications given for the rejection of the above-mentioned papers concealed other, less respectable motivations. These would be sufficient reasons for many people to react with angry indignation, and Heaviside was understandably upset by the rejection of three papers on what he saw as extraneous grounds. However, the "dead stop" actually addresses a far more fundamental concern. It reflects the practically irreparable disruption of the publication scheme that Heaviside began to conceive in February of 1886.

To study these developments in greater detail, the rest of this chapter will be organized as follows. Section 2 will trace the development of "E.M.I.& P." until April of 1886. Section 3 will describe Heaviside's reaction to Hughes's discovery and its effects on the course of "E.M.I.& P." The emergence of Heaviside's plan of presenting Maxwell's theory through the analogy it bears to transmission-line theory will be discussed in sections 4 to 6 together with the distortionless condition and its central role in this plan. This will be followed in section 7 by a more detailed discussion of Heaviside's clash with

Preece. With the details of sections 3 to 7 in mind, it will be shown that the rejection of the above-mentioned papers amounted to the disruption of Heaviside's grand publication plan, and that he actually did come to a temporary dead stop in 1887. When he resumed publication, it was along very different lines than the ones he began to develop from April of 1886. Heaviside did eventually publish most of the ideas and discoveries that he planned for the prematurely terminated series. But while it remained implicit in practically all of his subsequent work, he never explicitly returned to the unified framework that he tried to erect between "E.M.I.&P." and "On the Self-Induction of Wires." The emergence, crystallization and disintegration of this thematic framework are crucial to a proper appreciation of the unique position Heaviside eventually occupied among both his scientific and engineering contemporaries. All this will put us in position to better understand how Heaviside emerged from obscurity only to remain shrouded in enigma.

2. "Electromagnetic Induction and its Propagation" until April, 1886

In his rough sketch of Maxwell's theory, Heaviside showed how to determine the continuous transfer of usable electromagnetic energy given the electromagnetic field everywhere in space. This enabled him to pronounce the electromagnetic field system defined by Maxwell's equations "dynamically complete." However, he still had one problem to address in this connection. The system is really complete only when energy transfer is specified throughout time as well as everywhere in space. Heaviside still had to show that given the electric and magnetic field distributions at one moment in time, the two circuital equations completely determine the fields' distributions at all other times. Mathematically, this translates into proving that there always exists a solution to the two circuital equations, and that this solution is unique. The rough sketch was also incomplete in another sense. Heaviside began to introduce there a comprehensive dynamical structure for electromagnetic-field theory based on the principle of activity. This had to be developed further in two interconnected ways. In the derivation of the energy flux, Heaviside assumed the electromagnetic medium permeated by the fields to be isotropic and motionless. He still had to generalize this discussion to anisotropic media in relative motion. Furthermore, an electromagnet does not attract a piece of iron by a generalized force, but by a real mechanical force. Therefore, Heaviside needed to show how an electromagnetically stressed medium could give rise to real mechanical stresses. Thus, Heaviside had two themes to address: (1) the mathematical and dynamical completeness of the Maxwellian system, and (2) the development of its dynamical structure.

Until April of 1886 Heaviside basically followed these two general guidelines. In his usual manner, he proceeded from particular cases and slowly generalized them to include a wider scope. Along the way, he pointed out many important results of the theory. Thus, he developed the general framework along with the particular details of the electromagnetic picture. Naturally, this approach necessitated many small digressions along the way. As a result, Heaviside's discussions tend to give at first the impression of a loosely connected series of articles. However, once the two general themes of discussion are discerned one can never lose the argument's thread.

Proceeding past the "Rough Sketch of Maxwell's Theory," Heaviside developed the two main lines of analysis more or less in parallel. He began with theme (2) by extracting certain theoretical restrictions on the properties of specific capacity and inductivity from an examination of the activity of the electric and magnetic forces.[11] Then he showed that the expression for heat dissipation by the conductive current implies no such restrictions on electric conductivity. Immediately following this, he embarked upon theme (1) by proving the existence and uniqueness of a solution to Maxwell's equations in

---

11. This is a simple and instructive example of how Heaviside used the dynamical principle of activity to extract information concerning the physical properties of the electromagnetic medium. Using Heaviside's rational units to avoid lugging along a host of 4π's, the expression for the energy per unit volume of the magnetic field is $(1/2)\,\mathbf{H}\cdot\mathbf{B}$. In the most general case, **H** and **B** are related by a vector operator, μ, defined by up to nine independent coefficients, being a general 3x3 matrix. Thus, the energy per unit volume is $(1/2)\,\mathbf{H}\mu\mathbf{H}$. The rate of change of this amount is:

$$\frac{1}{2}\frac{d}{dt}(\mathbf{H}\mu\mathbf{H}) = \frac{1}{2}\left(\frac{d\mathbf{H}}{dt}\mu\mathbf{H} + \mathbf{H}\mu\frac{d\mathbf{H}}{dt}\right).$$

This change of energy per unit volume must be equal to the activity of the magnetic force, $\mathbf{H}\cdot\dot{\mathbf{B}}$. Hence, we have:

$$\frac{1}{2}\left(\frac{d\mathbf{H}}{dt}\mu\mathbf{H} + \mathbf{H}\mu\frac{d\mathbf{H}}{dt}\right) = \mathbf{H}\mu\frac{d\mathbf{H}}{dt}.$$

This is generally true only if $\dot{\mathbf{H}}\mu\mathbf{H} = \mathbf{H}\mu\dot{\mathbf{H}}$, which implies that the off-diagonal coefficients of μ must satisfy $\mu_{i,j}=\mu_{j,i}$. In other words, the physically meaningful inductivity μ is fully determined by no more than six independent coefficients. (*Electrical Papers*, Vol. I, p. 451).

the case of a steady magnetic field in space free of conduction currents.[12] Next, he switched to theme (2) with a discussion of the electromagnetic forces and their associated energies and activities. A little later, Heaviside resumed the topic of existence and uniqueness with a proof of the uniquely determined solution for the steady electric field.[13]

Heaviside could have done without the proofs of unique determinacy in the special cases above. He could have proceeded immediately to the general question of dynamical completeness with the following simple argument: first, let **E** and **H**, the electric and magnetic fields, be given throughout space at one specific moment in time. The impressed forces **e** and **h**, if they exist at all, must be known separately. Also, the electromagnetic properties of the relevant space must be given, namely the conductivity, inductivity and capacity. Finally, the manner in which **E** and **H** vary in time is everywhere uniquely determined by the circuital equations. We may, therefore, determine the rate of variation of **E** and **H** from their given values at an arbitrary initial moment. Using this rate of variation and the initial values, we may calculate the quantity for the next moment in time. We may now repeat the procedure using this new value. With this iterative procedure, the quantity may be calculated to any desired degree of accuracy by diminishing the time intervals taken while increasing the number of iterations. In the case before us, the two circuital equations provide the algorithm for calculating the variation of **E** and **H** from their value at a given moment. This necessarily establishes **E** and **H** at any other time, before and after the initially given state. Such an approach would have made Heaviside's discussion both concise and thematically coherent. It would have launched him directly onto the next logical step, namely, the outline of a general approach to the solution of Maxwell's equations.

Heaviside eventually got to this argument, as well as to the general solution of Maxwell's equations.[14] However, he obtained many interesting results about the nature of the electromagnetic system by examining the special cases on the way. Heaviside did not conceive of "E.M.I.& P." as a treatise for Maxwellian physicists, aimed solely at probing the unexplored regions of the theory.[15] As a continuation of his earlier introduction to field thinking he must have envisioned "E.M.I.& P." as education for the uninitiated but strongly mo-

12. *Electrical Papers*, Vol. I, pp. 453-454.
13. *Ibid.*, p. 499.
14. *Ibid.*, p. 520; Vol. II, pp. 468-483.

tivated telegraph or telephone engineer. Therefore, he must have attached great importance to the special results. They provided him with priceless educational tools that hammered in the distinguishing marks of the new vision of electromagnetism far more clearly than general treatments. Consider, for example, the steady electric field in the presence of steady conduction currents. Under these assumptions, Maxwell's equations yield the following:

$$\mathbf{C} = k\mathbf{E}, \qquad \operatorname{div}\mathbf{C} = 0; \tag{4-1}$$

$$\mathbf{D} = c\mathbf{E}, \qquad \operatorname{div}\mathbf{D} = \sigma; \tag{4-2}$$

$$\operatorname{curl}(\mathbf{e} - \mathbf{E}) = 0. \tag{4-3}$$

(**C** is the density of conduction current, **E** the electric force, **D** the electric displacement, **e** is any externally impressed force, σ is the volume distribution of electric charge, k is the conductivity, and c the capacity). As stated above, the main part of Heaviside's discussion is devoted to demonstrating that these conditions uniquely determine **E** within, as well as outside conductors. However, an additional, fascinating piece of information can be extracted here, as he proceeded to show.[16] Consider eqs. (4-1) and (4-2) within a part of the medium which possesses both conductivity and capacity. Being a steady situation, the transient displacement current no longer exists, and the non-divergence of the conduction current becomes the steadfast condition that demands fulfillment. This means that the state of the system must be made to conform to equations (4-1) and (4-3) first. Heaviside's discussion shows that these equations are perfectly sufficient by themselves to determine **E** completely within this region. Then, with **E** determined, **D** and σ may be derived from eq. (4-2). It becomes immediately evident that if the capacity c is a constant multiple of k, then σ is necessarily zero (because then div(**D**) reduces to div(**C**) multiplied by a constant, but div(**C**) is necessarily zero). In other words, there cannot be any electric charge as long as the ratio of conductivity to capacity is a constant throughout the space. Conversely, if electric charge exists at all under steady circumstances, it can do so only where the ratio of conductivity and capacity varies from place to place. Thus, the over-determination of the field in this

15. Whenever he considered he had such a message to convey, he usually published it in a separate article. Such is his paper on the electromagnetic wave-front, (*Electrical Papers*, Vol. II, Art. 31) published in *The Philosophical Magazine* at the same time that he completed the "rough sketch of Maxwell's theory" in *The Electrician*.

16. *Electrical Papers*, Vol. I, pp. 496-500.

case leads to the important conclusion that there cannot be charge without conduction and consequently without matter, since the conductivity of ether is everywhere assumed to be zero.

In this manner many small digressions in Heaviside's work can be seen as the result of specific questions that naturally arose in the course of examining the internal structure of Maxwell's theory. Such digressions explain why it took Heaviside several months and some 60 pages to arrive at the general demonstration of the dynamical completeness of the theory. There exist, however, more drastic shifts of emphasis that cannot be accounted for in this way. In particular, it is a rather different matter to explain why he addressed the general solution of Maxwell's equations nearly three years later, in August of 1888. A similar and parallel break occurred at just the same time in Heaviside's discussion of the theory's dynamical structure. By January of 1886, after an equally meandering tour of several specific dynamical issues, Heaviside began to address the general dynamical structure of the theory.[17] He began the analysis with a short, minimal survey of the mechanical theory of stress to establish the guiding analogy. The basic idea Heaviside wished his readers to take in is the following:

> A simple stress is either a tension or a pressure acting in a certain line. It implies the existence of mutual force between contiguous parts of the substance in which it resides, and of a corresponding state of strain, with storage of energy in the potential form, i.e., depending upon configuration, though perhaps ultimately resolvable into kinetic energy.[18]

Note, first, that with the last sentence Heaviside deliberately avoided the attempt to reduce the dynamical picture thus produced into the proverbial matter and motion. This means that at least for now, the mutual forces must be considered fundamental and irreducible. Next, the implication of the existence of mutual force between contiguous parts of the substance means that stress analysis does away with action at a distance.

What really interested Heaviside is the depiction of energy transfer in terms of the stress. He illustrated the basic idea with the case of the horse-drawn barge:

---

17. Just like his digressions along the road to dynamical completeness, the special issues Heaviside addresses on the way to the general dynamical structure are fascinating and illuminating in themselves. See, for example, his discussion of thermal resistance and the connection between electrodynamics and thermodynamics. (*Electrical Papers*, Vol. I, pp. 481- 488).

18. *Electrical Papers*, Vol. I, p. 542.

> ... when a horse tugs a barge along a canal, there is, besides the transfer of energy through space by the onward motion of the horse, rope, barge, and dragged water, carrying their kinetic energy with them, a transfer of energy through the rope from the horse to the barge, and through the strained barge to the water, where it is wasted in friction. The rate of transfer per second equals the product of the tension of the rope into its speed, and the direction of transfer is against the direction of motion.[19]

The upshot is that as long as the tension, or more generally, the stress in a medium is constant, no net force exists. Where the stress varies—at the end of a rope, or where a driving belt makes contact with a resisting pulley—a net force appears, and energy is transferred into or out of the stressed medium. A constant state of stress implies continuously stored potential energy within the stressed medium. Note, however, that this does not negate energy transfer in the medium. It implies that the energy current, whatever it really is, does not diverge within any region of constant stress, and that the constant potential energy there is maintained in a state of dynamic rather than static equilibrium.

The analysis of stress structures like the above may be used in three basic ways:

> The first is in the dynamical theory of elastic bodies; the second is, after Faraday and Maxwell, in the explanation of forces of unknown origin by means of stress

19. *Electrical Papers*, Vol. I, pp. 542-543. Once again, the abstract nature of the picture becomes clearly discernible when one realizes that the transfer of energy takes place against the direction of motion. The ease with which Heaviside glances over this observation is truly remarkable, considering that it presented a source of endless frustration to other physicists at the time (see J.Z. Buchwald, *From Maxwell to Microphysics*, [1985], pp. 42- 43). Hertz, for example, queried in reference to the belt-driven dynamo that powers an arc-light: "But is there any clear physical meaning in asserting that the energy travels from point to point along the stretched strap in a direction opposite to that in which the strap itself moves? And if not, can there be any more clear meaning in saying that the energy travels from point to point along the wires, or—as Poynting says—in the space between the wires? There are difficulties here which badly need clearing up." These difficulties are rooted either in the desire to see energy as possessing the sort of individual identity that material corpuscles possess in classical mechanics, or in the need to specify an explicit molecular mechanism for energy transfer. Heaviside, however, never subscribed to the notion of identifiable energy units, as his criticism of Lodge's pronouncements in this respect clearly show (see chapter III, section *4.12*). He was also quite skeptical with regard to molecular theories, and maintained that for the time being, Maxwell's theory should be developed as a purely dynamical scheme. The remarkable thing is that from the dynamical point of view, energy transfer is an analytical concept that can be endowed with physical meaning in terms of the activity of generalized forces. Taking this as the fundamental view, Heaviside effectively resolved Hertz's difficulty by simply sidestepping it.

> in a medium; and the third application consists in the use of the stresses, not for explanation, but for purposes of investigation.[20]

Naturally, one first derives the expression for the electromagnetic stress. This Heaviside did (for a stationary, isotropic medium) in section 22 of "E.M.I.&P." It was published in January of 1886. One would then expect the next logical step to be a generalization of the above to moving, heterogeneous media. Indeed, while section 24—which appeared in March of 1886—is yet another Heavisidean digression,[21] he had this generalization clearly in mind, for he says:

> ... the last section [23] being of a perfectly abstract nature, the one to follow this [24] will be on the magnetic stress in general, as modified by differences of permeability and other causes.[22]

In fulfillment of the above, section 25 (published in April) is a first step in this direction. Here Heaviside discussed fundamental problems of magnetism in connection with the view of magnetization as a generalized stress system. He concluded the section with an appended comment on the experimental difficulties of measuring the inductivity of various conductors by using the Wheatstone bridge as an induction balance:

> Of course, it will be understood by scientific electricians that it is necessary, if we are to get results of scientific definiteness, to have true balances, both of resistance and of induction, and not to employ an arrangement giving neither one nor the other. He will also understand that, quite apart from the question of experimental ability, the theorist sometimes labours under great disadvantages from which the pure experimentalist is free. For whereas the latter may not be bound by theoretical requirements, and can employ himself in making discoveries, and can put down numbers, really standing for complex quantities, as representing the specific this or that, the former is hampered by his theoretical restrictions, and

---

20. *Electrical Papers*, Vol. I, p. 544.

21. The discussion is devoted primarily to the use of the magnetic stress expression for a derivation of the forces experienced by magnets and currents immersed in the field. Heaviside used this discussion to show that equating time variations in the electric displacement to legitimate electric currents is a matter of definition. This arises from the observation, very clearly displayed in the extrication of the force from the stress, that the conduction current is not measured directly, but defined as a particular measurement of the mechanical force experienced by a magnet immersed in a magnetic field. This was obviously an important high point for Heaviside, whose desire had been to provide a comprehensive theory for the phenomena associated with the electric current.

22. *Electrical Papers*, Vol I, p. 556.

> is employed, in the best part of his time, in the poor work of making mere verifications.[23]

In the context of the section to which this comment was appended, it appears as no more than an isolated remark on the specific problems of experimenting in order to elucidate aspects of advanced theory. However, it would take Heaviside the better part of three years to return to the themes that guided "E.M.I.& P." until this point. In fact, when Heaviside wrote the comment his interest had actually shifted from the field theory of generalized magnetic stresses to the theory of the Wheatstone bridge as a linear circuit. With section 26 the discussion outlined in sections 23, 24, and 25 abruptly ceases. Eventually, Heaviside delivered on the promise of the latter three; but he did so only in June of 1890 when he communicated "On the Forces, Stresses, and Fluxes of Energy in the Electromagnetic Field" to the Royal Society.[24] Thus, just like the mathematical analysis of Maxwell's equations, the general study of Maxwellian dynamics seems to have been postponed in favor of a different topic. Heaviside devoted all of section 26 to the analysis of the inductive Wheatstone Bridge, and continued to develop this discussion in the five sections that followed it. The electromagnetic stress was nowhere to be seen and Heaviside obviously embarked on a new course. To see how that came about, we must turn our attention to David E. Hughes's discovery of the skin effect.

3. Emergence of a New Theme: The Skin Effect.

3.1 David E. Hughes's Discovery

In 1886, David E. Hughes was elected to preside over the Society of Telegraph Engineers and Electricians. In his inaugural address, Hughes chose to inform the Society about a series of experiments he had been carrying out.[25] In these experiments, Hughes set out to determine the electrical properties of various conductors in circuit with a source of intermittent current. He concluded that the results stood unaccounted for in terms of accepted electrical

---

23. *Electrical Papers*, Vol. II, p. 44.

24. *Electrical Papers*, Vol. II, Article 52.

25. D.E. Hughes, "The Self-Induction of an Electric Current in Relation to the Nature and Form of its Conductor," *The Electrician*, **16** (Feb. 5, 1886): 255-258.

knowledge. In particular, under pulsed currents Hughes measured variations in the self-inductance of various conductors that he did not expect, and that seemed to him potentially incongruous with some of Maxwell's conclusions. His interpretation of these experimental results seemed to call for a major revision of circuit theory.

Hughes's experimental procedure consisted of placing a segment of conducting material in one arm of a Wheatstone Bridge, and then balancing the bridge twice: first under a steady current, and then under pulse trains of variable frequency. During the first, steady stage, the self-inductance of the bridge arms was not expected to contribute to the balance, for it had been well established by that time that inductance manifests itself only during current variations. With steady balance achieved, Hughes anticipated that any imbalance during the alternating current stage would be due exclusively to unbalanced distribution of inductance. He found, however, that in general he could not reestablish balance without also redistributing the resistance around the bridge. Furthermore, proceeding to the manner in which the self-inductance varied with the cross section of the wires, Hughes intimated that the measured variations might not be in keeping with Maxwell's theoretical observations.[26] Both findings attracted the close attention of the scientists who attended the lecture,

26. Hughes began the crucial paragraph in the summary of his results with a direct quotation from Maxwell (*Treatise on Electricity and Magnetism*, Vol. 2, p. 291): "'The electromotive force arising from the induction of the current on itself is different in different parts of the section of the wire, being in general a function of the distance from the axis of the wire as well as time.' From this I expected that the increase of electromotive force [needed to restore balance in the variable stage] by an increased section would not increase directly as its sectional increase; but I was not prepared to find, as my experiments prove, that after a certain maximum diameter of wire has been reached a marked decrease in electromotive force takes place with each further sectional increase, and that each maximum is variable with each metal." (D.E. Hughes, "The Self- Induction of an Electric Current in Relation to the Nature and Form of its Conductor," *The Electrician*, **16** (Feb. 5, 1886): 256). Note that Hughes never claimed to have refuted Maxwell. He did not pretend to analyze Maxwell's mathematical presentation, and clearly considered only the qualitative message that variation in the EMF distribution in different layers of the wire would not allow a simple linear increase of self-induction with increasing section. He did not expect the maxima he discovered, but he did not suggest that Maxwell disallowed their existence. Far more curious, however, is that he did not comment on the inhomogeneous current densities that should be associated with the uneven EMF distributions in Maxwell's paragraph. Hughes needed nothing more than the paragraph he quoted from Maxwell to conclude that his experiments demonstrated inhomogeneous current densities during the transient stage. It appears that he missed that observation altogether.

and of Heaviside who read in *The Electrician* both the text of Hughes's lecture and the detailed report on the discussion that followed it.

Hughes quickly discovered that the physicists in his audience received his dramatic conclusions with considerable skepticism. As it turned out, Hughes's interpretation of his experimental results was severely compromised by his inconsistent application of linear circuit theory to the analysis of the circuit design and to the measuring procedure he employed.[27] In a short letter to *The Electrician*, Heaviside explained that the anomalous results Hughes obtained were all due to improper application of standard circuit theory. Reading Hughes's analysis, Heaviside observed, one would think that the science of electricity was in its infancy, with the laws that define resistance, inductance, and capacitance still awaiting discovery:

> When Prof. Hughes speaks of the resistance of a wire, he does not *always* mean what common men, men of ohms, volts and farads, mean by the resistance of a wire—only sometimes. He does not exactly define what it is to be when the accepted meaning is departed from. But by a study of the context we may arrive at some notion of its new meaning. It is not a definite quantity, and must be varied to suit circumstances. Again, there is his "inductive capacity" of a wire. We can only find roughly what that means by putting together this, that, and the other. It, too, is not a definite quantity, but must be varied to suit circumstances. ... Owing to the mention of discoveries, apparently of the most revolutionary kind, I took great pains in translating Prof. Hughes's language into my own, trying to imagine that I have made the same experiments *in the same manner* (which could not have happened), and then asking what are their interpretations? The discoveries I looked for vanished for the most part into thin air. ... I venture to think that Prof. Hughes does not do himself justice in thus deceiving us, however unwittingly, and that possibly there has been also some misapprehension on his part as to what the laws of self-induction are generally supposed to be. ... For consider what the mere existence of ohms, volts and farads means? It means that, even before they [Hughes' discoveries] were made, the laws of induction in linear circuits were known, and very precisely. To get, then, at new discoveries requires very accurate comparison of experiment with theory, by methods which enable us to see what we are doing and measuring, in terms of the known electromag-

27. For an analysis of Hughes's experiment and Heaviside's response see I. Yavetz, "Oliver Heaviside and the Significance of the British Electrical Debate," *Annals of Science*, **50** (1993): 135-173, appendix 1. For a detailed account of Hughes's experiment and the discovery of the skin effect, see D.W. Jordan, "D.E. Hughes, Self-Induction and the Skin-Effect," *Centaurus*, **26** (1982): 123-153.

> netic quantities. This is practically impossible, on the basis of Prof. Hughes's papers.[28]

At the same time, Heaviside showed that variations in current distribution over the wire's cross section should be expected under alternating currents of high enough frequency. Significantly enough, he showed that prediction of such variations would emerge regardless of whether the circuit is analyzed in Maxwellian field terms or in terms of electrical fluids interacting at a distance:

> As a general assistance to those who go by old methods—a rising current inducting an opposite current in itself and in parallel conductors—this may be useful. Parallel currents are said to attract or repel, according as the currents are together or opposed. This is, however, mechanical force on the conductors. The distribution of current is not affected by it. But when currents are increasing or decreasing, there is an apparent attraction or repulsion between them. Oppositely going currents repel when they are decreasing, and attract when they are increasing. Thus, send a current into a loop, one wire the return to the other, both being close together. During the rise of the current it will be denser on the sides of the wires nearest one another than on the remote sides. It is an apparent force, not between currents (on the distance-action and real motion of electricity views), but between their accelerations.[29]

Such variations in the distribution of current would entail changes in both resistance and inductance, and Heaviside proceeded to show how an experiment should be designed to illustrate clear-cut departures from the current-independent resistance and inductance that characterized traditional linear circuit theory. In order to bring out these departures, the design should strictly conform to the principles of linear circuits. Hughes, however, failed to comply with these principles in the design of his experiments. It was therefore difficult to assess which, if any, of his results really exhibited phenomena that stood outside the scope of linear circuit theory. Obviously then, the comment that Heaviside appended to section 25 of "E.M.I.& P." expresses much more than a passing remark on scientific experimentation. It reflects the intense interest Heaviside took in Hughes' work, and introduces the most profound change in his publication plans since he began to expose Maxwell's theory in 1882.

---

28. O. Heaviside, "Self-Induction of Wires," *The Electrician*, **16** (Apr. 23, 1886): 471-472; or *Electrical Papers*, Vol. II, pp. 28-29.

29. *Ibid.*, p. 31.

In the initial parts of "E.M.I.&P." Heaviside already took a crucial step toward delineating the phenomena that mark the domain of linear circuit theory. We have seen how he described the energy flow into the conducting wire from the surrounding dielectric. This description implies that there must be an intermediate stage between an inactive conductor and a conductor in which electromagnetic energy is dissipated homogeneously throughout its cross section. Since energy comes in from the outside, it must be that initially only the outer layers of the wire will dissipate energy, and that this dissipative process will gradually diffuse inward. However, energy dissipation in a conductor can occur only in the presence of electric current. It immediately follows that during the initial stages of current onset the current is not distributed homogeneously throughout the conductor, but that it begins in the outer layers and diffuses inward until a homogeneous current density is established. This is today referred to as skin conduction. Heaviside tried to give his readers an intuitive sense of how such conduction could come about in terms of water flow in a pipe. If the entire pipe, filled with water, is suddenly jerked into steady motion along its axis, the water layer directly in contact with the pipe's inner surface will join the motion immediately. However, inner layers will pick up the motion only gradually, so that at the beginning of the pipe's motion, water flow will be confined to a thin "skin" at the boundary.[30]

What Heaviside failed to do when he first outlined this process in 1885 was to take the analysis to its conclusion, and show that during the initial variable state, the resistance and inductance of the wire will not be the same as during the eventual steady state. In other words, he described skin conduction, but not its effects. To see how these effects come about, consider that the resistance of a conducting body is inversely proportional to the conducting cross

30. The skin effect usually manifests itself in conductors carrying oscillating currents, when oscillation periods are considerably shorter than the time it takes for the current to distribute itself across the entire cross section. As Heaviside observed, under such circumstances the current oscillations on the skin have no time to penetrate inward, and most of the current remains confined to a thin boundary near the conductor's exterior: "If the steady state is not fully set up before the impressed force is removed, we see that the central part of the wire is less useful as a conductor than the outer part, as the current is there the least. If there are short contacts, as sufficiently rapid reversals, or intermittences, the central part of the wire is practically inoperative, and might be removed, so far as conducting the current is concerned." (*Electrical Papers*, Vol. II, pp. 49-50). However, alternating currents are not in any way essential to the existence of the skin effect. Indeed, the alternating current aspect of the skin effect is an obvious special manifestation of the more fundamental considerations that Heaviside illustrated by the water-pipe analogy.

section. This must now be examined in light of the notion that current is confined to a thin boundary layer at the very beginning of a pulse. In effect the wire may be regarded as a thin-walled conducting tube surrounding a nonconducting core. Hence, during initial current onset, the conducting cross section is effectively reduced, and the overall resistance is greater than that of the same conductor under a uniform current distribution. By the same token, the current-free inner core is also free of the magnetic field associated with the current. As a result, the core does not contribute to the overall self-inductance of the wire. Thus, since during the initial stage of current onset the current distribution over the cross section is varying, one can no longer use the terms "resistance" and "inductance" as independent properties of the wire. They become functions of the varying current distribution.

In the papers he published following Hughes's work, Heaviside indicated an alternative way of showing even more sharply the incompatibility of the field and linear views of current onset in a homogeneous conducting wire.[31] Consider first the linear view under the assumption that resistance and inductance are properties of the conducting material independent of the current in it. As already mentioned, under this view the wire may be regarded as a line characterized by resistance and inductance per unit length. When such a linear conductor creates a closed circuit with an electromotive force, a current will arise in it. It will take some time for the current to attain its steady state, and during the initial stages its strength will increase at a rate determined by the ratio of resistance and inductance. Linear circuit theory, however, always considers integral current and assumes that the wire's inductance and resistance are independent of the current. Therefore, if the wire is a real wire with significant thickness, the current density throughout its cross section must be assumed constant during the entire process (recall that variations in current density generally entail variations of both resistance and inductance). In particular, the current density may be taken as evenly distributed throughout the cross section, everywhere growing in strength at the same rate until the steady state is reached. Now consider with Heaviside whether this description of current onset can be maintained together with Maxwell's circuital equations. A homogeneous current density, **C**, necessarily implies that the electric force is also homogeneously distributed throughout the conductor, because $\mathbf{C} = k\mathbf{E}$, and k, the conductivity (not to be confused with the conductance), is con-

31. *Electrical Papers*, Vol. II, p. 83.

stant throughout the homogenous wire. Use this with Maxwell's second circuital law, -curl(**E**) = d**B**/dt. According to our description thus far, **E** is homogeneous throughout the wire, its curl is therefore zero and consequently **B** does not vary in time inside the wire. In this particular case, we started out with no current, no electric force, and no magnetic flux. Therefore, a magnetic flux that starts out at zero and does not vary in time will necessarily remain zero. Unfortunately, Maxwell's first circuital law requires that curl(**H**) = **C**.[32] We assumed that the current grows homogeneously throughout the cross section, which requires that once current has started, the value of curl(**H**) must differ from zero everywhere in the wire. That, however, cannot happen when **H**, which is related to **B** by a constant, is uniformly zero throughout the cross section. We could avoid this contradiction and still satisfy the assumed constancy of resistance and inductance by starting with an uneven current distribution that retains a constant proportionality in time as the current intensifies toward the final steady state. This will allow a nonzero curl(**E**), and enable a growing magnetic field (both **H** and **B**) in the wire. However, both linear and field theories must accommodate the requirement that at some point the current has to stop increasing. In the field theory this means that at some point the time variation of **B**—and consequently **H**—in the wire must drop to zero, which requires that **E**—and consequently **C**—must become homogeneously distributed. This implies that the uneven current distribution we began with must evolve into a homogeneous one. The linear theory, however, does not allow for that, because such a process generally implies a variation of both the resistance and the inductance in keeping with the changing current distribution. Either way, then, we reach a contradiction which implies that Maxwell's theory is fundamentally incompatible with the linear theory of current onset in a circuit.

All this explains the excitement that Hughes's ill-designed experiments raised. The apparently anomalous variations in inductance that they indicated could be accounted for in terms of Hughes's inconsistent use of linear circuit theory. At the same time, the prospect of measurable departures from constant resistance and inductance during current onset suggested an opportunity to

---

32. Maxwell's first circuital law actually requires that curl(**H**) = **C** + d**D**/dt. However, there is no way to insure that the sum on the right amounts to zero in general (only when $D = cE$; $C = -kE$; and $E = E_0 e^{kt/c}$ can this requirement be satisfied). Therefore we may assume that the conductor is characterized by zero capacity without undermining the generality of the above argument.

highlight the usefulness of more comprehensive electromagnetic theories—Maxwell's field theory in particular. The basic assumption underlying the linear theory of electric circuits is that under all circumstances one only needs to take into account the integral current in the wire. The electrical properties of the wire in the most general version of this theory are then taken per unit length, as if the wire had no thickness. Maxwell's theory, on the other hand, by virtue of being a field theory, provides a description of the electromagnetic state in any minute volume of space. As such, it enables one to endow the electric wire with thickness, and to investigate, for example, the current distribution over its cross section. As we have seen, the Maxwellian analysis suggests that the current begins on the wire's surface and then diffuses inward much like heat would diffuse laterally into a long metal cylinder. Hughes's experiments indicated the possibility of associating this abstract prediction with measurable effects in practical situations, hence the excited reaction to his work.

At the same time, this explains why Heaviside and a few others, notably Lord Rayleigh, found much to criticize in Hughes's work. They realized that what Hughes saw as a challenge to Maxwell's theory actually reflected his own misunderstanding of an already well-established circuit theory. All this is well expressed in Heaviside's comment on Hughes's objectionable use of induction and resistance balances in the course of his experiments:

> In Prof. Hughes's researches, which led him to such remarkable conclusions, the method of balancing was not such as to ensure, save exceptionally, either a true resistance or a true induction balance. Hence, the complete mixing up of resistance and induction effects, due to false balances. And hidden away in the mixture was what I termed the "thick-wire effect," causing a true change in resistance and inductance. In fact, if I had not, in my experiments on cores and similar things, been already familiar with real changes in resistance and inductance, and had not already worked out the theory of the phenomenon of approximation to surface conduction, on which these effects in a wire with the current longitudinal depend, it is quite likely that I should have put down all anomalous results to the false balances.[33]

Hughes's work created intense interest that went beyond Britain's borders. It motivated subsequent work that eventually led to experimental corroboration of the skin effect, partially justifying the accreditation of Hughes with the effect's discovery.[34] Aside from this influence, Hughes's work had a far-

33. *Electrical Papers*, Vol. II, p. 100.

reaching effect on Heaviside by providing him with the first component of a new and ambitious publication plan. To this effect of Hughes's work we must now turn our attention.

Heaviside was drawn to Hughes's work and the skin effect by two different, though related sentiments. Heaviside's early correspondence with Oliver Lodge, as well as his statement that he had already been experimentally "familiar with real changes in resistance and inductance," reveal that he considered himself as deserving at least some credit for the discovery of the skin effect. More important than this personal interest, Heaviside saw together with others that Hughes's work could be used to highlight the practical relevance of higher theoretical concepts. In particular, he realized that through the effect he could develop his previous work in fulfillment of his old desire to form a comprehensive view of electric circuits. As we shall presently see, these two considerations motivated Heaviside to reorganize his publication scheme, and to return to the subject he abandoned in 1882, namely, the theory of electric circuits, and transmission lines in particular.

3.2 A Questionable Priority Claim

Heaviside's interest in establishing a priority claim for the discovery of the skin effect is worth examining if only because it demonstrates his incomplete acquaintance with the scientific literature of the time. The skin effect as described here consists of two components: the observation that a rapidly oscillating current tends to concentrate in the conductor's outer layers, and the associated prediction that under such circumstances the resistance and inductance would differ from their values during steady conduction. Heaviside was anticipated by Lord Rayleigh, Horace Lamb, and Joseph Larmor with respect to the first component, and by Maxwell himself with respect to the second. As early as 1882, Rayleigh published a short note in which he solved the problem of the decay of eddy currents in a cylindrical core. In this short, single-page discussion, Rayleigh only outlined the solution and did not explicitly interpret the mathematical expressions to conclude that the decay begins at the boundary and diffuses inward.[35] In the same year, Rayleigh also noted the

34. W.D. Jordan, "D.E. Hughes, Self-Induction, and the Skin Effect," *Centaurus*, **26** (1982): 141-142.

35. Rayleigh, "On the Duration of Free Electric Currents in an Infinite Conducting Cylinder," *Report of the British Association for the Advancement of Science—1882*, pp. 446-447.

screening effect that good conductors have on electromagnetic disturbances.[36] He showed experimentally that a copper board interposed between two coils can effectively screen them from mutual inductive interference at sufficiently high frequencies. He accounted for this by the inability of a rapidly oscillating field to penetrate the conductive board, thus anticipating Heaviside's often-quoted observation that "the prefect conductor is the perfect obstructor."[37] In 1884, Joseph Larmor published a detailed mathematical elaboration of these observations by Rayleigh, in which he calculated the penetration of a flat conducting board by a varying electromagnetic field.[38] In the course of his analysis, Larmor observed that:

> Thus, if at each instant we suppose a system of currents to start in the superficial layer of the body which neutralizes for internal space the effect of the outside changes, the actual state of the body is that produced by these currents soaking into it and decaying by their own mutual action. The equation of decay ... is the same as the equation for the diffusion of heat from the surface into the body.... We see that to have penetration into the solid *of the same order* in both cases the oscillations must be about 100 times quicker for the electrical case.[39]

In January of 1884, less than two weeks before Larmor's paper appeared, Horace Lamb published another study of current onset based on Maxwell's *Treatise*.[40] In the first five pages of his paper Lamb discussed the skin effect in long cylindrical conductors, namely, wires, and then proceeded to the case of a spherical conductor. The first part of Lamb's analysis addresses the case of a rapidly varying magnetic field *parallel* to the axis of the wire, in which case the induced currents are circles centered on the wire's axis. This is the case of a conductive core surrounded by a coil carrying an alternating current—exactly the case that Heaviside began to study four months later in "The Induction of Currents in Cores." In the course of his analysis, Lamb observed that:

---

36. Rayleigh, "Acoustical Observations - IV", *The Philosophical Magazine*, **8** (May, 1882): 340-347, esp. p. 344.

37. As Rayleigh noted in his paper, the basic observation that a perfect conductor would perfectly block magnetic fields was made by Maxwell, (*Treatise on Electricity and Magnetism*, 3rd edition, Vol. II, article 655).

38. Joseph Larmor, "Electromagnetic Induction in Conducting Sheets and Solid Bodies", *The Philosophical Magazine*, **18** (Jan. 1884): 1-23, esp. 6-7.

39. *Ibid.* p. 11.

40. Horace Lamb, "On the Induction of Electric Currents in Cylindrical and Spherical Conductors," *Proceedings of the London Mathematical Society*, **15** (1884): 139-149.

> ... when the oscillations in the intensity of the field are rapid in comparison with the decay of free currents, the induced currents extend only to a small depth beneath the surface of the cylinder, the inner strata (so to speak) being almost free of electromotive force by the outer layers.[41]

Heaviside discovered the theoretical possibility of skin conduction in his study of "The Induction of Currents in Cores" under oscillating fields. He began to publish this study on 3 May 1884. On 10 May 1884, he published his first explicit interpretation of the mathematical analysis as indicating the initial concentration of current at the core's boundary. In July he called attention to the analogy between the magnetic field in the core and the communication of motion to water contained in a pipe set into motion parallel to its axis.[42] Lamb, then, clearly anticipated Heaviside's efforts in this particular direction even if Heaviside's study is considerably more detailed and in many ways more satisfying from the physical point of view. To some extent, Lamb himself had been anticipated by Rayleigh, who solved the equation of current diffusion into a cylindrical core in 1882, and whose experimental study indicated that electromagnetic fields permeate good conductors only through a relatively slow process of diffusion.

In one important respect, however, the above investigations of the skin effect, including Heaviside's, fall short of addressing the problem Hughes encountered. The resistance of a wire to currents that circulate around its axis is not the same as its resistance to axial currents. Consequently, variations of resistance such as those observed by Hughes cannot be calculated directly from the case Lamb and Heaviside investigated in 1884. In January of 1885, Heaviside noted that it must take time for the energy flux to penetrate good conductors, and proceeded to comment that as a result "it may be expected" that the electric current would start on the boundary.[43] Here Heaviside was already discussing skin conduction in the case of axial currents, but he said nothing at all about observable effects of skin conduction, notably variations in inductance and resistance. These, after all, and not skin conduction per se, had been the subject of Hughes's experiments, and Heaviside could not be credited with predicting them in 1885, or at any other time before Hughes described his work. However, the effects of such variable currents had already been discussed in generalized form by Maxwell, when he investigated the general

41. *Ibid.*, p. 141.

42. *Electrical Papers*, Vol. I, pp. 353-416, esp. 360-361; 382-383.

43. *Ibid.*, p. 440.

problem of inhomogeneous current distributions in cylindrical conductors.[44] In an appended footnote to Maxwell's discussion, J.J. Thomson showed in four straightforward steps how Maxwell's general eq. (18) could be applied to the specific case of harmonically oscillating currents.[45] However, Maxwell did not associate these changes in resistance and inductance with the explicit notion of skin conduction, and his equations need further elucidation to bring out this feature. It appears that the first explicit association of skin conduction with its measurable effects was due to George Forbes during the discussion following Hughes's presentation. Forbes concurred with Rayleigh's observation that the manner in which Hughes used the concepts of resistance and inductance prevented meaningful comparison with Maxwell's discussion. He then proceeded to add that the theory Maxwell developed actually implied more, and showed that:

> ... the E.M.F. of induction must be greater in the centres than in the outside of a wire, and consequently there may be sometimes a current in the centre of the wire contrary in direction to the main current, the outer part of the wire forming a short circuit to this induced current. The energy consumed by such an eddy current, or by the main current being confined to the outer parts of the wire, is in the nature of a resistance, and does not change sign with an increase or decrease of the current, as self-induction is generally supposed to do. This explains how Prof. Hughes obtained a different value for the resistance of a wire in the variable state, and in the stable state. He [Forbes] had applied this theory combined with the views put forward by Lord Rayleigh, to all the experiments contained in Prof. Hughes' paper, and found that in every case the experiment was in accord with theory.[46]

Forbes's observation prompted Rayleigh to review Maxwell's discussion. He studied it carefully, and produced a detailed analysis with several case studies that demonstrated various aspects of skin conduction. Rayleigh communicated his paper to *The Philosophical Magazine* on April 3, 1886. It was published in May of the same year.[47] In fact, J.J. Thomson's footnote on p. 322 of Maxwell's third edition is practically a reproduction of steps (15)-(20)

---

44. J.C. Maxwell, *Treatise on Electricity and Magnetism*, (3rd edition, edited by J.J. Thomson), Vol. II, Art. 689, pp. 320-323.

45. *Ibid*., p. 322.

46. Report of the discussion following Hughes' presentation, in "The Society of Telegraph Engineers and Electricians" *The Electrician*, **16** (Feb. 19, 1886): 289.

47. Rayleigh, "On the Self Induction and Resistance of Straight Conductors", *The Philosophical Magazine*, 5th Series, **21** (May, 1886): 381-394.

of Rayleigh's original analysis.[48] Immediately following that analysis, Rayleigh wrote:

> At slow rates of alternation, the distribution of current, being such as to make the resistance a minimum, is uniform over the section; and this distribution, since it involves magnetization of the outer parts of the cylinder, leads to considerable self-induction.... On the other hand, when the rate of alternation is very rapid, the endeavour is to make the self-induction a minimum irrespective of resistance. This object is attained by concentration of the current into the outer layers.[49]

Strictly speaking, the impedance is minimized; but Rayleigh's statement is a fair approximation for both very low and very high rates of alternation. Heaviside's first detailed discussion of the association between skin conduction and variations in resistance and inductance was published by *The Electrician* on May 14, 1886.[50] We do not know when Heaviside communicated his paper to the editor of *The Electrician*, so that he may have had a hair's breadth priority over Rayleigh's 1886 study. But even if Heaviside did have priority over Rayleigh in 1886, it seems clear that between Maxwell, Rayleigh (in 1882), Lamb, Larmor, and Forbes he had been anticipated on all essential aspects of the skin effect. Of course, when Heaviside first became aware of Hughes's work in 1886 and realized its relevance to his own work from 1884 and 1885, he almost certainly did not know of the studies by Rayleigh, Lamb, and Larmor.

Heaviside never claimed priority for providing experimental evidence in support of skin conduction, although in his initial critique of Hughes's experiment he did indicate that he obtained such results from his own experimental work on the induction of current in cores. The surviving fragments of Heaviside's correspondence with his brother Arthur, as well as notes in one of his notebooks, clearly indicate that as early as 1881 he had been carrying out considerable experimental work on Wheatstone-bridge arrangements with capacitors and inductors.[51] However, the notes do not leave any clear indication that the purpose of this work was to obtain evidence of variations in resistance and inductance under alternating currents, or that it inadvertently revealed such evidence. At any rate, in his letters to Lodge, Heaviside clearly credited Hughes with the experimental corroboration of skin conduction. Heaviside rightly pointed out that Hughes did not intend his experiments as tests for the measur-

48. *Ibid.*, p. 387.
49. *Ibid.*, p. 388.
50. *Electrical Papers*, Vol. II, pp. 44-50, esp. p. 49.

able effects of skin conduction, and that he failed initially to appreciate the significance of his results. For himself, Heaviside limited his priority claim to the proper theoretical establishment of skin conduction from the Maxwellian point of view. From his correspondence with Lodge, Heaviside learned that he would have a difficult time convincing others to accept even this limited claim without further qualifications.[52] Heaviside's struggle to preserve something for himself despite the prior work that Lodge cited to him clearly shows how intensely he desired to associate himself with the discovery of skin conduction. Note first that the wording in section 26 of "E.M.I.& P." gives the impression that Heaviside considered the skin effect in straight wires to be an almost self-evident result of his 1884 study of the diffusion of currents into cores. In reference to the effect in straight conductors with the current longitudinal (relevant to the skin effect in Hughes's work), Heaviside wrote:

> The investigations are almost identical with those given in my paper on "The Induction of Currents in Cores," in *The Electrician* for 1884.... The magnetic force was then longitudinal, the current circular; now it is the current that is longitudinal, and the magnetic force circular.... The application to round wires with the current longitudinal was made by me in *The Electrician* for Jan. 10, 1885, so far as a general description of the phenomenon is concerned."[53]

Indeed, considering the elementary exercises of changing the roles of current and field that Heaviside presented at the beginning of his introduction to field thinking (see chapter III), his comment seems reasonable. Heaviside wrote this comment in 1886, just after he became aware of Hughes's work. Two

---

51. See Notebook 1A, Heaviside collection, IEE, London. This notebook, like many of the notebooks in the collection, contains digested accounts of original notes that have not been preserved. It begins with a note: "Abridged Acc^{t} of Exp^{ts} in May 1886. London. Many of these were mere repetitions of Whitely expts of Oct. 18__ elsewhere described, so that only a condensed account is necessary". Pages 10 to 25 contain descriptions of experiments that were almost certainly motivated by Hughes's work. The descriptions contain the same nomenclature (e.g. p. 23, "M6,4 method") he used in his analysis of Hughes's experiment, published 30 April 1886 in *The Electrician*. Pages 26-29, under the title "Whitley Experiments (Oct. 1881)" describe the earlier experiments.

52. Heaviside to Lodge, 27 June, 5 July, and 17 Sept. 1888, Lodge Collection, University College, London. The second letter in particular shows most clearly that Heaviside did not claim priority for the experimental discovery of variable resistance, nor even for predicting skin conduction. He claimed priority for being the first to arrive at skin conduction by considering the way energy diffuses into a conductor from the surrounding dielectric. In this qualified sense his claim is actually justified.

53. *Electrical Papers*, Vol. II, p. 45.

years later, in his correspondence with Lodge, Heaviside expressed a very different sentiment:

> You presented me with a formidable array of "anticipations". But I cannot see that they are to the point. ... Lamb worked out the problem of vibrating magnetic force in a core longitudinally magnetised sinusoidally. But is it at all obvious that the problem of current in a wire is mathematically similar? I didn't find it so. It required original investigation to discover it.[54]

It is impossible to decide which of these two statements should be taken more seriously. Analogies seem self-evident only after they have been seen. The existence of the analogy between the two cases does not mean that Heaviside saw it before 1886, or that it became instantaneously evident to him following Hughes's work. Then again, Heaviside was aware in principle of such analogies as his discussions from 1882 show. In 1886, when he wrote the comment in section 26 of "E.M.I.& P.", Heaviside was evidently unaware of the prior studies by Rayleigh, Larmor, and especially Lamb. Heaviside's statement on the longitudinal skin effect in 1885 was no more than an inconclusive side comment. However, his 1884 study of the induction of currents in cores provided an exhaustive mathematical demonstration of skin conduction. His overriding concern in 1886 could have been, therefore, to show that while his 1884 discussion did not directly apply to Hughes's work, it required no additional insight to establish the connection between the two.

In 1888 Heaviside learned from Lodge that Lamb actually anticipated his study on the induction of currents in cores. In light of that, it could have now become imperative for Heaviside to argue that the case in cores with the current circular did in fact require original insight before it could be made relevant to Hughes's experiments. He now emphasized the 1885 study, and claimed priority for being the first to establish skin conduction from considerations of the energy flux. The incongruence between Heaviside's two statements can be partially mediated by interpreting his letter to Lodge as suggesting that Lamb never discussed the longitudinal skin effect and never showed the interchangeability of current and magnetic field as Heaviside did in 1882. Therefore Lamb did not do the "required original investigation." In the final analysis, however, the two conflicting statements betray one dominant motivation, namely, Heaviside's desire to establish at least a partial priority claim in the discovery of the skin effect. The fact that Heaviside never pushed his

---

54. Heaviside to Lodge, 17 September 1888, Lodge Collection, University College, London.

claim with regard to the skin effect further than his private correspondence with Lodge suggests that he considered Lodge's arguments to be rather weighty. The important thing is that in 1886 Heaviside was not aware of all the anticipations, and that his desire to be associated with the discovery of an important effect might have contributed significantly to his decision to reorient "E.M.I.& P." This decision led to a real innovation; for as we are about to see, no one incorporated the effect into the kind of systematic and comprehensive revision of circuit theory that Heaviside produced.

3.3 Circuit Theory, Field Theory, and the Skin Effect

Heaviside was undoubtedly interested in the possibility of associating himself with the discovery of the skin effect, but the importance of this motivation should not be overestimated. Of far greater consequence was his realization that the interest that Hughes's work attracted provided an unprecedented opportunity for linking the theoretical work he had been doing since 1882 to important practical issues. It must be emphasized that Heaviside never pretended that the skin effect was in any way an exclusive Maxwellian field effect. Indeed, we have already seen (page 194) that it was Heaviside himself who showed how to consider the effect in terms of action at a distance between accelerating electrical fluids in conductors—a point of view that violates Heaviside's Maxwellian field view both by positing non-local interactions and by regarding the electric current as a material flow. In fact, neither this non-local interaction nor Maxwell's field theory is needed to deduce skin conduction from the measurement of frequency-dependent variations in resistance and inductance. It is enough to keep in mind the phenomenological laws of linear circuits together with the assumption that unlike integral resistance and inductance, both *specific* resistance and *specific* inductance (or resist*ivity* and induct*ivity* in the nomenclature scheme Heaviside was soon to develop) are frequency *in*dependent. Under these conditions, variations in resistance and inductance necessarily imply variations in the current distribution over the conductor's cross section. In particular, as Rayleigh's argument shows, concentration of the current in the outer layers accounts for the increased resistance and reduced inductance that accompany a rising current. Hence, Heaviside's interest in Hughes's work stemmed from his wish to emphasize the usefulness, rather than the exclusive necessity, of a field view that naturally incorporated skin conduction.[55] In 1889 he expressed that in a letter to Hertz:

> I was formerly accused of dogmatism, especially in regard to surface-conduction; but I was experimentally *sure* of the accuracy of the laws on which it is based; it has nothing to do with Maxwell's theory of a dielectric; I mean that it is true independently thereof; but how admirably Maxwell's theory of the dielectric explains the meaning of it.[56]

It appears that Heaviside considered this observation important enough to postpone his study of the dynamical foundations of Maxwell's theory and the electromagnetic wave-front. He set out instead to systematically place the theory of circuits and transmission lines in the wider context of Maxwell's field theory. In six successive sections (26-31) of "E.M.I.& P.", published in *The Electrician* between May and August of 1886, he developed this theme. Heaviside showed through a set of carefully designed case studies how the linear circuit equations should be modified to yield adequate approximations to various circuit configurations when resistance and inductance could no longer be considered constant. Then, in September of 1886, without formally completing the picture of linear circuit theory as a practical approximation of field theory, Heaviside began to focus his attention on transmission-line theory.

On first examination, it seems as though in section 32 of "E.M.I.& P." Heaviside regressed to the sort of transmission-line analysis that he developed in the 1870s. With hardly a hint of the sophisticated field view he developed in the previous sections, Heaviside now presented a purely linear formulation of the telegraph equation, with a discussion of its components. He continued the discussion in the same spirit in section 33. In section 34, however, he already extended the treatment "...to include the propagation of current into a wire from its boundary." In section 36 he was back to the "Resistance and Inductance of a Round Wire with Current Longitudinal...," in which he examined the effects of variable current distributions. However, the short digression in sections 32 and 33 is not merely a summary of linear circuit theory. It represents the crystallization in Heaviside's mind of his ambitious plan to devise a comprehensive framework for the case studies that he had been elabo-

55. "I contend that from the fact of increased resistance (and reduced inductance) Wh. Prof. H found ... and the fact which follows immediately of variable current density, we cannot conclude anything at all about the *way* it comes about. There might be 20 theories showing variable c[t] density & increased resistance." (Heaviside to Lodge, 5 July 1888, Lodge Collection, University College, London).

56. Heaviside to Hertz, 14 Aug. 1889, quoted in J.G. O'Hara and W. Pricha, *Hertz and the Maxwellians*, (London: Peter Peregrinus Ltd., 1987), p. 72.

rating since May of 1886. His plan involved simultaneous publication in two journals, *The Electrician* and *The Philosophical Magazine*. In *The Electrician*, Heaviside intended to continue his May-August discussion in "E.M.I.& P." by showing how linear circuit theory could be generalized into electromagnetic field theory. At the same time, he began a new series, "On the Self-Induction of Wires," in the *Philosophical Magazine*. Here he pursued the reverse course: starting from his duplex field equations he showed how, through judicious approximations, they could be made to yield the principles of linear circuits. One need not speculate that this was indeed Heaviside's plan, for he openly declared his intentions at the beginning of the apparent digression in section 32 of "E.M.I.& P.":

> In another place (*Phil. Mag.*, Aug., 1886, and later) the method adopted by me in establishing the equations of V and C, Section XXIX, was to work down from a system exactly fulfilling the conditions involved in Maxwell's scheme, to simpler systems nearly equivalent, but more easily worked. Remembering that Maxwell's is the only complete scheme in existence that will work, there is some advantage in this; also, we can see the degree of approximation when a change is made. In the following I adopt the reverse plan of rising from the first rough representation of fact up to the more complete. This plan has, of course, the advantage of greater intelligibility to those who have not studied Maxwell's scheme in its complete form; besides being, from an educational point of view, the more natural plan.[57]

The common thread that ran through both series was the analysis of transmission lines. As we have seen, Hughes's inaugural address to the Society of Telegraph Engineers and Electricians gave Heaviside the initial motivation for turning his attention to circuit theory. However, neither the address nor the skin effect suggested to Heaviside the idea of organizing his new double-sided exposition of Maxwell's theory around the analysis of transmission lines. To trace the development of this idea we must turn to the long and steady collaboration between Oliver Heaviside and his older brother, Arthur West Heaviside.

57. *Electrical Papers*, Vol. II, p. 76.

4. The Bridge System of Telephony and the Distortionless Condition.

As we have seen in the previous chapter, Heaviside devoted most of his publications between 1882 and 1886 to the basic concepts of electromagnetic field theory. This should not be interpreted to mean that he lost interest in the theory of circuits and transmission lines, or, for that matter, in practical problems of telegraphy and telephony. Since his departure from the telegraph service he kept in touch with telephone and telegraph engineering through his brother. I have already noted (see chapter I) that this connection provided Heaviside with more than electrical components for his own experimental studies. A.W. Heaviside, who valued his younger brother's command of advanced theory, used him as a scientific adviser on several occasions. Their collaboration provided Heaviside with a constant source of problems that complemented his interest in basic electromagnetic theory. Thus, the notes in one of Heaviside's notebooks corroborate the claim he later made in a letter to Hertz that he obtained experimental results that could have been refined into indications of electromagnetic wave detection.[58] It seems that the two brothers were experimenting in the early 1880s with long-distance inductive coupling of telephones for the purpose of obtaining wireless communications. This may have been the subject of the patent application they discussed in their correspondence from the same period. Nothing of practical value came from this, and Heaviside freely acknowledged that he recognized the relevance of this work to Hertz's discovery only in retrospect, with Hertz's achievements well in hand.[59] However, this clearly demonstrates that while Heaviside abandoned the publication of transmission-line studies in favor of introducing the elements of field theory, he kept in touch with the subject through its practical side.

In 1886, the collaboration of the two Heaviside brothers finally exerted a crucial effect on Oliver Heaviside's publications. Toward the end of that year, Arthur Heaviside masterminded a sweeping redesign of Newcastle's telephone network. Among his innovations was the idea of placing the telephones in par-

58. Notebook 1A: 77-82 (pp. 80-81 are missing) under the title "Tel. Exp^{ts} [A.W.H.'s]". These are all experiments dating from 1881 on inductive telephony, using circuits measuring dozens of miles in length.

59. Heaviside to Hertz, 14 Aug. 1889, in J.G. O'Hara and W. Pricha, *Hertz and the Maxwellians*, (1987), p. 72.

allel, as bridges in a matrix of telephone lines. The purpose of this design was to remove as much series impedance from the lines as possible, so as to reduce the attenuation of signals to the farther points of the network. When the arrangement was tested, it was found that in addition to increased volume, the transmitted voice actually came in more crisply than before. This reminded Oliver Heaviside of his work ten years earlier. The bridged telephones actually formed discrete leaks in the transmission lines, and in 1876 Heaviside gave a theoretical account of the experimental knowledge that leaks in a telegraph line could actually sharpen the signals and contribute to faster signalling. What followed next is best described in Heaviside's own words:

> Considering that the bridge system consists electrically of a circuit with a number of leaks in it, and bearing in mind my investigations concerning leaks, I inquired whether the intermediates were not actually beneficial to through communication .... Was not the articulation better, not merely than if the intermediates were in series in the circuit, but better than if the shunts or bridges were removed altogether, leaving insulation? My brother's answer was Yes, certainly, in certain cases but doubtful in others. It appeared, however, to be only a minor effect, of not much importance compared with the major one. Besides, the bridges caused a weakening of the intensity of the speech received when many bridges were passed.[60]

To avoid the weakening of intensity, the bridges were purposefully endowed with the highest possible resistance and inductance, so as to minimize the leak current and approach the state of perfect insulation. In other words, they were not designed with the idea of finding out precisely how much leakage would lead to the best clarity of speech, because obtaining higher volumes was the overriding concern. To Heaviside, however, this circuit suggested a novel outlook on the propagation of current pulses in a transmission line. In his old work, he was concerned with the detailed description of the pulses in terms of their harmonic content. In that investigation Heaviside noted that resistance had the effect of rounding off the sharp edges of a pulse. Arthur Heaviside's bridge system made Oliver Heaviside focus specifically on the encounter between a pulse and a leakage point, and he quickly realized that it would be counterproductive to bring the full power of the mathematical machinery to bear on this question. What he did instead was to consider first of all that in an ideal line, characterized only by inductance and capacitance, the

60. *Electromagnetic Theory*, Vol. I, p. 435.

current pulses would be purely elastic, namely, they would retain their shape and size perfectly. He then proceeded to examine the encounter between such pulses and a small discrete series resistance coupled to a small leak. This simplifying approximation enabled him to study resistance and leakance as contributing second-order effects to a primarily elastic pulse. The investigation led to a remarkable conclusion that is not immediately obvious from the bridge system itself, namely, that when leakance and resistance are related in a definite way, current pulses will traverse the line with attenuation, but without distortion. Heaviside's own description of how he arrived at that conclusion is again illuminating:

> The investigation of the bridge system of telephony suggested to me the distortionless circuit, as I have before acknowledged (*Electrical Papers*, Vol. II, p. 402). It may be of interest to state how it came about, since it is not presented to one by the above described experiences with the bridge system. The question was, What was the effect of a bridge, or of a succession of bridges, when the self-induction of the line was not negligible? Now the influence of the bridges can be very readily settled for steady currents, of course. There is no difficulty either in finding the full formulae required to express the voltage and current and their variations in any part of the system when the inductance and permittance are operative, as well as a set of leaks, when one knows how to do it. But the general formulae thus obtained are altogether too complicated to be readily interpretable. I was, therefore, led to examine the effects produced by a leak on a wave passing it of the elastic type. This implies that the resistance of the line is sufficiently low for waves of the frequency concerned to retain that type approximately. ... Now this was comparatively an easy problem, and the result was to show that a leak (or conductance in bridge) had the opposite effect on a wave passing it along the line to that of a resistance inserted in the circuit at the same place in this sense:- A resistance in the line reflects the charge positively and the current negatively, or increases the charge and reduces the current behind the resistance. On the other hand, a leak reflects the current positively and the charge negatively, or increases the current and reduces the charge behind the leak.... A resistance and a leak together at the same place therefore tend to counteract one another in producing a reflected wave, leaving only an attenuating effect on the transmitted wave. This led to the distortionless circuit. For although the compensation of the resistance in the circuit and the conductance across it is imperfect when they are finite, it becomes perfect when they are infinitely small.... This occurs when R and K, the resistance and conductance per unit length of circuit of the line and of the leakage respectively, are taken as the same ratio of the other constants, L and S. Or, when L/R = S/K, indicating equality of the electric and magnetic time-constants.[61]

There are several things to observe about the emergence of this discovery. We shall soon see that Heaviside's attempts to publish it and call attention to its practical significance met with stiff resistance. Perhaps this troubled reception gave rise to later claims that the discovery required the use of novel, even revolutionary, knowledge. At any rate, there is a long-standing and fairly widespread notion that Maxwell's theory played a crucial role in Heaviside's discovery of the distortionless condition. One of the earliest, and possibly the most blatant of these claims was made by Behrend, in an appreciation of Heaviside's work:

> We were concerned particularly with the conception of the distortionless circuit and its meaning in connection with alternating-current power transmission or speech transmission. ... It must be remembered that here was a problem baffling the ablest mathematicians, physicists, and engineers and that a novel form of approach had to be designed before the problem would yield to analysis.[62]

Behrend never bothered to explain what exactly baffled all the experts. He did, however, suggest that Heaviside derived the telegraph equation from his duplex form of Maxwell's equations.[63] This idea was developed in more refined observations to the effect that formulation of the distortionless condition required analysis of the transmission line in Maxwellian terms of propagating electromagnetic fields.[64] Notice, however, that throughout the discussion of the condition for distortionless transmission Heaviside used only the words resistance, conductance, inductance, and permittance. In Heaviside's nomenclature scheme (see ahead, page 236), these constitute the terminology of circuit theory rather than of field theory with its corresponding terms of resistivity, conductivity, inductivity, and permittivity. Accordingly, he performed the analysis of partial reflections that led him to the condition in terms of integral charge and current propagating along the line, not in terms of a Maxwellian field scheme propagating in the dielectric space surrounding the metallic circuit. The remarkable view of the transmission line as an electromagnetic waveguide that Heaviside developed is totally invisible in this dis-

---

61. *Electromagnetic Theory*, Vol. I, pp. 435-436. See also *Electromagnetic Theory*, Vol. I, p. 2 for another statement to the effect that the distortionless condition grew out of considering this practical telephony problem.

62. B.A. Behrend, "The Work of Oliver Heaviside," reprinted in O. Heaviside, *Electromagnetic Theory*, Vol. I, pp. 469-504, esp. 481-482.

63. *Ibid.*, p. 477.

64. Lately, this view has been reinforced in B.J. Hunt, *The Maxwellians* (1991), pp. 131-136.

cussion. The skin effect that featured so prominently in Heaviside's work since Hughes's discoveries is totally ignored. The entire discussion depends on the assumption that penetration of current into the wire is instantaneous—a view that clearly prevents the consistent application of the Maxwellian field view to the problem (see pp. 196-197). Heaviside was able to reach the distortionless condition precisely because he reduced the discussion to an approximation that enabled him to avoid the complexities of field theory.

More important to the formulation of the distortionless condition was the availability of mathematical theory. The condition is practically inconceivable without the mathematical relations between current, voltage, resistance, capacitance, and inductance that form the foundations of linear circuit theory. However, this hardly supports Behrend's suggestion that the formulation of the distortionless condition required analysis in terms of novel—perhaps even revolutionary—mathematical techniques. Furthermore, the availability of a mathematical circuit theory did not by itself suggest distortionless transmission. In other words, the possibility of distortionless transmission did not emerge out of a mathematical investigation of the telegraph equation. Rather, mathematical circuit theory enabled precise formulation of the condition for such transmission once experience suggested the counterbalancing effects of inductance, capacitance, and resistance in a leaky transmission line. We have seen (chapter II) that the distortionless condition can be extracted from the Fourier series solution of the telegraph equation. We have also seen that without foreknowledge of what we are after, the likelihood of performing this particular mathematical exercise is not very high. Both this point and the fact that the Maxwellian field view is not essential to the derivation of the distortionless condition are reflected in the work of the Danish physicist Ludvig Lorenz. He formulated the full telegraph equation without reference to Maxwell's theory, and may have obtained the distortionless condition a year or two before Heaviside. Lorenz apparently obtained the distortionless condition by examining the Fourier solution of the equation. Significantly, however, it appears that he did not begin to examine the Fourier solution with the elimination of distortion in mind until his attention had been focused on the possibly beneficial effects of self-induction by the experimental field work of Van Rysselberghe in the mid-1880s.[65]

Arthur West Heaviside's bridge system of telephony did for Oliver Heaviside what Van Rysselberghe's work did for Lorenz; it provided him with experience that suggested the possibility of distortionless transmission. Unlike

Lorenz, Heaviside did not proceed directly from that to a mathematical derivation of the distortionless condition from the telegraph equation. Instead, he went through an intermediate stage involving a sort of thought experiment in which he manipulated the basic electrical components of a transmission line and studied the ensuing effects. Noting the opposite polarities of reflections from resistances and leaks, he considered in detail the propagation of a square current pulse through several such reflective points. At each point he let the pulse split into two not necessarily equal parts, combining the reflective contributions of both the resistance and the leak. The backward-traveling reflections themselves split at every reflection point that the original pulse had already traversed. For each moment in time, Heaviside added up the components of the pattern of forward and backward reflections that the advancing pulse left behind. The emerging picture formed a sketch of an advancing pulse, gradually attenuated by reflections that combined behind it to form a diminishing "tail." Heaviside gave a three-page outline of this method of investigation in part 8 of "On the Self-Induction of Wires" and intended to devote part 9 to a detailed account of it.[66] Part 9 was never published, but we may get a sense of what stood behind the three pages in part 8 from some loose manuscript leaves that have been preserved. They show how Heaviside "propagated" a current pulse along a wire, calculating and adding up the forward and backward-traveling reflections to obtain a rough picture of a pulse with a diminishing tail behind it. As Heaviside clearly pointed out, this procedure had no recommendation as a means for making precise design calculations; but it suggested a general visual idea of traveling current pulses:

> In a somewhat similar manner to that in which we have roughly followed the growth of tails, we may follow the progress of signals through a circuit, and obtain the arrival-curves of the current at the distant end, or rather, we may obtain curves resembling the real ones somewhat by drawing curves through the zigzags which result. The method has no recommendation whatever in point of accuracy: its real recommendation lies in the facility with which a general

---

65. Helge Kragh, "Ludvig Lorenz and the Early Theory of Long-distance Telephony," *Centaurus*, **35** (1992): 305-324, esp. 311-317. Lorenz's manuscripts pertaining to his investigation of telephone communication have mostly been lost, and our knowledge regarding his anticipation of Heaviside is based on secondhand testimony.

66. "It was my intention to have given the equations of the tails, positive or negative, or mixed, but as the investigation would unduly extend the length of the present communication, I propose to consider the tails in the next Part IX." (*Electrical Papers*, Vol. II, p. 322).

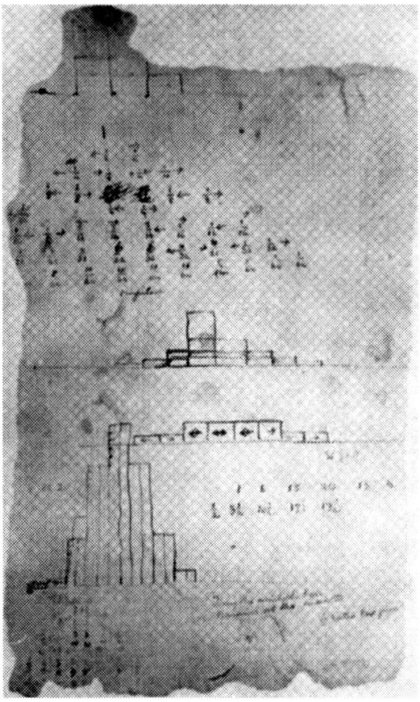
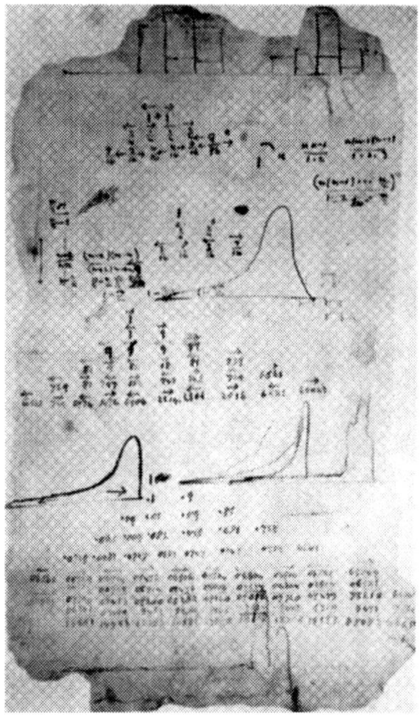
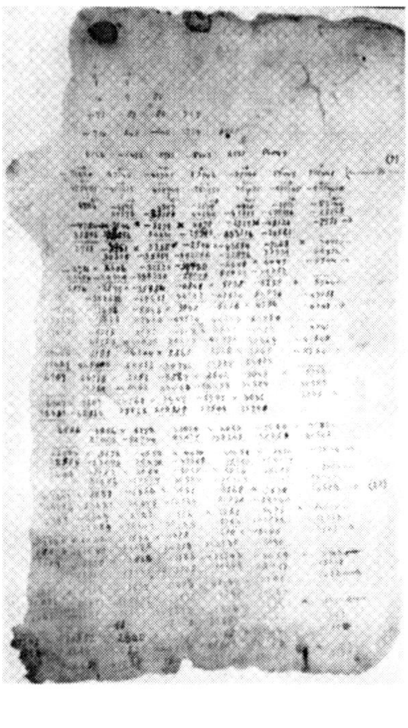

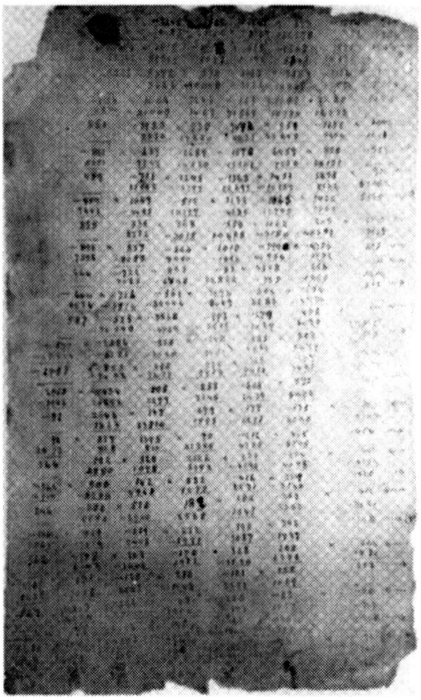

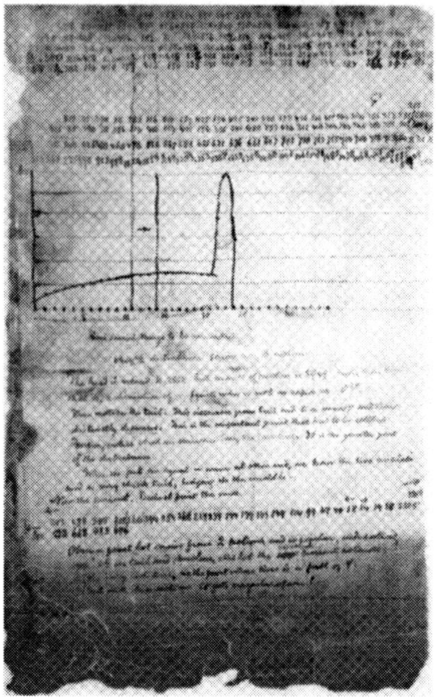

knowledge of the whole course of events may be obtained, and I dare say some people think that of not insignificant moment.

In particular, this investigation suggested to Heaviside the possibility of obtaining a diminishing pulse without a tail behind it by adjusting the resistance and leakance of every reflection point. In other words, the method of partial reflections avoided the difficulty of interpreting long Fourier series and directly brought up the possibility of attenuated but distortionless transmission. For the first time since Heaviside began to study the propagation of current pulses in wires, he was led to ask himself in a precisely focused manner what condition must be satisfied so that a pulse would traverse the line without losing its original shape save for a proportional attenuation. Once the question is formulated in this manner, the distortionless condition follows almost trivially. Mathematically, a pulse of arbitrary form that satisfies this requirement is described as $g(t)f(x\text{-}vt)$. The function $f(x\text{-}vt)$ describes a form that moves without any change along the x-axis with a constant velocity v. The multiplicative factor $g(t)$ makes the *entire* form diminish (or increase) with the passage of time. All that remained for Heaviside to do now was to ask under what circumstances the equation:

$$\frac{\partial^2 V}{\partial x^2} = LS\frac{\partial^2 V}{\partial t^2} + (RS + KL)\frac{\partial V}{\partial t} + RKV \tag{4-4}$$

would accommodate a solution of the form $g(t)f(x\text{-}vt)$. Indeed, if we simply substitute this into the equation in place of V, it turns out that the required condition is $R/L = S/K$. There is no need to actually solve the equation, and the complicated derivation of the condition from the Fourier series solution is completely circumvented.[67] Heaviside's first published account of the distortionless condition reflects a similar line of reasoning.[68] Clearly then, he arrived at the distortionless condition by considering the bridge system in

67. To see how that comes about, carry out the partial derivatives on $g(t)f(v\text{-}xt)$ using the chain rule. Then write pf and p^2f for the first and second derivatives of f to get a second degree polynomial in p multiplied by f, and require the polynomial to be zero, to satisfy the original telegraph equation. This means setting the coefficients of each power of p to zero separately, which yields three conditions: $v = (LS)^{-1/2}$, $g(t) = \exp[-(R/2L{+}K/2S)t]$ and $RS = LK$. The first two determine the propagation velocity and the attenuation rate. The third requirement restricts the allowable values of the electrical coefficients, and denotes the distortionless condition. In other words, with $RS = LK$, the telegraph equation will accommodate a solution of the form $\exp[-(R/2L{+}K/2S)t]f[x\text{-}(LS)^{-1/2}t]$, which describes an attenuating but non-deforming pulse.

terms of engineering building blocks (resistances, inductances, etc.) and linear circuit theory, without once recurring to the circuit's description in Maxwellian field terms. Painstaking analysis of partial reflections was the specific means that enabled him to develop an immediate, intuitive grasp of wave propagation along a transmission line. The tangible picture of square current pulses propagating from reflection point to reflection point led to the precise question that Heaviside failed to formulate when he appeared poised to derive the distortionless condition in 1881.

The distortionless condition was obviously interesting from the practical point of view. Distortion had first become highly noticeable in the operation of submarine telegraph cables during the 1850's. The phenomenon was strikingly documented by Bain's chemical recorder, which was capable of indicating both the intensity and duration of an electric pulse. The instrument displayed a signal comprised of several clearly demarcated dots as a single continuous "smear" once transmission exceeded a certain distance.[69] In 1854, this experimental effect found theoretical expression in the first sophisticated analysis of telegraphic transmission due to Kelvin. He had laid out this theory as a result of his interest in the problem of signal distortion or "retardation" in long-range submarine telegraphy.[70] Kelvin narrowed the scope of his theory to the requirements of the practical problem at hand. In particular, he took into account only the effects of capacitance and resistance. He justified the omission of self-induction and leakage on the basis of their negligible effect in the submarine cable. Distortion much like the kind observed by practical telegraphists emerges as an unavoidable effect in Kelvin's theory. As far as Kelvin's theory goes, there is no way of eliminating the distorting effects altogether.

---

68. *Electrical Papers*, Vol. II, pp. 122-123. Heaviside actually followed the reverse order and showed that subjecting the telegraph equation to the distortionless condition constrains the acceptable solutions to the form $g(t)f(x-vt)$. This suggests the possibility that he actually obtained the distortionless condition from considerations of partial reflections and used the telegraph equation only to verify the condition's generality.

69. W.H. Preece and J. Sivewright, *Telegraphy*, (1876), pp. 126-127. Bain's chemical recorder was invented in 1846. The basic principle of operation involved passing an electric current through a moving paper strip soaked in an electrolyte. The point of contact between the moving paper strip and the electrode darkened in proportion to the strength of the current.

70. Kelvin (William Thomson), "On the Theory of the Electric Telegraph," in *Mathematical and Physical Papers*, (1882-1911), Vol. II, pp. 61-76. See also S.P. Thompson, *Life of Lord Kelvin*, (1910), Vol. I, p. 328.

<u>Distortion over telegraph lines</u>

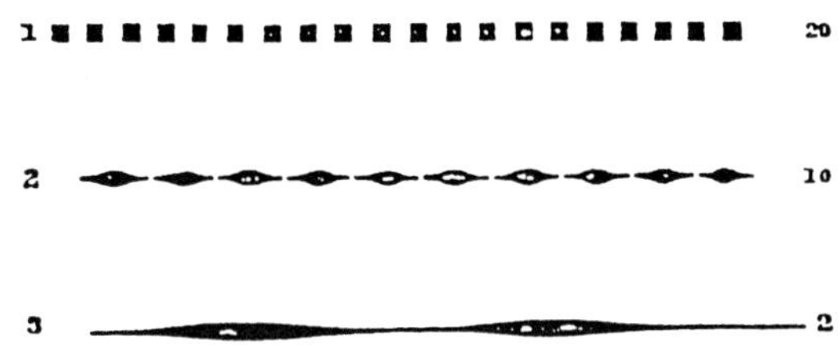

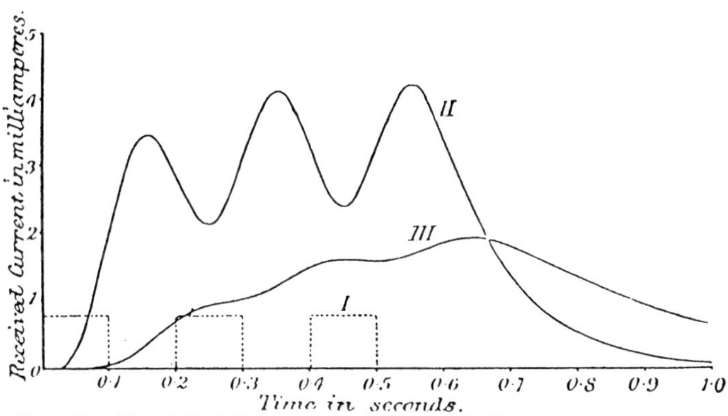

FIG. 10.—The dotted line represents the "S" Signal as sent, and the firm lines as received on Cables of various CR values, and lengths. For Curve II. length = 1,000 miles, $CR = 1{\cdot}0$, and for Curve III., length = 1,581 miles, $CR = 2{\cdot}5$.

The top three traces are reproduced from Preece & Sivewright's Telegraphy. *They show three series of dots as received by a Bain's recorder from cables that had progressively greater distorting effects. Note how signaling speed must be reduced as distortion grows. The bottom figure is from J.A. Fleming,* The Propagation of Electric Currents in Telephone and Telegraph Conductors, *4th edition, (London: Constable & Company, LTD, 1927), p. 182, depicting distortion of three consecutive dots. The figure overleaf, also from Fleming, p. 189, shows very clearly how distortion increases with the value of KR (Fleming's CR is the same as Preece's KR, on which see pp. 243-244 and Appendix 4.1).*

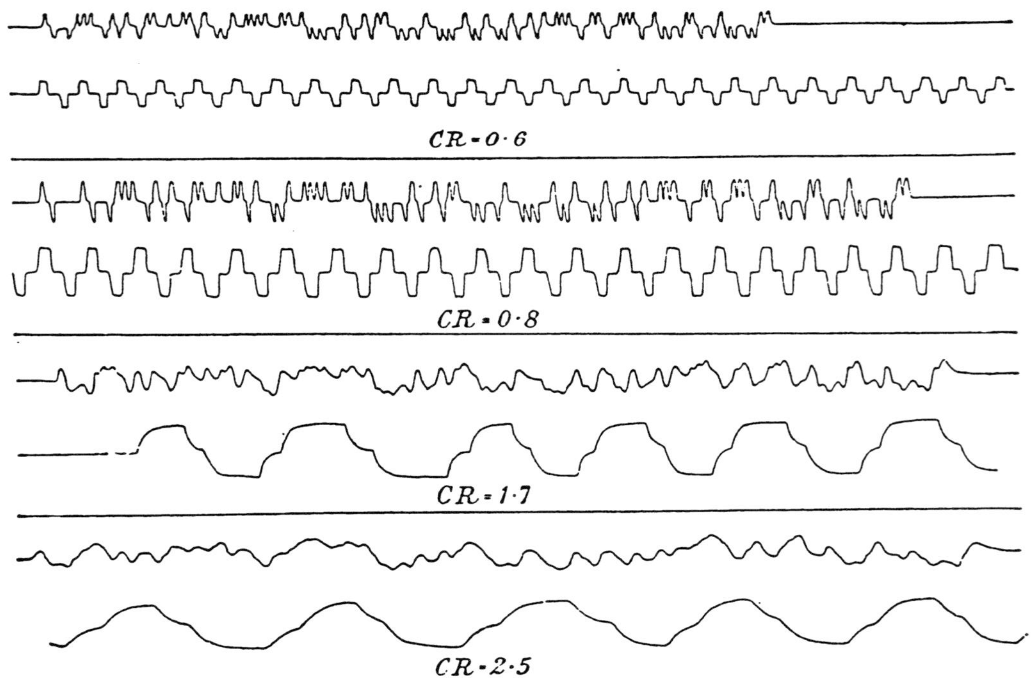

FIG. 13.—The above are Syphon Recorder Signals taken by Mr. H. Tinsley as received on Cables having values of *CR* as given below each line. The upper line is an alphabet and the lower line a succession of *dot dash* to show the rounding effect on the signals of gradually increasing the value of *CR*.

Long-distance telegraphy could still be effected under this constraint, albeit with a reduced signaling rate to ensure sufficient spacing between consecutive pulses. In telephony, frequencies are orders of magnitude higher to begin with, and they cannot be reduced at will, for they represent the electrical profile of sound signals. Uncritical generalization of Kelvin's 1854 theory would therefore impose a theoretical limitation on the range of long-distance telephony. Now, however, Heaviside had incontestable proof in terms of traditional circuit theory that distortion in telephone and telegraph lines could be avoided. The question was how to implement the distortionless ratio between the four line constants in the most practical way. One could always introduce large enough leakage to obtain the necessary ratio. Large leaks, however, necessarily imply large current losses that would limit communication range not by distortion, but by inaudibly weak signals. Here the self-inductance, L, receives its overriding importance:

> Given R and S then, if L should be so small that the introduction of K of a proper amount to reach the distortionless condition should produce unreasonable attenuation, then one way of curing this is to increase L. For then a smaller amount of leakage will serve. The bigger L, the smaller need K be....[71]

The conclusion is therefore clear: the practical way to render telephone lines distortionless is to artificially increase their inductance until the required ratio is achieved.

We shall soon see that the distortionless condition and the notion of a distortionless wave train along a wire played a central role in Heaviside's new publication scheme. At this point it should just be observed that by its own limitations, the distortionless condition could highlight the practical importance of the Maxwellian view. Just like the analysis of the inductive Wheatstone bridge, the distortionless condition leads to the possibility of measurable departures from linear circuit theory. In both cases the skin effect intervenes to spoil the perfection of the linear theory. The distortionless condition is derived under the assumption that the resistance and the inductance are constants, independent of the current. However, if the variations in the form of the pulses are quicker than the time it takes to achieve nearly homogeneous current distribution over the wire's cross section, there will be significant deviations from this assumed independence, and the analysis that led to the distortionless condition will no longer be valid. In other words, from the full

71. *Electromagnetic Theory*, Vol. I, p. 436.

Maxwellian point of view the distortionless condition is only a first-order approximation rather than a precise law. Heaviside repeatedly called attention to this observation.[72] This was one way in which transmission-line theory and special emphasis on distortionless transmission helped Heaviside to present electromagnetic field theory as the basic theoretical framework for practical telephone engineering. Recall how he emphasized the importance of keeping the field view in mind, because it enables one to "see the degree of approximation when a change is made" (see page 208).

The distortionless condition catapulted Heaviside and his work into the center of scientific and engineering interest through a bitter dispute with William Henry Preece, the Post Office's chief electrician. However, before attending to this affair, we must examine the nature of Heaviside's publication scheme in greater detail. This will be done in the next two sections.

5. Self-Induction and the Nature of Heaviside's Publication Scheme

It should be clear by now that the proper treatment of self-induction was crucial to both the controversy surrounding D.E. Hughes's experiments and Heaviside's distortionless transmission line. To better understand the nature of Heaviside's contribution, we must take a brief look at the state of engineering knowledge regarding self-induction at the time. William Henry Preece, the senior electrician of the British Post Office and later its chief engineer, exercised considerable influence on Heaviside's publications following the discovery of the distortionless condition. Therefore his opinions regarding the question of self-induction provide a good starting point.

---

72. Heaviside clearly stated that the linear-theory-based derivation of the distortionless condition completely neglects the effects that require the field-theoretical view of the wire as a thick cylinder: "Now a telegraph circuit, when reduced to its simplest elements, ignoring all interferences, and some corrections due to the diffusion of current in the wires in time [namely the skin effect], still has no less than four electrical constants, which may be most conveniently reckoned per unit length of circuit—viz, its resistance, inductance, permittance, or electrostatic capacity, and leakage-conductance. These connect together the two electric variables, the potential-difference and the current, in a certain way, so as to constitute a complete dynamical system, which is, be it remembered, not the real but a simpler one, copying the essential features of the real." (*Electrical Papers*, Vol. II, p. 120).

At the British Association meeting in Manchester (September 1887), Preece described how a capacitor shunted to the electromagnet in a telegraph receiver entirely eliminated the coil's retardation. He proceeded to observe that:

> ... now the only cause of slow working is the mechanical efficiency of the apparatus and the condition of the circuit. I have pointed out in another paper how copper wire has removed electro-magnetic inertia [meaning self-induction] from the wire, and there remains only the retardation ... of the circuit to check speed of working.[73]

Thus, on the one hand Preece explained that the shunted capacitor *cancelled* the coil's delay, and left only mechanical efficiency and the condition of the circuit to determine the signalling speed. On the other hand, he proceeded to observe that the use of copper wire removed the inductive effects, leaving "only the retardation" of the circuit. The second statement, in contradiction to the first, makes magnetic effects seem *additive* to the capacitive retardation. Preece's campaign for the use of copper wire only adds to the confusing implications of his views. Having recommended copper over iron for telegraph circuits owing to the latter's greater self-induction, Preece went on to attempt an explanation of how the inductive (magnetic) "retardation" works by way of analogy. Consider, he suggested, a water pipe whose walls contract when cooled. As water starts flowing through, the pipe cools, contracts, and offers greater resistance to the flow:

> This is analogous to the effect of self-induction in a circuit conveying rapidly alternating currents of electricity; the whole passage becomes choked throughout its length, ... it is as though an opposing EMF were set up in the wire diminishing the value of the original driving power. ... The effect of friction on the flow of fluids is by no means dissimilar to that of self-induction. Anyhow, this self-induction, or as it is better known in its general sense *electro-magnetic inertia*, is a serious hindrance to rapid telegraphy and to long-distance speaking on telephones.[74]

Preece must have felt somewhat uncomfortable with this analogy, for he introduced it with a warning:

---

73. W.H. Preece, "Fast Speed Telegraphy," *The Electrician*, **19** (Sept. 23, 1887): 425.
74. W.H. Preece, "On Copper Wire," *The Electrician*, **19** (Sept. 9, 1887): 373.

> This influence of self-induction is extremely difficult to explain. An analogy sometimes assists explanation, but it more often makes confusion worse confounded.[75]

It should be said to Preece's credit that he always strove to keep abreast of the latest developments. Unfortunately, occasional incomplete mastery of such developments never prevented him from incorporating them into his public discussions of technical matters. His choking water pipe analogy was clearly influenced by the recent excitement around Hughes's experiments and the discovery of the skin effect. However, the analogy actually reflects his failure to comprehend the roles that resistivity, inductivity, resistance, and inductance play in the description of the effect. The skin-depth, namely, the depth to which alternating current penetrates a conductor, is inversely proportional (in first approximation) to the product of the conductivity and the inductivity of the conductor. However, this is not as if the self-inductance of a wire has anything like the effect of friction. A strong skin effect reduces the effective conducting cross section by crowding the current out of the core. As a result, the integral resistance of the conductor increases (the resistivity, it must be stressed, remains unchanged). For the very same reason, the elimination of current from the core removes the magnetic field from there, resulting in a reduced integral inductance (although the inductivity, just like the resistivity, remains unchanged). Throughout these changes, resistance remains the equivalent of a frictional, heat-dissipating agent, while inductance remains equivalent to inertial mass. To say, as Preece did, that self-induction has an effect not dissimilar to that of friction in a water pipe is to hopelessly misconstrue the significance of the skin effect as well as to confuse the dynamical roles of inductance and resistance in linear circuit theory. This seems all the more mystifying considering that Preece explicitly referred to self-induction as electromagnetic inertia, but failed to use the water's mass in his analogy.[76]

---

75. *Ibid.*

76. Remarkably enough, with proper understanding of self-induction as an analogue of mass in fluid flow, Preece's analogy of the narrowing water pipe can be used to some advantage. The opposed inductive electromotive force associated with a varying current should be compared to the opposed force associated with a varying velocity in the case of a moving mass. As we have seen, Preece failed to do that altogether. However, with this properly understood, the "choking" effect described by Preece could act as an inaccurate counterpart to the increased resistance associated with the skin effect, which restricts the size of the conducting cross-section. The resemblance is inaccurate, because Preece's analogy cannot reproduce the approximation to surface conduction that the skin effect involves.

Indeed, lack of clarity regarding these dynamical roles was at the core of Preece's inability to make sense of the skin effect. In this, his analogy reflects the same state of knowledge that prevented Hughes from properly interpreting his own experiments. Preece, as we have seen, was aware that a shunted capacitor tends to eliminate the distortion induced by a coil in the receiving apparatus. It seems only natural for one to consider at least the possibility that if capacitance can compensate for too much inductance, then inductance should be able to compensate for too much capacitance. Preece, however, seems to have neglected this possibility altogether. Apparently, his general notions concerning self-induction prevented him from extending this observation to the entire transmission line. We shall soon see that for the same reason Preece failed to comprehend how resistance, inductance, leakance, and capacitance could combine to eliminate distortion from telegraph and telephone lines.

Preece's failure to consider the use of inductance to counteract capacitive effects along a transmission line does not fully represent the extent of engineering knowledge at the time. In the fall of 1879, Willoughby Smith, who had been deeply involved with the Trans-Atlantic telegraph cable since the early 1860s, read a paper before the Society of Telegraph Engineers on the subject of "The Working of Long Submarine Cables."[77] In the course of his discussion, Willoughby Smith recalled an experiment that seems to have been carried out in the 1860s, wherein:

> ... eighty-five knots of gutta percha covered wire in coils were immersed in water in iron tanks and joined in one continuous length. The absolute resistance of the conductor was 626 ohms and the absolute electrostatic capacity twenty-seven micro-farads. When the distant end was put to earth through the wire of an electro-magnet the resistance of which was 850 ohms, and an ordinary resistance coil of 1600 ohms, the charge became neutralized by the electro-magnetic current, and consequently the discharge was *nil.* But by removing the keeper from the magnet, or varying the resistance of the resistance coil, the true or false discharge predominated. *From the results of these and similar experiments, I was under the impression that by a judicious distribution of electro-magnets on subterranean lines greater speed might by obtained.*[78] (my italics).

The emphasized lines are most striking considering that Heaviside's idea of the loading coil seems to enjoy in the eyes of many the privileged status of an unanticipated discovery. Willoughby Smith's astute observation appears to be

77. *The Electrician*, **3** (Sept. 6, 1879): 188-189.
78. *The Electrician*, **3** (Nov. 15, 1879): 304-305.

rather exceptional. No contemporary publications by other engineers (in *The Electrician*, at least) echo it. However, the general observation that self-inductance and capacitance are antagonistic in their effects was far more commonplace. F.C. Webb, for example, observed about the current in a circuit overloaded with self-inductance:

> This current is that described by Faraday as the induction of a current on itself, and is said by the higher electricians to be due to an electro-magnetic effect. It is also called the extra current. ... As the effect is the reverse to the discharge it has also been termed *false discharge*. It is evident, that as the effect is reversed to the discharge, that we may, if the wire is coiled and has great electrostatic capacity, have the two effects in antagonism.[79]

Given such observations, it may seem surprising that the inductively "loaded" transmission line had to wait for George A. Campbell's work in 1900. However, this very fact, together with the difficulties surrounding Hughes's experiments, should indicate that the situation could not have been as simple as the selected quotations above suggest. Indeed, in the same paper where he observed the antagonistic effect of capacitance and self-inductance, F.C. Webb also remarked that:

> We have lately had discussions on this subject, and it seems clear that the relative effects of electrostatic induction, and of what has been termed electromagnetic induction, are much disputed, and, indeed, it seems that the conditions must be often such as to give one or the other the preponderance, while in other cases they may be very difficult to unravel from one another.[80]

Willoughby Smith, who so clearly saw the possibility of improving the quality of telegraphic communication by a judicious addition of inductors to the line, also revealed the source of Preece's observations on the beneficial effects of capacitors shunted to the receivers' coils. These originated with a circuit design attributed to Cromwell Varley, in which a shunted capacitor proved capable of "sharpening" the transmitted pulse's profile, the capacitor annulling the coil's "retardation."[81] Here, however, Smith referred to a report by Richard S. Culley in which the latter explained that these circuits were designed for the sole purpose of filtering out the disturbances of random earth currents. The notes on Smith's lecture acknowledge the general observation that condensers

79. F.C. Webb, "Momentary Currents in Wires," *The Electrician*, **2** (April 26, 1879): 286-287.
80. *Ibid*., p. 275.
81. *The Electrician* **3** (Oct. 11, 1879): 243-244.

appreciably increase signalling speed, but:

> ... it is to be observed that no theory explanatory of such advantages has yet been brought forward; and, as we have seen, their existence has been denied on very respectable and generally-accepted authority.[82]

Thus, Preece's confident pronouncements on the ill effects of self-induction create a false sense of consensus, and his authoritative style is simply misleading. Only inadvertently, through the confused "explanation" of how self-inductive retardation works, do Preece's statements betray the disarray that characterized his colleagues' ideas on the subject.

As the sample of quotations from various engineering authorities at the time indicates, the problem did not stem from ignorance of an inductor's fundamental property. Everyone seems to have known that self-induction manifests itself by opposing an electromotive force to any change in the intensity of the current. The greater the rate of current change, the greater the opposing induced electromotive force, or, as encapsuled in the defining equation of self-inductance:

$$V_L = -L\frac{dC}{dt}.$$

This property was constantly utilized by engineers in telegraph and telephone circuits as well as in the rapidly developing dynamo. It should be noted that dynamically considered, there is nothing more complicated, mathematically or physically, about self-induction than about capacitance in a linear circuit. However, when a question arose concerning the basic role of capacitance in a general telegraph circuit, the telegraph engineer could call upon Kelvin's submarine cable theory for guidance. This could not be done in the case of self-induction. Self-induction was strictly outside the range encompassed by Kelvin's theory, and telegraph engineers recognized no alternative framework for guidance. Until 1885, save for a remarkably advanced paper by Kirchhoff of 1857,[83] and Heaviside's rarely consulted work from the late 1870s, no theoretical framework existed that tied self-induction to other circuit elements in an unambiguous and comprehensive manner. This point was explicitly observed by Thomas H. Blakesley, in the introduction to a series of articles he

82. *The Electrician*, **3** (Oct. 18, 1879): 259-260.

83. G. Kirchhoff, "On the Motion of Electricity in Wires", *The Philosophical Magazine*, **13** (1857): 393-412.

published beginning January of 1885 in an attempt to address this need:

> It is often taken for granted that the simple form of Ohm's law—total E.M.F. / total resistance = total current, is true for alternating currents. ... That there are causes which modify the value of the current as deduced from this simple equation, such as mutual or self-induction, or the action of condensers, is often acknowledged in text books, and the values and laws of variation of the current correctly stated for certain cases of instantaneous contact and breaking of circuit. But the effect of an alternating E.M.F. upon a circuit affected by self-induction, mutual induction, and condensing action, has not been, as far as I know, put into a tangible working form.[84]

As a result, telegraph engineers did not know how to consistently assess the effects of self-induction combined with those of other circuit elements. This is the situation so sharply expressed in the statements of F.C. Webb and Willoughby Smith as well as in Hughes's experiments. They show prolific experience with coils and electromagnets side by side with bewilderment as to the place of such circuit elements in the general scheme of things.

This state of affairs must have been on Blakesley's mind when he composed his series of articles on alternating currents. Accordingly, he showed how to calculate the admittance of a section of a discrete network containing a resistance and inductance in series and a capacitance in parallel. In this context he explicitly noted the mutually counteracting effects of capacitance and self-inductance:

> ...there is one important point to be noticed, viz., that self-induction in the sections by no means necessarily diminishes the currents in them, but up to a certain point may be actually beneficial. This cannot be the case when there is no capacity in the circuit. Under such circumstances the self-induction must invariably diminish the current produced by fixed electromotive force....[85]

He did not, however, consider leakage, and it appears that at this stage he did not see how his analysis could be made to yield the clue to long distance telephony. All he did was to relate the above to the case of a single circuit consisting of a resistance, inductance and capacitance in series and to reiterate more explicitly the point on the mutually canceling effects of capacitance and inductance:

---

84. Thomas H. Blakesley, "Alternating Currents," *The Electrician*, **14** (Jan. 15, 1885): 199.
85. *The Electrician*, **14** (Apr. 18, 1885): 470-471.

> When an alternating generator operates upon a circuit which is closed by a condenser without leakage, and which possesses a coefficient of self-induction, then there is a certain period of alternation which may be given to the generator, at which the condenser might be replaced by a junction introducing no additional resistance into the circuit, the coefficient of self-induction being also removed, without disturbing the current. The condenser, in fact, in conjunction with the coefficient of self-induction, will obliterate the effects of the breach of continuity in the conductivity caused by the infinite resistance of the condenser itself. Moreover, this state of things is quite independent of the resistance of the circuit itself, which will then simply regulate the current in the same manner as with continuous uniform electromotive force.[86]

When he proceeded to the case of the continuously distributed coefficients characteristic of transmission lines, Blakesley limited his discussion to the noninductive, non-leaking problem solved by Kelvin thirty years earlier.[87] Accordingly, his observations concerning long-distance telephony reflect the received engineering wisdom of the time regarding the range limitation on telephone communication:

> Thus, at the end of a cable of any considerable length and capacity the various tones of the voice would be received in a state of degradation depending upon their pitch. If this were not the case, if all the tones were reduced in strength in the same proportion, a relay might be employed to restore the various currents to their original intensity, or to one in which the ear would readily appreciate the meaning of the tone. But the ear has not the synthetic power of reconstructing a composite tone from the wreck of variously degraded components. In this consideration reside the limits of telephony. Until this consideration is more clearly understood than it seems to be at present, people will fail to understand the exquisite nonsense to which they are often now content to listen about the possibilities of being able to listen to the minutest modulations of voice of a transoceanic prima donna, and so on. [88]

On May 12, 1887, W.E. Sumpner read a report to the Society of Telegraph Engineers and Electricians on "The Measurement of Self-Induction and Capacity." Sumpner's report indicates that by 1886 Blakesley was already thinking along lines similar to Willoughby Smith's about the possibility of using self-induction to obtain improved telephony:

---

86. *Ibid.*
87. *The Electrician*, **14** (May 2, 1885): 510-511.
88. *The Electrician*, **15** (June 5, 1885): 58.

> Prof. Silvanus Thompson told us (*Journal* [*STE&E*], May 13, 1886) that Mr. Black in 1878 showed that one difficulty of long distance telephoning could be effectively got over by putting a condenser in the circuit as a bridge to any electro-magnet in the line. The reason of this is possibly that the condenser diminishes the effective self-induction of the line. Mr. Blakesley remarked on the same occasion that the working of a telephone circuit completed through a condenser could be improved by adding self-induction to the circuit up to a certain amount. After that maximum had been reached, an increase of the self-induction of the circuit or a diminution of its capacity would be a disadvantage. Now possibly the reason for this is to be found in the fact that capacity is like a negative self-induction. Capacity in a circuit helps change of current. Self-induction in a circuit hinders change of current. Just as self-induction in a telephone circuit retards some of the waves of current more than others, and thereby renders impure the sounds given out, so capacity in a telephone circuit accelerates some of the waves more than others, and produces a similar result. The ideal telephone circuit should have no effective self-induction. What the resistance of it is does not matter so much. When Mr. Blakesley added self-induction to his telephone circuit the improvement he noticed in the working was probably not because the mere addition of self-induction was an improvement, but because the effect of the capacity was diminished more and more as self-induction was added, until a point was reached when the effect of the latter counterbalanced that of the former. It was then that the line worked at its best. It was then that the waves of current travelled along the line without changing much in form.[89]

Remarkably, both Sumpner and Blakesley considered the problem of telephonic distortion from the wrong end, so to speak. Both saw the elimination of distortion as a balancing act between inductive and capacitive effects; both considered the resistance of the line to be of no great importance. Actually, an ideal line, containing any finite amounts of capacitance and inductance but no resistance and no leakage, would be a perfect telephone line. All signals would travel along it without distortion regardless of the ratio of inductance to capacitance. The capacitance and inductance merely determine the propagation speed of these undistorted signals. Add to such a line any finite amount of resistance, and the result would be distortion that no combination of inductance and capacitance could eliminate. Dynamically speaking, Sumpner and Blakesley failed to see clearly that a medium must possess both inertia and elasticity to support propagating waves. In a transmission line, capacitance

89. W.E. Sumpner, "The Measurement of Self-Induction and Capacity," *The Electrician*, **19** (June 17, 1887): 127-128.

and inductance are the dynamical equivalents of elasticity and inertia, hence both must be present if the line is to propagate current waves. In such a line, *resistance* should be regarded as the cause of distortion. This much was clear to Heaviside as early as 1878 when he noted that resistance rounds out the sharp edges of square pulses communicated to a line possessing both capacitance and inductance. However, only sometime between late 1886 and early 1887, prompted by his brother's experience with the bridge system of telephony, did he find the missing ingredient that could counteract the distorting effects of resistance, namely, leakage.

With this in mind, we may begin to assess Oliver Heaviside's contributions since his critique of Hughes's experiments. He traced their inadequacy to Hughes's failure to design his circuit and interpret its behavior in terms of an unambiguous circuit theory encompassing the combined effects of resistance and self-inductance. In particular, this failure prevented Hughes from isolating the effects of skin conduction. Within a few months of the controversy's eruption Heaviside published his "Notes on the Self-Induction of Wires." Here he first showed in outline how a Wheatstone bridge containing both inductive and resistive elements should be analyzed. With the ground rules thus laid down, Heaviside proceeded to indicate the sort of departures from the predictions of this analysis that one should look for as manifestations of the skin effect.

No new theoretical ground needed to be broken to accomplish these tasks. The definition of self-induction was as old as Faraday, and one merely needed to show how to take it into account in conformity with Kirchhoff's circuit laws. Maxwell's work could have been consulted for examples of the Wheatstone bridge properly adapted to the measurement of self-inductance.[90] The non-revolutionary nature of this undertaking is attested in two publications in *The Electrician*, one by W.E. Sumpner, the other by his teachers, Professors William Edward Ayrton and John Perry. Like Heaviside, Sumpner produced an analysis of the general Wheatstone bridge with resistance, self-inductance,

90. James C. Maxwell, *A Treatise on Electricity and Magnetism*, Vol. II, (1954), Arts. 756-757, pp. 397-398. Maxwell wrote a more detailed discussion of the above in his earlier (1864) exposition of "A Dynamical Theory of the Electromagnetic Field," (*The Scientific Papers of James Clerk Maxwell*, Ed. W.D. Niven, [New York: Dover publications, Inc., 1965], Vol. I, pp. 526-597). Here (Arts. 35-46, pp. 543-550) he analyzed self-inductance in a single circuit, extended the discussion to two circuits interacting by mutual inductance, and then proceeded to apply the results to the measurement of self-inductance with an adapted Wheatstone bridge.

and capacitance all taken into account.[91] Unlike Heaviside, Sumpner spelled out every little detail of the analysis, providing a host of fully solved particular examples involving different arrangements. Save for a few general remarks, however, Sumpner concentrated on current and voltage in discrete networks, not on current and voltage waves propagating along transmission lines. As the quotation from his series indicates (see page 226), he did not significantly add to the question of distortion beyond the observations of Willoughby Smith and Thomas Blakesley.

Sumpner's series was actually a long technical appendix to the work of Ayrton and Perry on "Modes of Measuring the Co-Efficients of Self and Mutual Induction."[92] In sharp contrast to the spirit of Hughes's paper, Ayrton and Perry made it clear that their purpose was not to produce remarkable new effects. Instead, they wished to "help the practical electrician to obtain as clear a conception of self-induction as he now has of resistance...." To this end, Ayrton and Perry designed an instrument dedicated to the measurement of self and mutual inductance. They christened it the "secohmeter," after the practical unit of inductance, the secohm, which was later replaced by the "Henry." The user was merely required to follow a mechanical procedure, the most demanding part of which involved the all too familiar process of balancing a purely resistive Wheatstone bridge. It would seem easy to forget, Ayrton and Perry observed, that a scant thirty years earlier resistance was not at all the simple and unproblematic proposition it seemed to be in the 1880s. Only the practical measurement of millions of resistors by a standard procedure—balancing a Wheatstone bridge—rendered it a household concept among practical engineers. Ayrton and Perry asserted that in the same way, self-inductance would become unproblematic only after practical men had familiarized themselves with it by measuring millions of inductors using a standard procedure.

Contrary to the analysis of linear circuits containing self-inductance, the experimental detection of the effects associated with skin conduction was undoubtedly a new development, even if, as Heaviside pointed out, it did not require a Maxwellian field outlook to be understood. As we have seen, however,

91. W.E. Sumpner, "The Measurement of Self-Induction, and Capacity," *The Electrician*, **19** (June 17, 1887): 127-128; **19** (June 24, 1887): 149-150; **19** (July 1, 1887): 170-172; **19** (July 8 1887): 189-190; **19** (July 15, 1887): 212-213; **19** (July 22, 1887): 231-232.

92. W.E. Ayrton and John Perry, "Modes of Measuring the Co-Efficients of Self and Mutual Induction," *The Electrician*, **19** (May 13, 1887): 17-21; **19** (May 20, 1887): 39-41; **19** (May 27, 1887): 58-60; **19** (June 3, 1887): 83-85.

Heaviside was not the first to establish the effects of skin conduction on Maxwellian grounds. Thus, neither in elucidating the place of self-inductance in linear circuit theory, nor in explaining the skin effect did Heaviside's work distinguish itself from other publications on these subjects at the time.

Two distinct aspects, besides his uncompromising individualistic style and conventions, set Heaviside's contribution apart. The first involves his subject of discussion; the second involves the manner in which he discussed it. Motivated by his brother's bridge system of telephony and by the discovery of the distortionless condition, Heaviside returned to his old subject of the late 1870s, namely, transmission-line analysis. The distortionless condition suggested a scientifically justified solution to one of the main obstacles on the way to long-distance telephony. Unlike the detached guesses of Willoughby Smith, Heaviside's recommendation to artificially load transmission lines with extra inductance was grounded in a solid and noncontroversial linear circuit theory. It was only natural for Heaviside to develop this theme in great detail. Indeed, while others were busily studying self-induction as a lumped circuit element, Heaviside turned his attention to the linearly distributed coefficients that characterize the transmission-line problem. Here Heaviside found the case of distortionless transmission particularly useful as a means of introducing his transmission-line theory to practical engineers. Using ideal lines and the method of reflections, he could communicate at least a qualitative sense of how current pulses propagate along a line without reliance on the complicated Fourier series solution of the telegraph equation. In particular, he could crisply demonstrate the difference between the diffusion picture created by Kelvin's old theory, and the propagation of square pulses that occurs when all four elements of the transmission line are taken into account. Heaviside expressed all this in the final sections of "E.M.I.& P." with oblique references to Preece's misconceptions concerning the propagation speed of telegraph pulses (see page 245):

> The solution of the above problem [pulse propagation along a distortionless line terminated by a resistance] by means of Fourier-series is extremely difficult. It expresses the whole history of the variable period by a single formula. But this exceedingly remarkable property of comprehensiveness, which is also possessed by an infinite number of other kinds of series, has its disadvantages. The analysis of the formula into its finite representatives ... is trying work. And the getting of the formula itself is not child's play. Considering this, and also the fact that a large number of other cases besides the above can be fully solved by common algebra (with a little common-sense added), the importance of a full study of the

> distortionless system will, I think, be readily admitted by all who are dissatisfied with official views on the subject of the speed of the current. The important thing is to let in the daylight on a subject which it was difficult to believe could ever by freed from mathematical complications.[93]

In addition to this first aspect of his work, Heaviside's contribution possessed an outstanding general characteristic. Unlike other commentators, who concentrated on pinpointed accounts of the skin effect or the measurement of self-inductance in linear circuits, Heaviside sought to explicitly place all such analyses within one general scheme. The majority of his articles in *The Philosophical Magazine* and in *The Electrician* in the two years following D.E. Hughes's work were meant to accomplish this ambitious task. Thus, Heaviside did not simply describe the essence of the skin effect. He examined a conducting wire of circular cross-section. He envisioned the wire as enveloped by a cylindrical conductor insulated from the central wire by an intermediate dielectric layer. Using the thickness of the intermediate layer and of the coaxial conductors as guides, Heaviside distinguished particular cases from one another. He derived a practical approximation for the case of a very close return conductor, and demonstrated its radical deviation from the linear theory. Then he outlined the way to alternative approximations, appropriate for thicker insulating layers. In the same way, he developed practical approximate theories for circuits wherein the capacitive element predominated; others addressed situations in which the inductive element predominated. He showed how a circuit with a strong skin effect could still be "linearized" as long as it carried a harmonically oscillating current.[94] Each of these approximations is less general than Maxwell's theory. Strictly speaking, they are all false. They may be thought of as "limited-reference theories," following Heaviside's observation that:

> ... we shall never know the most general theory of anything in Nature; but we may at least take the general theory so far as it is known, and work with that, finding out in special cases whether a more limited theory will not be sufficient, and keeping within bounds accordingly. In any case, the boundaries of the general theory are not unlimited themselves, as our knowledge of Nature only extends through a limited part of a much greater possible range.[95]

---

93. *Electrical Papers*, Vol. II, p. 134.
94. *Ibid.*, pp. 44-76.
95. *Ibid.*, p. 120.

However, while clearly limited in the scope of their application, each of these limited theories possesses an enormous advantage over the full, unadulterated Maxwellian field scheme: they provide practical problem solving environments for real engineering situations.

It should be noted that W.E. Sumpner's illuminating account of Heaviside's work could be misleading on this point. Perhaps out of a sincere desire to exalt Heaviside's contributions, Sumpner suggested that from 1882 on, Heaviside thought strictly in terms of electromagnetic waves traveling in the dielectric.[96] There is no doubt that this is indeed a distinctive mark of Maxwell's theory; but it is equally certain that in most cases any attempt to analyze electric circuits directly in these terms is simply unthinkable. Every little twist and turn of the wire, every detail in the geometry of capacitors and inductors, influences the electromagnetic field and energy flow associated with a circuit. In the majority of cases, the most powerful mathematical tools available will fail to provide the full analytical solution. It will be found no less difficult to form an intuitive mental picture of the required fields. Almost invariably, any intelligible observation on such circuits will emerge from limited-reference theories such as linear circuit theory. For a simple example, consider a coil carrying an alternating current. As the frequency increases, the capacitance between neighboring wire loops becomes more noticeable. Under such circumstances, the integral inductance of the coil no longer represents it properly. However, this failure of the linear characterization does not mean that the engineer must immediately fall back to Maxwell's equations. It would be easier to characterize the coil as a compound linear circuit in which a series of small coils are shunted by small capacitors. As the frequency gets even higher, the capacitance between more remote loops may be added. Only when such progressively refined linear modeling fails or grows too complex does it become useful to reformulate the problem in explicit field terms.[97] The great advantage of linear circuit theory, which makes it so well adapted to practical application, is that within its framework the three-dimensional electromagnetic

96. W.E. Sumpner, "The Work of Oliver Heaviside," *I.E.E. Journal*, **71** (Dec. 1932): 837-851, (837).

97. S. Ramo, J.R. Whinnery and T. Van Duzer, *Fields and Waves in Communication Electronics*, 2nd edition, (1984), pp. 196-197. This is not to suggest that field analysis has no role in practical engineering. Most advanced electrical engineering texts contain a chapter on field and flux plotting. For an early text dedicated to these techniques, see L.V. Bewley, *Two-Dimensional Fields in Electrical Engineering*, (1948).

fields completely disappear. The theory focuses on the conducting wire, and folds up, in a manner of speaking, all of the field effects into lumped or continuously distributed linear coefficients. This extremely useful simplification of the problem comes at a price, however. As Heaviside himself noted, linear circuit theory is but a first approximation. Its range of reference is limited to a subset of the electromagnetic phenomena spanned by field theory. If one is not judicious in its application, it is liable to be as misleading as it is useful.

The numerous approximations Heaviside drew against the background of Maxwell's theory outline a general approach that characterizes much of his work. Instead of applying basic field theory mechanically with the brute force of mathematical manipulation, Heaviside first considered the facts of the practical situation at hand. From this, he judiciously gave precedence to some of the available theoretical concepts, while intentionally diminishing the importance of others. These considerations led to the approximations that in turn guided the derivation of each limited-reference theory. This procedure is not fully defined by the fundamental theory—in this case Maxwell's electromagnetic field theory. Only intimate acquaintance with practical experience coupled with a healthy dose of common sense can guide the approximation process. The importance Heaviside attached to this point may be gathered from both his correspondence and published work. In 1889, Hertz asked Heaviside about the propagation of electromagnetic waves around a coiled conductor.[98] Heaviside replied:

> The question you asked about the spiral has occurred to me, but the theory (exact) is hopelessly difficult. It is hopeless to follow in full detail the theory of such cases.[99]

When Hertz ventured to disagree,[100] Heaviside reiterated the point even more forcefully:

> My remarks referred to an *exact* theory, without simplifying assumptions introduced; a theory giving the form of the waves, and all about them. *That* is immensely difficult.[101]

In 1891 Heaviside repeated and elaborated this point in his published work:

---

98. J.G. O'Hara and W. Pricha, *Hertz and the Maxwellians*, (1987), p. 60.

99. Heaviside to Hertz, 14 February 1889, *Ibid.*

100. Hertz to Heaviside, 21 March 1889, *Ibid.*, p. 63.

101. Heaviside to Hertz, 1 April 1889, *Ibid.*, p. 65.

> But we must always be careful to distinguish between theory and the application thereof. The advantage of a precise theory is its definiteness. If it be dynamically sound, we may elaborate it as far as we please, and be always in contact with a *possible* state of things. But in making applications it is another matter. It requires the exercise of judgment and *knowledge of things as they are*, to be able to decide whether this or that influence is negligible or paramount (my italics).[102]

It is instructive to compare Heaviside's style of drawing up limited-reference theories to the one outlined in a modern textbook on circuit analysis:

> Just how many effects must be taken into account in representing a system by equivalent parameters? We can answer our questions only by asking another: Just how good do we expect the results to be? The accuracy of our results will be determined by how many separate electrical effects we can take into account by a parameter. We must stop somewhere. We must, at some point, make an *approximation*. Approximation requires engineering judgement. An approximation which is valid in one case will not be in another. ... In the discussions to follow in other chapters, we will assume that when a schematic of a system is given, all significant parameters have been taken into account. Engineering judgement has been exercised by the individual who made up the problem. But when the student finally applies the techniques of analysis to a problem that *he* makes up, these questions associated with approximation must be answered. It is difficult to write answers to such questions in textbooks; experience is usually the best teacher. ... Approximation and analysis are bound together. To ignore the problem of approximation is to lack understanding of the results of analysis.[103]

To sum up, the remarkable aspect here about Heaviside's work is that each of his limited-reference theories creates a link between a general theory and an independently existing body of practical "hands-on" experience. Of crucial importance is to constantly keep in mind that the construction of a limited-reference theory in relationship to a general comprehensive theory involves two kinds of approximations that must be made simultaneously to harmonize with each other. On the one hand, certain simplifying assumptions enable the derivation of relationships (say, the laws of linear circuits) that approximate to the basic principles of the fundamental theory (say, Maxwell's field equations). On the other hand, the idealization symbolized by a circuit diagram is really an approximation-based representation of a far more complex physical system. Thus, each limited-reference theory is bounded and defined on one side

102. *Electromagnetic Theory*, Vol. I, p. 403.
103. M.E. Van Valkenburg, *Network Analysis*, (1974), pp. 21-22.

by the general, fundamental electromagnetic theory of Maxwell, and on the other by the pertinent facts of the particular situation at hand. Consequently, each limited-reference theory provides more than a practical approach to a specific engineering problem. The very manner of its derivation immediately underscores the circumstances under which it ceases to be applicable, and suggests the general approach to deriving an alternative by focusing on those elements that stand outside its limited scope.[104]

Limited-reference theories based on particular approximations are by no means an exclusively technological phenomenon. "Pure" scientists employ them constantly, and probably have been doing so before the practice appeared in engineering contexts.[105] The difference is in what these approximation-based theories refer to. As Heaviside noted, intimate familiarity with "things as they are" is indispensable to a useful approximation; but the relevant "things as they are" in the pure scientist's laboratory may not be the relevant ones in the engineer's realm. Approximations carefully formulated with regard to a particular experimental setup could yield a limited-reference theory of great value on the forefront of pure scientific research; but that does not necessarily make it useful in other contexts, in particular regarding application to engineering problems. In other words, to be usefully applied to engineering, science must become an engineering science. For Heaviside, specifically, the transmission-line was far more than a remote source of abstract problems. His six years as a telegraph operator instilled in him a strong interest in genuine engineering problems. He continuously nourished that interest through his collaboration with A.W. Heaviside. The approximations he constructed were carefully guided by case studies rooted in the basic properties of the transmis-

104. Technologists can manufacture limited-reference theories even in the absence of explicit guidance by basic theory similar to that supplied by Maxwell's theory to linear circuit theory (see Walter G. Vincenti, *What Engineers Know and How They Know It*, [1990], pp. 137-169). However, in the absence of boundaries outlined by such a basic theory, the search for limited-reference theories may become very difficult. Harry Ricardo's failed attempt to unravel the causes of gasoline engine knock may be a case in point (Walter G. Vincenti, "The Air Propeller Tests of W.F. Durand and E.P. Lesley: A Case Study in Technological Methodology," *Technology and Culture*, **20** [1979]: 745). The security and efficiency provided by a general theory which, in Heaviside's words, allows one to "be always in contact with a possible state of things" can hardly be overestimated.

105. In Germany, for example, the art of drawing limited-reference theories on the basis of carefully formulated approximations was a cornerstone of F.E. Neumann's seminar for physics. See Kathryn M. Olesko, *Physics as a Calling: Discipline and Practice in the Koenigsberg Seminar for Physics*, (1991), pp. 165-166.

sion line as a real engineering system. Thus, by developing his limited-reference theories Heaviside made fundamental contributions to the basic engineering theory of electrical communications by wire.

In conclusion, Heaviside's work from February 1886 to December 1887, which constitutes nearly two-thirds of the second volume of his *Electrical Papers*, was more a contribution to engineering science than to the frontiers of field-theory. This contribution anticipates the sort of generalized-machine theories that Charles Steinmetz and his co-workers developed in General Electric's "Calculating Department" during the 1890s.[106] From the field-theoretical point of view, Heaviside's 1886 to 1887 studies introduced nothing fundamentally new to his studies of 1882 to 1886. Instead, Heaviside took the basic concepts that he had developed in the earlier period and discussed them in the context of telephone and telegraph engineering. Elaboration of the relationship between these engineering systems, electromagnetic field theory, and the circuit theory that mediates between them permeates his work following Hughes's experiments and A.W. Heaviside's bridge system. In the course of this work he forged the basic structure for the interconnected presentation of the theories of circuits and fields. This structure will be examined in the next section.

6. The "Royal Road" to Maxwell's Theory

In his initial, elementary review of "The Equation of Propagation along Wires," Heaviside wrote:

> It is far more difficult to obtain a satisfying mental representation of the electric force of inertia $-L\dot{C}$ than of that due to the potential. The water pipe analogy is ... simple enough. ... It is, however, certainly wrong, as we find by carrying it out more fully into detail. Remark, however, that, as $1/2LC^2$ is the magnetic energy per unit length, LC is the *generalised* momentum corresponding to C as a generalised velocity, $L\dot{C}$ the generalised externally applied force, an electric force, of course, and $-L\dot{C}$ the force of reaction—that is, electric force of inertia. This is by the simple principles of dynamics, disconnected from any hypothesis as to the mechanism concerned.[107]

106. Ronald R. Kline, *Steinmetz: Engineer and Socialist*, (Baltimore: The Johns Hopkins University Press, 1992), pp. 106-120.

107. *Electrical Papers*, Vol. II, pp. 83-84.

In a similar dynamical way, the voltage across a capacitor becomes the electric equivalent of elastic reaction, as in a strained spring: V is the generalized force, SV is the generalized displacement, $S V$ is the generalized velocity, and $1/2SV^2$ is the energy of the configuration. The energy can therefore be stored in a generalized elastic tension and in a generalized motion. Using the same sort of reasoning, it is possible to draw further dynamical analogies between the field formulation and the theory of linear circuits. As the quotation above indicates, Heaviside was well aware of the difficulties that beset mechanical analogies to the Maxwellian field scheme. In his theory of circuits Heaviside perceived a way of avoiding all these difficulties while still availing himself of an analogy that an experienced electrical engineer could use as a stepping stone to the field view. A crucial added advantage of this analogy is that the linear theory of circuits is directly related to the field scheme as a useful first approximation—a distinction that no mechanical analogy could claim for itself.

Heaviside's nomenclature scheme for electromagnetism provides the simplest, yet one of the most striking demonstrations of this dual connection between the field and linear theories. On one hand, the scheme distinguishes

Table 1: Heaviside's Nomenclature Scheme

Scheme	Force	Flux	Force/Flux	Flux/Force
Linear	E.M.F.	Current	Resistance	Conductance
Field	**E** Field	Current Density	Resistivity	Conductivity
Linear	M.M.F.	Mag. Flux	Reluctance	Inductance
Field	**H** Field	**B** Field	Reluctivity	Inductivity
Linear	E.M.F.	Charge	Elastance	Permittance
Field	**E** Field	Displacement (**D**)	Elastivity	Permittivity

between the terms of field analysis and those of circuit analysis, on the other, it reflects their shared dynamical structure. Resistance and resistivity, for example, are not terms that Heaviside chose merely to distinguish between related units of measurement for bulk material and per unit volume respectively. He chose them to maintain this distinction while pointing out the unifying dynamical framework of ratios between analogous forces and fluxes. Heaviside also strove to retain continuity with traditional names. He deleted only the term "capacitance," for which he substituted the word permittance. He objected to the implication of "capacitance" as a capability of storing electricity.[108] He found this suggestion totally contradictory to the Maxwellian view of capacitors as energy-storing dielectric springs rather than storage devices for electrical material. With all of these characteristics, Heaviside's nomenclature scheme testifies yet again to the engineering background he brought to his reading of Maxwell. It may be seen as a dynamically-minded engineer's interpretation of Maxwell's general desires regarding nomenclature:

> In forming the ideas and words relating to any science, which, like electricity, deals with forces and their effects, we must keep constantly in mind the ideas appropriate to the fundamental science of dynamics, so that we may, during the first development of the science, avoid inconsistency with what is already established, and also that when our views become clearer, the language we have adopted may be a help to us and not a hindrance.[109]

With his nomenclature scheme, Heaviside had a highly suggestive framework for the introduction of field concepts by analogy with circuit elements. However, to make this analogy really useful, he still needed a central link through which phenomena in linear circuits could be used as analogies to the far less familiar field phenomena. With the newly discovered concept of distortionless transmission lines, he found that link in the telegraph equation and in the fictitious notion of magnetic conductivity.

We already saw (chapter II) that a transmission line may be described by two simultaneous equations in the framework of linear circuit theory:

$$\left.\begin{aligned} -\frac{\partial V}{\partial x} &= RC + L\frac{\partial C}{\partial t} \\ -\frac{\partial C}{\partial x} &= KV + S\frac{\partial V}{\partial t} \end{aligned}\right\} \qquad (4\text{-}5)$$

Now consider a plane electromagnetic wave, advancing along the x-axis. Being a plane wave, it can vary only along the propagation axis. Therefore the spatial derivatives with respect to y and z become zero throughout. Further-

108. *Electrical Papers*, Vol. II, p. 328.

109. James C. Maxwell, *A Treatise on Electricity and Magnetism*, Vol. II, (1954), Art. 567, p. 210.

more, Heaviside already showed in his investigation of the electromagnetic wave front that the electric and magnetic forces must be perpendicular to the direction of propagation and to each other. In other words, the electromagnetic field $\mathbf{E} = (0,0,E)$ and $\mathbf{H} = (0,H,0)$ provides a legitimate description of a wave-front advancing along the x-axis. Application of Maxwell's circuital equations to these fields yields:

$$\left.\begin{aligned} -\frac{\partial E}{\partial x} &= gH + \mu\frac{\partial H}{\partial t} \\ -\frac{\partial H}{\partial x} &= kE + c\frac{\partial E}{\partial x} \end{aligned}\right\} \qquad (4\text{-}6)$$

Let the fictitious magnetic conductivity g be analogous to the resistance R. This renders the inductivity μ, conductivity k, and permittivity (capacity) c respectively analogous to the inductance L, leakance (leak-conductance) K, and permittance (capacitance) S.[110] At the same time, the voltage becomes analogous to the electric force, and the current becomes the analogue of the magnetic force.

Heaviside saw this analogy as a particularly useful gateway to Maxwellian electromagnetism. The entire subject of electromagnetic wave propagation could be introduced through the partial-reflection method and the transmission line. Telegraph and telephone engineers who could not or would not follow the mathematical treatment, could acquaint themselves with electromagnetic waves by relying upon their experience with transmission lines instead. The discussion of phenomena encountered in telegraph and telephone cables in terms of propagating and interfering current waves had been a common practice among electrical engineers in Heaviside's time. Such a manner of speaking should by no means imply a Maxwellian outlook. Consider that W.H. Preece, who could hardly be suspected of adopting this outlook, alluded to the analogy of interfering water waves to give an audience a sense of interfering telegraph signals. Having described the interfering ripples in a pond disturbed by a stone and a jumping fish, Preece continued:

> This super-position of wave on wave is called interference, and the interference of undulations plays a most important part in the phenomena of sound, of light,

110. Considering the intractable difficulties that the conduction current presented to the pre-electron Maxwellian theory, the analogy between the real resistance and the purely imaginary magnetic conductivity seems rather ironic in hindsight.

> and of electricity. In electricity wave upon wave can be superimposed, either in waves flowing in the same direction or in waves flowing in opposite directions. The usual indication of the presence of electric waves is either by the attraction of a magnet or by the deflection of a needle.[111]

Within this limited scope, while keeping to the discussion of signal propagation strictly in linear terms, Heaviside could now convey an intuitive idea of traveling current pulses using the partial-reflection method. The student could begin with the ideal line, say, a coaxial arrangement of two perfect conductors separated by a perfectly insulating layer. In such an arrangement signals would propagate undistorted and unattenuated. The student could start learning how to conceive of traveling current waves by considering simple cases of propagation and reflection along the ideal line. The practical distortionless line would involve a slight complication in the form of undistorted, but attenuated propagation. Having mastered propagation and reflection in this slightly more complicated case, the student may begin to think of less than perfect cases by considering small departures from these ideal situations. Throughout these exercises, only a good practical familiarity with transmission-line phenomena would be required together with basic arithmetic. In Heaviside's own words:

> The mathematics was reduced, in the main, to simple algebra, and the manner of transmission of disturbances could be examined in complete detail in an elementary manner. Nor was this all. The distortionless circuit could be itself employed to enable us to understand the inner meaning of the transcendental cases of propagation, when the distortion caused by the resistance of the circuit makes the mathematics more difficult of interpretation.[112]

Only a small step is required to pass from this to full-fledged Maxwellian electromagnetic waves. Without ever needing to actually solve Maxwell's equations for various cases, the student could still obtain an intuitive feeling for how electromagnetic waves would propagate through various media. To begin with, consider again the ideal transmission line, characterized by no leakance and no resistance. From the mathematical point of view, it is perfectly analogous to Maxwell's equations for the propagation of plane electromagnetic waves in a nonconducting dielectric:

111. W.H. Preece, "Multiple Telegraphy," *The Electrician*, **3** (June 7, 1879): 34-36.
112. *Electromagnetic Theory*, Vol. I, p. 2.

> The problem we have been considering is not merely mathematically similar to, but is identically the same problem as the propagation of plane waves of Light [according] to Maxwell's electromagnetic theory. V stands for the transverse electric disturbance and C for the transverse magnetic disturbance and their product VC for the transfer of energy. Consider only a ray of unit section, and for V take **E** and for C take **H** and we have a pencil of light.[113]

Furthermore, when extended to the full telegraph equation with both resistance and leakance taken into account, the analogy could be maintained through the real electric and fictitious magnetic conductivities. Without ever losing sight of the familiar telegraphic transmission line, the student would be provided with a direct route to a general Heaviside-flavored depiction of propagating electromagnetic fields. Nearly ten years after he first conceived of the distortionless transmission line as the "Royal Road" to Maxwell's theory, Heaviside was still stressing this point:

> ...I recommend every electrician to study in full detail [the theory of the distortionless circuit], as an introduction to electromagnetic waves in general, since it casts light in the most obscure places. It allows us to understand electromagnetic waves mathematically not merely as a collection of formulae, sometimes disagreeably complicated, but in terms of physical ideas of translation, attenuation, distortion, absorption, reflection, and so on.[114]

For Heaviside, the usefulness of transmission-line analysis did not end with this remarkable analogy. Still following the linear presentation, Heaviside could refer to the well-known fact that the self-inductance, L, actually depends on the properties of the dielectric surrounding the current-bearing conductor. This immediately exposed the main weakness of comparing the electric current to water flow in a pipe: in the latter, the inertia is associated with the mass of the water current within the pipe; in the former, it is associated with an environment outside the current. Heaviside repeatedly used this, and the closely related observation that energy transfer takes place in the dielectric outside the wire, to justify passing to a field-oriented view of the conduction current.[115] The "inertia" represented by L is the inertia of the magnetic field associated with the current rather than a direct property of the current itself. With this argument, Heaviside could offer an almost compelling motivation to review the transmission line with emphasis on the propagating elec-

---

113. Notebook 10, p. 202, Heaviside collection, IEE, London.

114. *Electromagnetic Theory*, Vol. II, p. 316, (Oct. 9, 1896).

115. *Electromagnetic Theory*, Vol. I, pp. 16-18.

tromagnetic fields in the dielectric. For Heaviside, the analogy between the propagation of current pulses along a transmission line and the propagation of electromagnetic plane waves was therefore doubly useful. It could provide the uninitiated with a way of developing an intuition for a novel theoretical outlook in familiar terms. Once this goal has been attained, drawing attention to the analogy's limitations could be a powerful way of illustrating the fine points of the new theory. Thus, Heaviside carefully noted the difference between the electromagnetic pencil of light and the case of propagation along wires:

> It is also well to remember that we are not exactly representing Maxwell's scheme, but a working simplification thereof. The lines of energy-transfer are not quite parallel to the conductors [while in the plane wave case they are always parallel to the direction of propagation], but converge upon them at a very acute angle on both sides of the dielectric. Only by having conductors to bound it of infinite conductivity can we make truly plane waves. Then they will be greatly distorted, unless we at the same time remove the leakage by making the dielectric a non-conductor instead of a feeble conductor; when we have undissipated waves without attenuation or distortion.[116]

Once again, then, the linear analysis of transmission lines provided Heaviside with a natural stepping stone to electromagnetic field theory. Not surprisingly, he regarded the ideal line as the "Royal Road" to Maxwell's theory:

> But that this matter of the distortionless circuit has, directly, important practical applications, is, from the purely scientific point of view, a mere accidental circumstance. Perhaps a more valuable property of the distortionless circuit is, that it is the Royal Road to electromagnetic waves in general, especially when the transmitting medium is a conductor as well as a dielectric.[117]

All of this could come to pass precisely because the distortionless condition and the analysis of current waves stand independent of Maxwell's theory. They are based on the principles of linear circuit theory and are fully compatible with the image of the electric current as an energy-transferring flow within the conductor. The distortionless line could hardly have provided Heaviside with his Royal Road to Maxwell's theory had this theory been a prerequisite to the establishment of the distortionless condition and the analysis of current waves in conductors.

116. *Electrical Paper*, Vol. II, p. 311.

117. *Electromagnetic Theory*, Vol. I, p. 2. See also *Electrical Papers*, Vol. I, p. x.

All told, the evolution of Heaviside's work since February of 1886 shows that by early 1887 he was prepared to weave practically his entire work from 1872 into one comprehensive scheme. He now had the terminology with which to discuss the natural science of electromagnetic fields and the engineering science of linear circuits in a dynamically interrelated way. The common thread through these themes was the transmission line considered as a linear circuit. It was indeed a Royal Road that led through two journals to a truly grand vision. All it required was decent comprehension of the principles of linear circuit theory. Unfortunately, this one requirement proved to be the scheme's undoing.

7. "But in the year 1887 I came, for a time, to a dead stop"

7.1 Prelude: W.H. Preece and S.P. Thompson on the Improvement of Telephone Communications

In the beginning of 1887, the senior electrician of the British Post Office, Mr. William Henry Preece, became involved in a dispute with professor Silvanus P. Thompson over the question of improving the quality of telephone communications. The theoretical background for this dispute was given by Kelvin's submarine cable theory of 1855. Until Heaviside's work of 1876 to 1878 this was the only detailed mathematical theory of electrical communication by wire. It has already been pointed out (see page 216) that the special characteristics of submarine cables convinced Kelvin that inductive effects in them may be ignored without affecting the practical value of the analysis.[118] Between 1876 and 1878 Heaviside showed how self-induction should be included in the transmission-line equation, and solved the problem for lines

118. In 1859 Kelvin reported on some measurements made on coiled cables. He explained that the observed secondary motions of the galvanometer needle in the reverse direction was due to "... mutual electro-magnetic induction between different parts of the coil and anticipated that no such reversal could ever be found in a submerged [uncoiled] cable." (*Mathematical and Physical Papers*, Vol. II, p. 129). Thus, by 1859 at the latest Kelvin seems to have been well aware of the general effects of magnetic induction. There seems to be no reason to assume that he had no notion of the above as early as 1855 when he actually worked out the theory of the submarine cable.

characterized by both electrostatic capacity and self-inductance. But he did not continue the analysis to include leaky cables, and did not pursue the particular question of distortionless transmission. At any rate, there is no indication that his extensions of Kelvin's telegraph theory had been studied—not to mention officially embraced—by any significant segment of the community of telegraph engineers at the time. The first aspect, then, to note about the Preece-Thompson debate, is that it was underscored by failure to properly incorporate the effects of self-induction.

Thompson seems to have argued from the point of view that practically speaking very little could be done about line-generated distortions. Indeed, he could have referred to Kelvin's telegraph theory in support of this point of view. Consequently, he advocated that attention should be paid to the end apparatus. Communication range could be increased, Thompson suggested, with more powerful transmitters and more sensitive receivers. Preece disagreed and stated that the question was not at all one of apparatus, but of the characteristics of the line as summarized by Kelvin's KR law (see below) that determines the rate of telegraphic signalling. He explained this point in a paper he read before the Royal Society on March 3, 1887, in which he called upon an impressive array of experience and important acquaintances to testify on his behalf:

> The law that determines the distance through which speaking by telephone on land lines is possible is just the same as that which determines the number of currents which can be transmitted through a submarine cable in a second. The experimental evidence on which this law is based was carried out in 1853 by Mr. Latimer Clark (whose assistant I then was). The experiments were made by me in the presence of Faraday; many were his own; he made them the subject of a Friday evening discourse at the Royal Institution, January 1854, and they are published in his 'Researches' (Vol. 3, p. 508). They received full mathematical development by Sir William Thomson in 1855 ... who determined the law, the accuracy of which was proved by Fleeming Jenkin and by Cromwell Varley, and the 110,000 miles of cable that now lie at the bottom of the ocean afford a constant proof in their daily working.[119]

The law referred to in this statement by Preece is the "KR law." Actually an approximation based on Kelvin's telegraph theory, the law states that the rate of rise of a signal's intensity at the receiving end of a cable is inversely propor-

119. W.H. Preece, "On the Limiting Distance of Speech by Telephone," *The Electrician*, **18** (Mar. 11, 1887): 395-397.

tional to the product of the cable's overall capacitance, K, and its overall resistance, R. Preece explained it as follows:

> ... as [the cable] absorbs the first portion of every current sent, it has the same effect as if it retarded or delayed the first appearance of the current at the distant end. Thus the apparent velocity of the current is diminished by the amount of induction present in the circuit. In a circuit of very low capacity there is practically no induction, and the current appears instantaneously at the distant end. In a circuit where there is capacity there is induction, and the first appearance of the current is retarded according to the amount of induction present.[120]

As we already saw, Preece believed that the self-inductance of a transmission line adds to the retardation caused by the capacitance. He therefore understandably asserted that owing to the very low self-induction of copper, communication ranges could be extended by the use of copper as opposed to iron wires:

> Copper wire being practically free from electro-magnetic inertia or self-induction, its time constant, or the amount of retardation it exercises on the rate of flow of electricity, is simply the product of its capacity K, and resistance R. KR for 400 lb of iron is .2116 per mile and for 150 lb. copper is .0786 per mile. But iron has electro-magnetic inertia, which still further retards the rate of working; and therefore the speed on a copper aerial line ought to be at least three times greater than that of an iron line. [121]

During the debate with Thompson as in the quotation above, Preece made it quite clear that as far as he was concerned magnetic induction and electrostatic induction had the same effect on signal propagation. Both retard the signal, both therefore cause distortion and both should be minimized. Thus, he was quite oblivious to the deeper significance of Kelvin's neglect of self-induction in his theory. A proper argument on the basis of Kelvin's theory should have referred to lower resistance or lower capacitance. Self-inductance, however, was neglected by the theory. Consequently it provided no clue as to how self-induction would change the signal-carrying capabilities of a transmission line already endowed with capacitance and resistance. Furthermore, Preece grossly underestimated the actual value of the inductance of copper wires. His reasoning was therefore demonstrably erroneous, but, as

120. W.H. Preece and J. Sivewright, *Telegraphy*, (1876), pp. 125-126.

121. W.H. Preece, "On Copper Wire," *The Electrician*, **19** (Sept. 9, 1887): 373. See also D.W. Jordan, "The Adoption of Self-Induction by Telephony, 1886-1889," *Annals of Science*, **39** (1982): 441-451.

Heaviside was to show soon thereafter, copper wires were still to be preferred due to their relatively high ratio of self-inductance to resistance. Practically speaking then, Preece's advocacy of copper wires turned out to be correct despite his reasoning.

To what extent did Preece understand the manner in which Kelvin's theory describes the evolution of a telegraph pulse in a submerged cable? The question remains without a clear answer. Preece's rendition of this description appears to imply that he conceived of the current as actually traveling more slowly along the line as the induction—magnetic or electrostatic—increases. The correct analogy, however, is that the current diffuses along the line in the same way as heat diffuses along a heat conductor.[122] The retardation, due exclusively to electrostatic capacity and resistance, is not the result of the pulse's slower motion, but of the longer time it takes for the essentially instantaneous current to increase beyond the receiver's detecting threshold. In principle, an infinitely sensitive receiver would detect a telegraph pulse instantaneously, and there is no definite way to speak of the speed of electricity in this case. (It should be noted that herein lies a great difference between a line theory that ignores self-induction and one that takes it into account. In a distortionless cable satisfying Heaviside's condition $R/L = K/S$, there is a very real velocity of propagation equal to $1/\sqrt{SL}$). If, however, the KR retardation in a noninductive cable is attributed to the current's velocity, then it turns out that the velocity is inversely proportional to the line's overall length. This should suffice to raise serious doubts with regard to the intelligibility of referring to the speed of the current in this case. How could the current know the line's length before it has actually traversed it? It seems, however, that the view did enjoy some popularity at the time, as Heaviside noted with yet another flare of his caustic sense of ridicule:

> Although the speed of the current is not quite so fast as the square of the line, yet, on the other hand, it is not quite so slow as the inverse-square of the length, as a writer in a contemporary (*Electrical Review*, June 17, 1887 p. 569) assures us has been proved by recent researches. However, if we strike a sort of mean, not an

122. E.C. Baker states that Preece learned from Kelvin that the analogy of heat diffusion is the correct one to apply. That Preece had indeed heard arguments concerning the indefiniteness of the velocity of electricity is without doubt. He had also heard arguments concerning the beneficial effects of self-induction in long-distance telephony, and of the absurdity of considering electricity to be a form of energy; but he never did assimilate these notions. Thus, it is not clear to what extent he assimilated the analogy of heat diffusion into his view of electric circuit theory.

> arithmetic mean, nor yet a harmonic mean, but what we may call a scienticulistic mean (whatever *that* may mean), and make the speed of the current altogether independent of the length of the line, we shall probably come as near to the truth as the present state of electromagnetic science will allow us to go. But, apart from this, there is some *a priori* evidence to be submitted. Is it possible to conceive that the current, when it first sets out to go, say, to Edinburgh, *knows* where it is going, how long a journey it has to make, and where it has to stop, so that it can adjust its speed (scienticulistic speed) accordingly? Of course not; it is infinitely more probable that the current has no choice at all in the matter, that it goes just as fast as the laws of Nature, preordained from time immemorial, will let it; and if the circuit be so constructed that the conditions prevailing are constant, there is every reason to expect that the speed will be constant, whether the line be long or short. Q.E.D.[123]

Ever since this sarcastic paragraph, Heaviside often referred to Preece as "the Eminent Scienticulist."

Preece, it must be noted, was not averse to theory in telegraphic matters. When he trained telegraph operators he tried to instill in them the importance of keeping theory in mind. He explicitly stated that in telegraphy, theory and practice could not be separated.[124] What the above demonstrates, however, is that whatever Preece considered to be Kelvin's theory of the submarine cable was in fact a distorted image of the real thing. Still, he invoked this theory in his debate with S.P. Thompson. While Thompson himself did not fully appreciate the role of self-induction until he was enlightened by Heaviside's work, he did not need Heaviside's guidance to realize that something was fundamentally wrong with Preece's reasoning. Thompson went on to emphasize the fallacious nature of Preece's theoretical analysis, which he correctly perceived to be the weak point of his argument.

Preece did not like to be wrong; especially not when it could imply that his professional expertise would thereby appear questionable. This is nicely exemplified by a previous incident. In 1873 Preece suggested a scheme for duplex telegraphy. It was severely criticized by David Lumsden, who had been the scientific advisor of Preece's superior, R.S. Culley. Preece's reaction indicates that his first concern was to preserve his dignity. While he quickly withdrew his plan, he denounced Lumsden's criticism, and complained that it was disrespectful. It seems he tried to persuade Culley to get rid of Lumsden, but

---

123. *Electrical Papers*, Vol. II, pp. 128-129.

124. E.C. Baker, *Sir William Preece, F.R.S. Victorian Engineer Extraordinary*, (1976), p. 86.

the latter refused. He wrote back to Preece, saying that impulsive comments like Lumsden's are "a pleasant habit at the time—but bad in result, yet better than never speaking your mind.... [Lumsden] is very persevering and hard working but a wee prejudiced. He seems loyal also and not flighty as some are."[125]

In 1887 Preece was the Post Office's senior electrician. His dignity was once again on the line, but there was no longer a Culley to enforce upon him the criticism of an expert. Thus, unlike the above affair, which was settled quietly behind the scenes and never blossomed into a public challenge, Preece's clash with Thompson took place for all to see. Instead of pointing out the positive, practical merits of his own point of view, Preece reacted savagely to Thompson's challenging criticism, and attacked mathematical reasoning in the most general terms. In so doing he exposed himself to further attacks from Thompson, who bluntly suggested that Preece should do himself a favor, and avoid all future interference in mathematical issues.[126]

7.2 Scientist vs. "Scienticulist"

In February of 1887, with Thompson and Preece at loggerheads, the Heaviside brothers communicated their joint paper on the bridge system of telephony to the Society of Telegraph Engineers and Electricians. Oliver Heaviside's appendices to Arthur's description of the system contained a revision of basic telegraph theory, with special emphasis on the role of self-induction and the interplay of resistance and leakance. The last of the appendices contained the first explicit statement of the distortionless condition.

For purposes of discussion, Heaviside distinguished between five basic classes of cables. The manner in which he introduced the last of them is particularly instructive:

> Distortionless circuits, now to be first described, in which, by means of a suitable amount of leakage, the distortion of waves is abolished. Though rather outside practice,... this class is very important in the comprehensive theory, because it supplies a sort of royal road to the more difficult parts of the subject.[127]

125. *Ibid.*, pp. 110-111.

126. D.W. Jordan, "The Adoption of Self-Induction by Telephony, 1886-1889," *Annals of Science*, **39** (1982): 445.

127. *Electrical Papers*, Vol. II, p. 340.

Note that in this very first introduction of the distortionless circuit it is a suitable amount of leakage, not of self-induction, that makes distortionless communication possible. It was, as we have seen (page 226), the inclusion of leakage that originally enabled Heaviside to see the possibility of a practical distortionless condition; hence the emphasis he gave it here.

Since the characteristic telephone cables of the time were well insulated, the value of *R*/*L* was considerably larger than the value of *K*/*S*. In certain copper cables, especially as designed on the Continent and in the U.S., the two ratios came closer to equality owing to the use of highly conductive copper wires with a relatively smaller *R*/*L*. This accounts for the advantages of using copper in terms of the distortionless condition. As we already saw, Heaviside showed that the preferable way to further approach undistorted transmission was to further reduce *R*/*L* and bring it to equality with *K*/*S* by increasing *L*. In other words, Heaviside's analysis supported Preece's practical suggestions regarding the use of copper but flatly contradicted his theoretical reasoning. This difference of theoretical opinion found expression in the conflicting recommendations concerning the use of self-induction. Accordingly, the strong emphasis on the role of inductance during the debate over the question of distortion emerged once it became clear that Preece did not understand how it is to be incorporated into transmission-line theory.

As an expert on telephonic matters, Preece was called upon to referee Arthur and Oliver Heaviside's paper prior to publication in the *Journal of the Society of Telegraph Engineers and Electricians*. Had Preece's interests been aimed purely at practical results, he could have used this paper to support his preference for copper wires and to justify his practical concentration on the line rather than the end apparatus. Naturally, this would have required of him to concede that he seriously misunderstood the basic theory behind the practical correctness of his suggestions. Unfortunately, it appears that by this time Preece's first priority had switched from the question of practical efficiency to the defense of his image as an expert on the fine points of telephony. He gave a very unfavorable review of the paper, and blocked its publication.[128]

Perhaps the reasons for Preece's drastic action were not limited to the immediate context of his dispute with Thompson. Other circumstances may have made it particularly difficult for Preece to allow himself to be educated by Heaviside. Preece had known Heaviside since the latter's days as a telegraph operator in Newcastle-On-Tyne. His opinion of Heaviside was not exactly amicable in the wake of events that closely coincided with Preece's abortive duplex system. In June of 1873, while Heaviside was still employed by the company, he intimated in a short comment appended to an analysis of duplex telegraphy that he had the key to practical quadruplex telegraphy:

> ... from experiments I have made, I find it is not at all a difficult matter to carry on *four* correspondences at the same time, namely, two in each direction....[129]

Understandably, Preece was annoyed by such a cryptic remark, especially considering that it coincided with Thomas Edison's visit to England at a time when the famous inventor was working on quadruplex telegraphy. Preece commented on this and other remarks in Heaviside's paper to his superior, R.S. Culley:

> Oliver Heaviside has written a most pretentious and impudent paper in the *Philosophical Magazine* for June. He claims to have done everything, even Wheatstone's automatic duplex! He must be met somehow.

Culley was quite agreeable to this suggestion. He wrote back:

> O. Heaviside shows what is to be done by cheek. This we see every day—look at Thomson [Kelvin] among the great, brings forward the tangent and sine scales on galvanometers as new. He does not read his *Handbook [of Practical Telegraphy]* it is evident. He claims or is supposed to have brought out lots of other things. We will try to pot Oliver, somehow.[130]

---

128. While it is certainly tempting as well as easy to cast Preece as the unmitigated villain of this whole affair, it seems that his behavior is more aptly described as that of a wounded animal. At the same time, attempts to justify his actions on objective, rational grounds are even less appealing. E.C. Baker writes: "Silvanus P. Thompson later wrote that Preece had been unable to appreciate Oliver Heaviside's work on transmission but he might equally well have written that Oliver had neglected opportunities to persuade Preece that Thomson's KR law was not enough. Instead of interjecting an occasional sarcastic remark into his erudite articles, to the alarm at times of editors to whom libel suits were not unknown, if he had emphasized his argument by reiterating their gist in relatively simple terms his purpose might have been better served. His prose did not possess that persuasive quality found in the writings of Clerk Maxwell, who wrote:... 'It is by the use of analogies ... that I have attempted to bring before the mind, in a convenient and manageable form those mathematical ideas which are necessary to the study of electricity...'". (E.C. Baker, *Sir William Preece, F.R.S. Victorian Engineer Extraordinary*, [1976], p. 206). The extremely abstruse nature of Maxwell's treatise notwithstanding, Baker seems quite oblivious to the many analogies and simple discussions Heaviside constantly interjected into his *Electrician* papers. True, Heaviside's part 8 of "On the Self-Induction of Wires" was not published until well after the debate ended. This was without a doubt the most illuminating discussion of distortion in simple terms. It would have benefitted the mathematically incompetent Preece more than any other discussion at the time. Ironically enough, it was in large measure Preece's personal intervention that blocked its publication.

129. *Electrical Papers*, Vol. I, p. 24.

Worthy of notice is that at the end of his 1873 paper on duplex telegraphy, Heaviside promised to give:

> ... the formulae necessary for calculating the proper proportions of the resistances, etc., to suit different lines and apparatus, so that the greatest possible amount of current may be driven through the receiving instruments, where alone it is of practical service.

The promised paper appeared only in 1876, nearly three years later.[131] The reason for the delay could not have been simply rooted in some unforeseen mathematical difficulty. The problem involved little more than adapting Heaviside's own analysis of the best arrangement of the Wheatstone bridge—a problem he had solved and published as early as February of 1873—to suit the requirements of duplex telegraphy. Consider, however, that Heaviside left the telegraph service in 1874 under less than amicable circumstances that may or may not have had something to do with his deafness. Perhaps these circumstances had more to do with him bringing upon himself the joint wrath of Culley and Preece, and perhaps the two tried to "pot Oliver" by interfering with his publications. We have no evidence, however, to suggest that Culley and Preece really were responsible for the delayed publication of Heaviside's paper, "On Duplex Telegraphy (Part II)". Furthermore, in the intervening years Heaviside did publish three other papers on subjects other than duplex systems.

Clearly, then, hard feelings existed between Heaviside and Preece long before the publication of the joint Heaviside paper on telephony. Thus, possibly

130. Like Preece, Culley may not have reacted solely on the basis of an objective, technical assessment of Heaviside's paper. Heaviside opened his analysis with the following observation: "Duplex telegraphy, the art of telegraphing simultaneously in opposite directions on the same wire,... until lately seemed never likely to be carried out in practice to any extent. According to the very practical author of *Practical Telegraphy*, 'this system has not been found of practical advantage'; and if we may believe another writer, the systems he describes 'must be looked upon as little more than feats of intellectual gymnastics—very beautiful in their way, but quite useless in a practical point of view.' However, notwithstanding these unfavourable reports as to the practicability of duplex telegraphy, the experience of the last year has negatived them in a striking manner, and made the so-called 'feats' very common-place affairs.... There seems little reason to doubt that this system will eventually be extended to all circuits of not too great a length, between the terminal points of which there is more than sufficient traffic for a single wire worked in the ordinary manner—that is to say, only one station working at a time." (*Electrical Papers*, Vol. I, p. 18). Significantly enough, the "very practical author of *Practical Telegraphy*" was no other than R.S. Culley himself.

131. *Electrical Papers*, Vol. I, pp. 24-34.

aided by past memories of his encounters with Heaviside and put on the defensive by Thompson's superior mastery of theory, an already pestered Preece turned to brute force. He blocked the one paper that he could have used to justify his practical suggestions.

By itself, the rejection of this paper had no effect on Heaviside's overall publication plans. It was a separate article, completely independent of his two-way exposition of Maxwell's theory. Furthermore, he was just coming to the analysis of transmission lines in the natural course of his series "On the Self-Induction of Wires," where he could discuss the issue in great detail; or so he thought. His long-term plans notwithstanding, Heaviside was not about to forgive Preece's excesses.[132] In part 7 of "On the Self-Induction of Wires", following a reference to the manner of reckoning the electrostatic capacity of overhead wires, Heaviside added a note:

> On the other hand, Mr. W.H. Preece, F.R.S., assures us that the capacity is half that of either wire (Proc. Roy. Soc. March 3, 1887, and Journal S. T. E. and E., Jan. 27 and Febr. 10, 1887). This is simple, but inaccurate. It is, however, a mere trifle in comparison with Mr. Preece's other errors; he does not fairly appreciate the theory of the transmission of signals, even keeping to the quite special case of a long and slowly worked submarine cable, whose theory, or what he imagines it to be, he applies, in the most confident manner possible, universally. There is hardly any resemblance between the manner of transmission of currents of great frequency and slow signals.[133]

Heaviside's footnote was certainly not a pleasant comment for Preece to read. Biting sarcasm aside, however, it demonstrates how clearly and precisely Heaviside pinpointed the source of Preece's difficulties. Preece did not understand Kelvin's theory within its proper range of application, and as a result of that he applied it to situations beyond its range. Preece, however, was beyond reason. All he saw in Heaviside's footnote was an unmitigated attack on

132. Heaviside suspected Preece of unwarranted handling of his papers before. In the private summary of his 1878 paper "On Electromagnets, etc." (*Electrical Papers*, Vol. I, pp. 95-112), Heaviside wrote: "I sent paper to Sec[retary] S.T.E. in the spring (about March) with request for publication in next number, or else its return. In the summer number was published; my paper *not* in it. Asked for its return, to make addition. Got it back, & found it had been in Preece's hands. Wrote paper for Phil Mag. ... When this was published, I was asked for paper through A.W.H. [Arthur West Heaviside], P., and Webb. I complained of the great delay as reason for writing Phil. Mag. paper. Webb got blame for it. Perhaps it wasn't *his* fault. Almost certain paper was kept by P. Have some reasons for that." (Notebook 3A, Art. 19, Heaviside collection, IEE, London).

133. *Electrical Papers*, Vol. II, p. 305.

his person. Preece called Kelvin's attention to the footnote. On July 12, 1887, the renowned scientist replied:

> Dear Preece,
>
> Thanks for calling my attention to the footnote on p. 82. I was very sorry to see it. It ought never to have been there and I have written a little editorial to say so which no doubt G. Francis [the editor of *The Philosophical Magazine*] will put into the July Number. Personalities tinctured however slightly with ill nature are utterly unsuited for a scientific journal, and this one was really outrageous. I think Oliver Heaviside is a good deal off his head. I have given Francis an absolutely general rule for Phil. Mag. to cut out in future anything tinctured in the [smallest] degree with personal ill nature.[134]

Whether Kelvin sent the editorial note to the magazine is unclear. It was not printed in the July number, or in the following ones. His disapproval of Heaviside's comment, however, must not be taken for an insincere attempt to appease an angry Preece. He disapproved of Heaviside's bruising sense of fun at least once before, in 1883, when Heaviside poked some acidic ridicule at church officials.[135] In 1889 Heaviside learned from Oliver Lodge of Kelvin's displeasure and recorded it in his private notes on the offensive letter:

> ... Present letter rather nonsensical, about the Archbishop. Heard from O.J.L. in 1889 that Sir W.T. was very much disgusted at my remarks about archbishops. Really, however, there was nothing to be disgusted about. It was simply a bit of fun.[136]

Significantly enough, Heaviside decided not to reproduce the contents of this letter in *Electrical Papers*.

Regardless of what Kelvin's opinion of Preece's complaint really was, Preece must have felt greatly fortified by the unqualified support he offered. What then transpired between himself, the editor of *The Philosophical Magazine*, and possibly Kelvin remains largely unknown. We do know that Heaviside sent the eighth installment of "On the Self-Induction of Wires" to *The Philosophical Magazine* around July of 1887. It was published for the first time only in 1892, in the second volume of the collected *Electrical Papers*. Part 9 never went beyond Heaviside's declared intention to devote it to a de-

---

134. Lord Kelvin to W.H. Preece, 12 July 1887, Preece collection, I.E.E., London.

135. P.J. Nahin, *Oliver Heaviside: Sage in Solitude*, (1988), pp. 107-108.

136. Notebook 3A, p. 50, Heaviside collection, IEE, London.

tailed elucidation of the partial-reflection method. In his own summary of part 8 of "On the Self-Induction of Wires," Heaviside wrote:

> Declined with thanks by Ed. *Phil. Mag.* Told him circumstances, paper blocked by Preece, and that it and Part IX on distorted waves were divested of their original telephonic applications, and contained entirely new results, and asked to be allowed to return it for reconsideration; or put them under separate titles, instead of being Parts VIII and IX of S.I. of Wires. No reply.[137]

As we have already seen (page 183), "On the Self Induction of Wires" was originally planned to run through four installments only, which the editor of *The Philosophical Magazine* used as an excuse to terminate the series. Of course, he could have invoked that excuse three installments earlier but elected not to do so. Furthermore, as Heaviside's last sentence shows, he was perfectly willing to change the contents and publish parts 8 and 9 under different titles; none of which made any difference.

Unlike the rejection of the three appendices to his brother's paper, this setback really hurt Heaviside's work plan. "On the Self-Induction of Wires" was intended as the counterpart of "Electromagnetic Induction and its Propagation." Heaviside was just coming to the series' high point, namely, the distortionless circuit, which would serve as the royal road not merely to distorted transmission, but to wave propagation in Maxwell's field theory. In the wake of these developments, Heaviside decided to salvage what was left of his grand plan by communicating the analysis of distortionless transmission in his ongoing "E.M.I.& P." At the time, the series was still in the midst of linear circuit analysis which was to build up to the transmission line and then to Maxwell's field theory. On June 3, 1887, Heaviside cut the process short with the following introduction to section 40 of the series:

> Although there is more to be said on the subject of induction-balances, I put the matter on the shelf now, on account of the pressure of a load of matter that has come back to me under rather curious circumstances. In the present section I shall take a brief survey of the question of long-distance telephony and its prospects, and of signalling in general. In a sense, it is an account of some of the investigations to follow.[138]

Heaviside's private summary of this paper makes the nature of the "rather curious circumstances" quite clear:

137. Notebook 9, p. 197, Heaviside collection, IEE, London.
138. *Electrical Papers*, Vol. II, p. 119.

> This article is abstract of results given in App. C. to my brother's paper ... which was blocked by the em. scienticulist, W.H. Preece, F.R.S. in spring of '87. He ordered all my work (20 pp.) in text to be omitted, on ground of irrelevancy & want of novelty.[139]

In section 40 of "E.M.I.&P." Heaviside finally put down the distortionless condition.[140] Given enough time, he could still complete the interrelated presentation of circuit and field theory along the lines he drew up in August of 1886. At this stage, however, Heaviside had clearly come to see Preece as an intellectual monster. Just as Preece lost sight of the positive side of his argument with Thompson, Heaviside was now out to even the score with his personal nemesis.

The confrontation between Heaviside and Preece came to a head over the measurement of self-inductance in copper wires. In September of 1887 Preece published the results of an extensive experimental undertaking whose purpose was to establish the self-inductance of iron and copper wires. Preece's values for the copper wires were hundreds of times lower than the known ones. Citing this, Heaviside set out to show that considering everything known about electromagnetism Preece's results should be highly suspicious. As already mentioned, by this time Heaviside was utterly enraged by Preece's treatment of his work. Therefore, instead of simply demonstrating the probable fallacy in Preece's work, he proceeded to discredit Preece's scientific ability in the most direct and biting manner. Considering Heaviside's extraordinary knack for sharp expression, one can imagine the severity of the punishment. That his criticism was not at all out of touch with reality only added to its bite. Heaviside opened his letter with the following tirade:

> A very remarkable paper "On the Coefficient of Self-Induction of Iron and Copper Telegraph Wires" was read at the recent meeting of the B. A. by William Henry Preece, F.R.S., the eminent electrician.... It contains an account of the latest researches of this scientist on this important subject, and of his conclusions therefrom. The fact that it emanates from one who is—as the *Daily News* happily expressed it in its preliminary announcement of Mr. Preece's papers—one of the acknowledged masters of his subject, would alone be sufficient to recommend this paper to the attention of all electricians. But there is an additional reason of even greater weight. The results and the reasoning are of so surprising a character that one of two things must follow. Either, firstly, the accepted theory of elec-

139. Notebook 3A, article [137], Heaviside collection, IEE, London.
140. *Electrical Papers*, Vol. II, p. 122-123.

> tromagnetism must be most profoundly modified; or secondly, the views expressed by Mr. Preece in his paper are profoundly erroneous. Which of these alternatives to adopt has been to me a matter of the most serious and even anxious consideration. I have been forced finally to the conclusion that electromagnetic theory is right, and consequently, that Mr. Preece is wrong, not merely in some points of detail, but radically wrong, generally speaking, in methods, reasoning, results, and conclusions. To show that this is the case, I propose to make a few remarks on the paper.[141]

After a detailed criticism of Preece's work, Heaviside concluded:

> In fact, I may remark that Mr. Preece employs such entirely novel and unintelligible methods, that it would surely be right that he should give some reason for the faith that is in him.[142]

Considering the past history of Heaviside's written attempts to discredit Preece's work, one should not find it surprising that the editor of *The Electrician* declined to publish this letter. To make matters worse, the unpublished text of the letter seems to have found its way into Preece's hands. In August, just prior to the events that prompted the above letter, Heaviside sent in part 47 of "E.M.I.& P." It was published only in December with the appended note declaring the series' premature termination (see page 182). In his private notes, Heaviside added these words to the paper's technical summary:

> Observe date. Dec. 30. Sent in in Augt. Just after B.A. meeting I wrote critique of P's paper "On the C. of S.I. of iron & Copper wires'. Ed. sent it to P., who made his marks on it. The Ed. refused to publish it, or to give reason for refusal. Then requested me to discontinue E.M.I. & P.[143]

Further explanation is provided by a note Heaviside appended to his handwritten copy of "On the Self-Induction of Wires," part 7. After noting the fate of parts 8 and 9, he added:

> Got portions published in the Electrician, especially on the non-distortional circuit, Arts XL to XLVII of E.M.I. & its P. The journal then suffered from a change of Editor, or in other ways. Requested to discontinue, Nov. 87, after keeping last article since August.[144]

---

141. *Electrical Papers*, Vol. II, p. 160.
142. *Ibid.*, p. 163.
143. Notebook 3A, Art. 147, Heaviside collection, IEE, London.
144. Notebook 9, p. 197, Heaviside Collection, IEE, London.

This is how Heaviside's grand plan of bringing Maxwellian electromagnetism to the attention of his past colleagues was put to rest. In February of 1888, just over a month after "Electromagnetic Induction and its Propagation" had been terminated, Heaviside began to publish a new series in *The Philosophical Magazine*. In a manner of speaking, it continued the disrupted theme of the two parallel series of 1887. In the first part, Heaviside discussed at length the analogy between signal propagation in the linear theory and the propagation of disturbances in the electromagnetic field. However, already the sophisticated wording of the new series' title makes it clear that it did not proceed in the spirit of his previous work: "On Electromagnetic Waves, Especially in Relation to the Vorticity of the Impressed Forces; and the Forced Vibrations of Electromagnetic Systems." Obviously, this was aimed at a different audience. The outlook is thoroughly Maxwellian from the beginning; the idea of leading to the field equations through a gradual analysis of electrical circuits is completely gone. Perhaps the saddest thing about the new series is the following statement:

> It seems to be imagined that self-induction is harmful to long-distance telephony. The precise contrary is the case. It is the very life and soul of it, as is proved both by practical experience in America and on the Continent on very long copper circuits, and by examining the theory of the matter. I have proved this in considerable detail; but they will not believe it. So far does the misconception extend that it has perhaps contributed to leading Mr. W.H. Preece to conclude that the coefficient of self-induction in copper circuits is negligible (several hundred times smaller than it can possibly be), on the basis of his recent remarkable experimental researches.[145]

Thus, in one short paragraph, with simple words and well disguised sarcastic innuendoes, Heaviside explained the reason behind Preece's extremely low estimate of self-induction in the clearest manner possible. Preece considered self-inductance harmful to long-distance telephony, and he knew that long-distance telephony was successfully implemented by the use of copper wires. Therefore, he had every reason to trust the exceedingly low values he obtained for the self-inductance of copper: they were in keeping both with his theory and with practical experience.

On a more basic level, Preece's misguided ideas on the role of self-inductance stemmed from his failure to assimilate the principles of linear circuit the-

145. *Electrical Papers*, Vol. II, p. 380.

ory. As we have seen he was not alone in this, and Heaviside's failure to dissuade him and other engineers from the notion "that self-induction is harmful to long-distance telephony" was actually a failure to teach the principles of linear circuit analysis. This, more than anything else boded ill for Heaviside's publication plan with its crucial dependence on proper understanding of these principles. It may have persuaded Heaviside that the audience simply was not there for the sort of presentation he designed. The unified development of circuit and field theory built around transmission-line analysis was intended primarily for telegraph and telephone engineers. Its reception, however, indicated that what seemed to Heaviside as a "Royal Road" was perceived by his intended readers as a rather perilous alleyway. Not surprisingly, the above quotation indicates Heaviside's resignation to the idea of giving up the grand plan he conceived in the wake of Hughes's work on variable currents.

Heaviside did something else with his new series that he had not done before: he sent a copy to Kelvin before submitting it to *The Philosophical Magazine*. A correspondence ensued, which led to a short, written discussion between them on the velocity of electricity. In its wake Heaviside appended three notes on the subject to the first part of the series. In his summary of the installment, Heaviside described the events:

> Ms. sent to Sir W.T. He made queries about J.J.T. view v. copper. Also referred me to his article on Velocity of El^{y} in Nicols' Cyclopaedia.
> Consequently, I wrote the three notes.
> Proof went to Sir W. and he modified my remarks upon inductance decreasing etc. and on P[reece] and also cancelled the (to be cont^{d}) saying "Better not say that, even at end of Notes".
> No doubt whatever that Sir W.T. was all at sea on the subject. Completely misunderstood the meaning of self-ind^{n} carrying on the waves.[146]

Heaviside went to even greater lengths with this series. He obtained fifty bound copies of it, and mailed them to scientists in England and abroad. In addition to this attempt to call wider attention to his work, Heaviside focused his efforts on two other potential allies aside from Kelvin: S.P. Thompson and Oliver Lodge. Thompson, who could naturally be expected to harbor little sympathy for Preece, wrote back using strong words to express his disapproval of Preece's scientific censorship. Publicly, however, he did little to help Heaviside receive the recognition he had been seeking.[147] In Lodge, however,

146. Notebook 3A, Art. 150, Heaviside Collection, IEE, London.

Heaviside found a more useful supporter. In February of 1888, Lodge delivered two controversial lectures on the theory of lightning discharges and on the means of protecting life and property from their effects. Some of Lodge's theoretical analysis was deficient, his critique of lightning protection practices at the time was unpersuasive, and his interpretation of certain experimental results was questionable. Still, Lodge relied on electromagnetic field theory for some of his analysis, correctly criticized the neglect of the inductive voltage drop along a struck lightning conductor, and Heaviside sensed that he could be helpful. He wrote to Lodge on June 5, 1888, requesting a copy of his lightning papers, about which "some rather sensational statements are being made."[148] The question of lightning, however, was quickly put aside. Once Lodge responded, Heaviside turned the correspondence into an introduction to his own work. He called Lodge's attention to his transmission-line studies and claimed priority for the theoretical prediction of the skin effect. Lodge found Heaviside's transmission-line work too laborious to follow (see chapter I), and was not convinced by Heaviside's priority claims, as we have already seen. Nevertheless, Heaviside and Lodge shared two sentiments that easily transcended such minor mismatches: they were both ardent admirers of Maxwell's electromagnetic field theory, and both were seriously at odds with Preece, who uncritically rejected everything Lodge had to say about lightning as well as Heaviside's observations regarding long-distance telephony. Lodge became the first scientist of significant standing to refer admiringly to Heaviside's work in both his publications and lectures.

For the purpose of calling public attention to his work, Heaviside could not have chosen a better mouthpiece than Lodge at that particular time. Lodge's lightning work became the focus of intense attention by both scientists and engineers in the course of the British Association meeting at Bath in September of 1888. The Bath meeting provided Preece and Lodge with a highly visible forum in which to air their differences. In Lodge, Preece found an opponent who was just as comfortable in front of a large audience as himself, and who matched his witticisms blow for blow. At the end of the day, Lodge had the better of the encounter as far as witty one-upmanship was concerned. Along the way, Lodge also handed Heaviside his first victory over Preece. No one

147. D.W. Jordan, "The adoption of Self-Induction by Telephony, 1886-1889," *Annals of Science*, **39** (1982): 451.

148. Heaviside to Lodge, 5 June 1888, Lodge Collection, University College, London.

who either attended the lively debate in Bath, or who followed its course in the pages of *The Electrician*, could have come away without encountering the name of Oliver Heaviside. He was finally beginning to emerge into the public eye. However, if Lodge was successful in his contest with Preece as far as sheer debating prowess was concerned, he was far less persuasive on the scientific and technical side. His critique of lightning-protection practices was not accepted by most of the scientists and engineers who witnessed the debate, while central aspects of his theoretical analysis and of his experimental work were severely criticized by Maxwellian authorities such as George Francis FitzGerald and Henry Augustus Rowland. Under such circumstances, Lodge could not really supply Heaviside's work with the sort of authoritative recommendation to command the respectful attention of most scientists and engineers.

In the end, Kelvin gave Heaviside the seal of approval he had been seeking. In January of 1889, the Institution of Electrical Engineers elected Kelvin to replace the outgoing President, Edward Graves (then Engineer-in-Chief of the British Post Office, and Preece's superior). Kelvin used his inaugural address to comment at length on several intense controversies that were raging in the British electrical world at the time. One of them, and the only one that he settled unequivocally, was the question of self-induction in telephony.

Kelvin's comments on this issue are particularly enlightening in view of Heaviside's private remarks above (page 257). Kelvin began with a short outline of how his own 1855 submarine cable theory came to be developed once he realized that the effects of self-inductance could be neglected. Then he proceeded to describe how the old question referring to these effects resurfaced and rekindled his interest in it. Kelvin's words and the response of his audience were recorded in *The Electrician*:

> Within the last forty days I have really worked it out to the uttermost merely for my own satisfaction. But in the meantime it had been worked out in a very complete manner by Mr. Oliver Heaviside (applause), and Mr. Heaviside has pointed out ... that electro-magnetic induction is a positive benefit, it helps to carry the current. It is the same kind of benefit that mass is to a body shoved along against a viscous resistance.... I am not doing justice, of course, I know, to his statement in one short sentence. The whole question is treated in the most complete mathematical way. The effect of electro-magnetic induction and electro-static induction taken together (and they cannot be separated) is fully worked out.... *Heaviside has included [leakage] along with electro-magnetic induction, and this point he has particularly accentuated....* Now, in the mathematical theory

> there are two things to be considered in respect to the distortion (as Heaviside called it) of the signals passing through the cable. One thing to be considered is the retardation of phase; another is the diminution of amplitude. *If the retardation of phase was the same for alternating current of all periods, then this retardation of phase would be of no consequence whatever—it could not diminish the distinctness at all. Again, if the diminution of amplitude was precisely in the same proportion for ... all periods, then when we come to make non-periodic signals we should find that the signals would be transmitted with perfect sharpness....* Heaviside points out that electro-magnetic induction causes a less great difference in the attenuation of different periods than there is without it; and that electro-magnetic induction (as we knew forty years ago) tends to reduce the retardation of phase to the same for all different notes—that is, to the retardation equal to what would depend on a velocity not very different from the velocity of light if the signals have but sufficient frequency [my italics].[149]

As the emphasized lines clearly show, Kelvin explained the problem of distortion in terms of an arbitrary signal's harmonic content. He undoubtedly understood that the inclusion of leakance was crucial to Heaviside's discovery of the distortionless condition and was aware that misconceptions regarding the role of self-induction stood in the way of the theory's acceptance. However, the most remarkable feature of this unqualified support for Heaviside's distortionless condition is its complete lack of Maxwellian undertones.[150] As already explained, linear distortionless analysis need not presuppose the Maxwellian field-theoretical view. It is difficult to say with certainty what prompted Kelvin to disassociate his praise for Heaviside's telegraph theory from the crucial role it played in Heaviside's exposition of electromagnetic field theory. He may have avoided such associations on purpose, to get the message across to the engineering audience in the most direct manner possible. An alternative view is indicated by Heaviside's impression in 1888 that "...Sir W.T. was all at sea on the subject" and that he "completely misunderstood the meaning of self-induction carrying on the waves". This, together with Kelvin's indisputable understanding of the linear principles behind the distortionless condition, suggests that Kelvin may have been uncomfortable with the Maxwellian view of the conductor as surrounded by an electromagnetic disturbance, propagating through the dielectric medium at near light speeds. This point is further reflected in the short correspondence between Heaviside and Kelvin in 1888. Kelvin pointed out in a note that Kirchhoff had

---

149. Sir William Thomson, *The Electrician*, **22** (Jan. 18, 1889): 305-306.

first shown in 1857 that electrical signals in wires propagate at about the speed of light. Heaviside responded:

> In Maxwell's theory, however, as I understand it, we are not at all concerned with the velocity of electricity in a wire (except the transverse velocity of lateral propagation). The velocity is that of the waves in the dielectric outside the wire.[151]

Kelvin, in fact, harbored grave doubts about the electromagnetic theory of light, and objected to the abstract dynamical nature of Maxwell's theory:

> ... it seems to me that it is rather a backward step from an absolutely definite mechanical motion that is put before us by Fresnel and his followers to take up the so-called electromagnetic theory of light in the way it has been taken up by several writers of late.[152]

---

150. The closest Kelvin came in his speech to making a direct association between Maxwell's field theory and Heaviside's transmission-line theory was when he said: "'Maxwell's electromagnetic theory of light' marks a stage of enormous importance in electro-magnetic doctrine, and I cannot doubt but that in electro-magnetic practice we shall derive great benefit from a pursuing of the theoretical ideas suggested by such considerations. In fact, Heaviside's way of looking at the submarine cable problem is just one instance of how the highest mathematical power of working and of judging as to physical applications helps on the doctrine, and directs it into a practical channel." (*The Electrician*, **22** [Jan. 18, 1889]: 306). This is a remarkable passage, considering that Kelvin never subscribed to the electromagnetic theory of light. If nothing else, it clearly attests to Kelvin's belief that a theory need not be true to have practical value. Under the immediate circumstances, however, this was not a surprising attitude on Kelvin's part: he must have had the successes of his own limited but highly useful submarine cable theory vividly in mind. In general, the passage lends itself to two interpretations. It may be taken for a somewhat noncommittal statement that Heaviside's distortionless condition is a practical triumph for Maxwell's electromagnetic theory of light. This interpretation, however, hardly tallies either with Kelvin's exposition of Heaviside's distortionless condition, or with the fact that Heaviside himself never suggested that Maxwell's theory is required for the condition's establishment. What Kelvin's endorsement of Heaviside's distortionless condition and his above-quoted praise of Maxwell's theory do have in common, is the emphasis on mathematical analysis. This, it seems, is the clue to the paragraph's more plausible interpretation, to wit: great practical value may be derived from the application of mathematical reasoning to electromagnetic doctrine (Maxwellian or otherwise). Heaviside's application of mathematical reasoning to the problem of distortionless transmission provides an excellent case in point. In the same way, mathematical reasoning applied to Maxwell's theory may also yield results of great practical value, especially since "Maxwell's electromagnetic theory of light marks a stage of enormous importance in electro-magnetic doctrine." Considering Preece's vehement attack on mathematics in practical matters as well as Kelvin's doubts regarding Maxwell's theory, it seems all the more likely that this was indeed the message Kelvin intended to get across.

151. *Electrical Papers*, Vol. II, p. 395.

152. Kelvin's first Baltimore Lecture, in R. Kargon and P. Achinstein, (eds.), *Kelvin's Baltimore Lectures and Modern Theoretical Physics*, (Cambridge, Mass.: MIT Press, 1987), p. 12.

Kelvin praised Maxwell's theory as a great scientific achievement during his inaugural address, and predicted that it would have a great future, but he also expressed his misgivings about the theory and never adopted its general outlook.[153] Some of this has been colorfully expressed by FitzGerald in a letter to Heaviside:

> ... nor does he [Kelvin], I think, even yet, understand Maxwell's notion of displacement currents being accompanied by magnetic force. I tried to get him to see that his own investigations of the penetration of alternating currents into conductors was only the viscous motion analogue of light propagation but he shied at it like a horse at a heap of stones which he is accustomed in another form to use for riding over.[154]

It is possible, as a result, that Kelvin was very sensitive to the knowledge that linear telegraph theory and the distortionless condition were basically independent of the Maxwellian field view, and considered that they must not be presented as triumphs of this view. At the same time, Kelvin regarded Heaviside's work as highly esoteric (see chapter I). He might not have considered it worthwhile to immerse himself in the time-consuming task of learning Heaviside's route to electromagnetism and electric circuits. Consequently, he may not have been at all aware of the role linear circuit theory played in Heaviside's larger scheme; indeed, he may not have been aware of the scheme in the first place. Regardless of his reasons, Kelvin clearly avoided presenting the issue of distortionless telephony as a vindication of Maxwell's theory. While he stressed the comprehensive nature of Maxwell's theory and suggested that it would be of future practical service, he carefully separated this general statement from the question of the distortionless transmission line.

Heaviside was understandably pleased with Kelvin's endorsement and quite appreciative of the service extended to him by the great scientist. A few months after Kelvin's address, Heaviside wrote to Hertz about the advantages of his conventions and nomenclature. "But," he added, "there is a very strong prejudice against me and all my work here; though Sir W. Thomson has done me a very good turn lately."[155] In one of his notebooks Heaviside celebrated

---

153. S.P. Thompson, *Life of William Thomson, Baron Kelvin of Largs*, (1910), Vol. II, pp. 1012-1085. C. Smith and M.N. Wise, *Energy and Empire: A Biographical Study of Lord Kelvin*, (1989), pp. 488-494. M.N. Wise and C. Smith, "The Practical Imperative: Kelvin Challenges the Maxwellians," in R. Kargon and P. Achinstein (eds.), *Kelvin's Baltimore Lectures and Modern Theoretical Physics*, (1987), pp. 323-348.

154. FitzGerald to Heaviside, 11 June 1896, Heaviside Collection, IEE, London.

his Kelvin-aided triumph over Preece with a victory rhyme that he annotated to make its meaning unmistakable:

Self-induction's "in the air" (W.T. and O.H.)
Everywhere, Everywhere; Bath
Waves are running to and fro
Here they are, there they go
Try to stop' em if you can, (W.H.P.)
You British Engineering man!

Conceive it (if you can)
the engineering man,
Docking & Blocking & Burking a paper,
Up in St. Martins-le-grand! (W.H.P.)[156]

Along with electromagnetic waves and self-induction, Heaviside's became a household name in wide circles of physicists and electrical engineers. While he maintained his solitary life-style, he never returned to the anonymity under which he composed the main body of his *Electrical Papers*.

8. Epilogue: The Making of a Riddle

Understandable as Heaviside's elated response to Kelvin's support may be, perhaps he should have been somewhat less enthusiastic about it. Kelvin completely ignored the key role that transmission-line analysis played in Heaviside's work. Therefore, Kelvin's inaugural address had a most peculiar effect on the reception of Heaviside's work. It certainly put the official seal of approval on the distortionless condition as a practical guide to long distance telephony; but at the same time it quite effectively hid from view how transmission-line analysis lent coherence to all of Heaviside's work since August of 1886. From this point of view, the confusion regarding both the character of Heaviside's work and his classification as engineer, physicist, or mathematician dates back to the same event that brought him out of the obscurity under which he worked until the winter of 1888-89.

155. Heaviside to Hertz, April 1, 1889, quoted in O'Hara and Pricha, *Hertz and the Maxwellians*, (1987), p. 64.

156. Notebook 7, p. 94, Heaviside collection, IEE, London. The last line refers to the location of W.H. Preece's office.

8.1 Out of Place with the Physicists...

Heaviside's lifelong interest in the general theory of electric circuits strongly conditioned his particular brand of electrodynamics. The latter, however, clashed with the standard analytical tools used by his peers. The same interest in electric circuits directed his main research efforts away from the forefront of electromagnetic research at the time. Furthermore, anyone who wished to study Heaviside's work had to master both his unorthodox operational techniques and his vector algebra . In this section we shall see how these closely interrelated issues prevented Heaviside from carving a comfortable niche for himself among the Maxwellian physicists of his day.

If 1887 could be described as the year in which the outside world slammed the door in Heaviside's face, then 1888 should be seen as the year in which the door was flung wide open. He published a number of papers on the motion of charged bodies that proved of great interest to other field theorists. In particular, he became the first to note that when a moving charge approaches the speed of light, its electric field component perpendicular to the direction of motion intensifies by a factor of:[157]

$$\sqrt{1 - v^2/c^2}.$$

Heaviside never deemed this result capable of justifying the famous Lorentz-FitzGerald contraction, which he considered to be as problematic as the difficulties it sought to resolve.[158] Nevertheless, he seemed well under way to joining the rank and file of the physicists' community.

---

157. *Electrical Papers*, Vol. II, p. 499.

158. Commenting on Lorentz's fast moving, oblate electron, Heaviside wrote: "It is attractive theoretically, on account of the simplicity of energy and mass formulas. It is also suggested by the 'explanation' given by FitzGerald and by Lorentz of the Michelson-Morely experiment, that the negative result could be accounted for by a certain lateral contraction, in the line of motion, of the bodies supporting the apparatus. Here the real difficulty is to explain the explanation." (*Electromagnetic Theory*, Vol. III, p. 475). B.J. Hunt suggested that Heaviside's discovery of the deformed fast moving electromagnetic field gave FitzGerald a theoretical basis for the contraction hypothesis, and that it was not, therefore, the "brilliant baseless guess of Irish genius" (see B.J. Hunt, "The Origins of the FitzGerald Contraction," *British Journal for the History of Science*, **21** (1988): 67-76). However, A. Warwick has shown that without a specific electromagnetic theory of molecular structure, which Heaviside did not supply and which FitzGerald did not have, the field deformity could not possibly lead to FitzGerald's hypothesis in a justifiable manner (A. Warwick, "On the Role of the FitzGerald-Lorentz Contraction Hypothesis in the Development of Joseph Larmor's Electronic Theory of Matter," *Archive for History of the Exact Sciences*, **43** (1991): 29-91, esp. pp. 43-45).

Undoubtedly, the success of his series on electromagnetic waves, which was aimed at a scientifically sophisticated readership, encouraged him to address such advanced topics. But Heaviside did not really need the interest of others to prod him into writing on the topic of moving charges. With the disintegration of his grand publication plan, it was only natural for him to resume the discussion in "E.M.I.& P." where he left off in 1886. The study of moving charges reflected his more general endeavor to extend the discussion of electrodynamics to the case of moving media. Perhaps that is why aside from the topics he discussed, he did little to endear his work to his newly found audience. In a way, his work from 1888 to 1890 only contributed more to the incomprehension that greeted his approach to Maxwell's field theory. On the face of it, his publications in this period might have given the impression of an attempt to join the world of theoretical physics. Heaviside, however, never abandoned his basic formulation of Maxwell's theory. As a result, while he addressed questions that would have been of considerable interest to other theoretical Maxwellians at the time, he did it in terms and methods that were quite foreign to them. Perhaps, had he stayed in London after Kelvin's endorsement and proceeded to join the activities of the scientific community that just discovered him, his work would have changed sufficiently to become more aligned with that of others. But in the middle of 1889 he moved to Paignton with his parents, so that his association with other scientists remained confined to letter writing. In 1891 he began to work on the first volume of *Electromagnetic Theory*, where he introduced for the third time in six years his unique force-oriented approach to electrodynamics. He never tried to repeat his two-pronged presentation of Maxwell's theory according to the guidelines he drew in August of 1886. However, his rendition of Maxwell's theory applies most naturally to electric circuits, and the analysis of circuits occupies the major part of the three volumes of *Electromagnetic Theory.*

Obviously then, the attention of his newly found colleagues never swayed Heaviside from the original motivations that drove him to study and reinterpret Maxwell's work. Like Heaviside, other Maxwellians were certainly aware of the intimate connection between electromagnetic field theory and circuit theory. Unlike Heaviside, they never turned this connection into the focal point of their research, and could not appreciate Heaviside's insistence on maintaining his particular brand of electrodynamics. This is particularly true of his closest Maxwellian friends, FitzGerald, Lodge, and Hertz. In 1889 Heaviside wrote to Hertz:

> My method of working electromagnetic problems is eccentric, and so is my notation. And so is your method of experimenting, on electromagnetic waves. It suits the waves, and is what was wanted. I venture to believe that my eccentricities too have the soundest foundations. They are deliberate and result from considerable experience. ... My methods and symbolism are strictly appropriate to the subject matter. Most writers here follow Maxwell slavishly, and repeat his faults and errors, even though they sometimes may disobey the spirit of his treatise by following the letter. I have struck out a path for myself, and I am quite sure that if Maxwell had lived, he would, because he was a progressive man, have recognised the superior simplicity of my methods.[159]

The most obvious expression of Heaviside's individualistic path to Maxwell's theory was his algebra of vectors, which he developed as the natural language of fields. Among the devoted Maxwellians who actively contributed to the theory's development, only Hertz was quick to adopt vector algebra. Lodge's forte was mostly experimental, and he never really contributed to the theory's mathematical structure. FitzGerald, who was a very capable mathematical physicist, never took to the vectorial language. In 1889, FitzGerald wrote to Heaviside:

> I am rather sorry you have not been content to work with the ordinary quaternions notation. It makes a very great difficulty to many people who want to look over and pick out the points in your work.[160]

Then, in 1892:

> I hope you will succeed in making the ordinary mathematical physicist think in vectors although I do not think your notation an improvement. You see I was very "big" on Tait and get very much [bothered] by your omission of S [in front of a scalar product] and when one gets bothered every time one naturally takes a dislike to the botheration.[161]

By 1893, FitzGerald realized that he could never resolve this difference with Heaviside. He still disagreed with him about the usefulness of vectors, but "Anyway, we differ too much to discuss it."[162] Similar sentiments characterize the two most influential British Maxwellians of the 1890s, J.J. Thomson and Joseph Larmor.

---

159. Heaviside to Hertz, 14 Feb. 1889, in O'Hara and Pricha, *Hertz and the Maxwellians*, (1987), p. 58.

160. FitzGerald to Heaviside, 4 Feb. 1889, Heaviside Collection, IEE, London.

161. FitzGerald to Heaviside, 26 Sept. 1892, Heaviside Collection, IEE, London.

162. FitzGerald to Heaviside,8 Aug. 1893, Heaviside Collection, IEE, London.

Important as the algebra of vectors was to Heaviside, it was not as crucial to the unique nature of his field theory as his special brand of electrodynamics. As we have seen (chapter III), he constructed all of his electromagnetic field theory on the dynamical principle of activity, which states, in accordance with Newton's third law, that the sum of all energy outputs by the various forces in a complete dynamical system must be zero. In 1891, Heaviside wrote a paper entitled "On the Forces, Stresses, and Fluxes of Energy in the Electromagnetic Field."[163] In this paper he completed the discussion that he interrupted in 1886 when he diverted his attention to Hughes's work. Heaviside showed how the principle of activity leads to the description of the energy flux and to expressions for the electromechanical stresses in the general case of anisotropic media in relative motion. This was Heaviside's last great contribution to the dynamical foundations of electromagnetic field theory, in which he presented significant extensions of his work from 1885 and 1886.

Berkson has noted that Heaviside's theory of ether stress had been "dropped without refutation by an independent test."[164] It is perhaps more accurate to say that Heaviside's analysis of the Maxwellian stresses was never picked up in the first place. His paper was greeted by general incomprehension, and not one of Heaviside's contemporaries ever attempted to further develop the dynamics of electromagnetic fields using his approach. The paper certainly requires patience and willingness to learn a set of novel conventions; it could present baffling pitfalls if these conventions were not properly assimilated together with Heaviside's use of the activity principle. However, it must be stressed that the paper is far from undecipherable. There were many talented mathematical physicists in Britain at the time who could have mastered it had they really wanted to. It is hard to believe that Heaviside's paper presented an insurmountable obstacle to the likes of Larmor, Rayleigh, J.J. Thomson, and FitzGerald. The paper was certainly clear enough for a cursory reading to show that it contributed nothing fundamentally new to the basic physical outlook of Maxwell's electromagnetic field theory. It did offer a novel way of formulating the dynamics of Maxwell's electromagnetic fields. Analytical mechanics, however, had been well developed by Heaviside's time, and the powerful formulations of Hamilton and Lagrange were widely used. It was,

163. *Electrical Papers*, Vol. II, pp. 521-574.

164. William Berkson, *Fields of Force: The Development of a World View from Faraday to Einstein*, (1974), p. xi.

therefore, quite natural to conclude that all of Heaviside's results could be obtained by more conventional means, such as the principle of least action.

This approach in particular was popular among British mathematical physicists. Without due appreciation of Heaviside's telegraphic background, his preference for the activity principle and the force formulation of electromagnetism must have seemed almost purely aesthetic and unduly dogmatic. He considered it more physical because it explicitly involved forces. The involved forces, however, were always *generalized* forces, not real mechanical ones, and it seems somewhat contrived to stress the greater physicality of an *analogy* to mechanical force over that of an extremum principle. Under these circumstances, it should not be too surprising that other mathematical physicists elected not to invest considerable time and effort merely to learn a novel formalism that promised no new results.

Heaviside, on his part, was constantly irritated by the frequent use of Hamilton's extremum principle and Lagrange's equations which usually replaced the principle of activity in the work of his colleagues. In 1899 he wrote to Lodge that for the most part he relegated the principle of least action to the kitchen, to be practiced with pots and pans:

> I have adopted the Principle of Least Action. It is a most clumsy machine in electromagnetics, but is splendid in the house; assisted by the old principle that prevention is better than cure. E.g., nasty job blacking boots. Don't black 'em; use tan boots. Fires is a most horrid nuisance, with the dirt and the work. Abolish them; use gas fires; no more trouble and labour.[165]

As usual, this private observation found its way into Heaviside's published work. Lumping it with the disappointing reaction of certain Cambridge mathematicians to his operational calculus, he declared in 1903:

> Whether good mathematicians, when they die, go to Cambridge, I do not know. But it is well known that a large number of men go there when they are young for the purpose of being converted into senior wranglers and Smith's prizemen. Now at Cambridge, or somewhere else, there is a golden or brazen idol called the Principle of Least Action. Its exact locality is kept secret, but numerous copies have been made and distributed amongst the mathematical tutors and lecturers at Cambridge, who make the young men fall down and worship the idol.
>
> I have nothing to say against the Principle. But I think a good deal may be said against the practice of the Principle. Truly, I have never practised it myself

165. Heaviside to Lodge, 30 Oct. 1899, quoted in G.F.C. Searle, *Oliver Heaviside, the Man*, (1987), p. 15.

> (except with pots and pans), but I have had many opportunities of seeing how the practice is done.[166]

With Heaviside, as always, one should look carefully beyond the sarcasm. A page later reveals that he had no objection to the principle's use on some occasions, but objected to giving it the status of an acceptable *physical* basis for dynamics. He also rejected the notion that Lagrange's equations and the principle of least action are somehow more powerful because one may use them to derive new equations of motion, while his own method always requires that these equations be given first:

> [The Principle of Least Action] is usually employed by dynamicians to investigate the properties of mediums transmitting waves, the elastic solid for example, or generalisations or modifications of the same. It is used to find equations of motion from energetic data. I observe that this is done, not by investigating the actual motion, but by investigating departures from it. ... Is not Newton's dynamics good enough? Or do not the Least Actionists know that Newton's dynamics, viz, his admirable Force = Counter-force and the connected Activity Principle, can be directly applied to construct the equations of motion in such cases as above referred to, without any of the *hocus-pocus* of departing from the real motion, or the time integration, or integration over all space, and with avoidance of much of the complicated work? It would seem not, for the claim is made for the P. of L. A. that it is a commanding general process, whereas the principle of energy is insufficient to determine the motion. This is wrong. But the P. of L. A. may perhaps be particularly suitable in special cases. It is against its misuse that I write.[167]

Heaviside proceeded to show that the information necessary for the derivation of new equations of motion using the principle of least action or Lagrange's equations also suffices for their derivation directly from the principle of activity.[168] The real basis for his criticism becomes clear yet another three pages later, after an outline of a simple and straightforward derivation of the

166. *Electromagnetic Theory*, Vol. III, p. 175.

167. *Electromagnetic Theory*, Vol. III, pp. 175-180, esp. pp. 175-176. As long as purely classical applications are concerned, it is difficult to discard Heaviside's point unequivocally—it really boils down to convenience and habit. With the benefit of modern hindsight, of course, the situation becomes dramatically different once one realizes that the "hocus pocus" departures from the real motion may be considered just as real, but less probable. See R.P. Feynman and A.R. Hibbs, *Quantum Mechanics and Path Integrals*, (New York: McGraw-Hill Book Company, 1965), pp. 28-30.

168. But see J.Z. Buchwald, "Oliver Heaviside, Maxwell's Apostle and Maxwellian Apostate," *Centaurus*, **28** (1985): 288-330.

Lagrangian generalized force by the activity principle. Referring to the derivation, Heaviside wrote:

> Some people who had worshipped the idol did not altogether see that the above contained the really essential part of the establishment of Lagrange's form, and that the use of the activity principle is proper, instead of *vice versa*. To all such, the advice can be given, Go back to Newton. There is nothing in the P. of L.A., or the P. of L. Curvature either, to compare with Newton for comprehensive intelligibility and straight correspondence with facts as seen in Nature.[169]

Thus, Heaviside wanted it made perfectly clear that the physics is in Newton, in the dynamical side of nature—not in an abstract mathematical Principle of Least Action that is derived from, but effectively hides the forces underlying the transference of energy. The Hamiltonian formulation may be a formidable problem-solving machine in certain cases, but it does everything directly in terms of the kinetic and potential energies of a system. It avoids explicit reference to energy-transferring forces. It had become all too mathematical. This is precisely what Heaviside would not accept given his approach to physics. In the Lagrangian and Hamiltonian mathematical machines he seems to have lost the physical mechanism that the force-based approach provided him with, even if it was only a mechanism by a dynamically sound analogy.

As we have seen (chapter III), the importance Heaviside attached to the principle of activity derived from two sources. Heaviside had been deeply influenced by Tyndall's exposition of heat not as a material substance, but as a state of matter.[170] However, unlike the case of heat, Heaviside did not treat electromagnetism as a mechanical interaction among the ultimate particles of matter. He was conscious of the difficulties associated with various attempts to create explicit electromagnetic mechanisms. He preferred instead to leave the mechanism latent and developed the dynamics in abstract terminology, independent of a particular mechanical setting. This, however, does not sufficiently explain the emphasis he laid on force. Heaviside never really explained why he considered the force formulation of the activity principle more physical than the available methods of analytical mechanics; but his work contains ample evidence to suggest that the reason goes back to his telegraphic career and persistent interest in electrical engineering. The reality he came into contact with through telegraphy was not mechanical, but electrical. It did not

169. *Electromagnetic Theory*, Vol. III, p. 178.

170. John Tyndall, *Heat: A Mode of Motion*, Fourth Edition (1870), p. 25.

consist of boilers, steam engines, shafts, pressure tanks, flywheels, and mechanical friction, but of batteries, dynamos, wires, capacitors, electromagnets, and electrical resistance. He did not constantly experience the pull of a piston, but the attraction of an electromagnet. In the principle of activity, Heaviside found a natural affinity to the tangible world of electrical circuits on one hand, and to the equally tangible world of physical forces on the other. The principle gave him a way of infusing Maxwell's theory with a sense of the reality he knew best. With the principle of activity, Heaviside could sketch the dynamical foundations of electromagnetism in terms most familiar to the practical telegraphist: voltage, current, resistance, inductance, and Ohm's and Ampère's laws. The principle expressed the relationship between these elements in the familiar terms of force, counter-force, resistance, inertia, energy, and power. Thus, with the principle of activity he could build the reality that he came to know as a telegraph operator into the very foundations of Maxwell's field theory. Conversely, he considered the adoption of an abstract principle of least action tantamount to divorcing electromagnetism from this tangible reality. Indeed, Heaviside opposed Maxwell's emphasis on the vector and scalar potentials because he realized that fields and fluxes, rather than potentials, related to voltage and current through the dynamical interaction of force and velocity.

The same emphasis on the need to distinguish between force and flux in a physical medium sharply distinguishes between Heaviside's and Hertz's views, despite the latter's strong preference for the field formulation and for vector algebra. In Hertz's system, the permittivity, resistivity, and inductivity of a medium are mere ratios, whose value is unity in the ether. To Heaviside, this was unacceptable. In December of 1890, he gave Hertz a piece of his mind:

> Your units of **E** and **H** have some advantages, manifestly; also some disadvantages,... what is, it seems to me, rather important is this. Can you conceive of a medium for el. mag. disturbances which has not at least *two* physical constants, *analogous to* density and elasticity? [The analogous constants are the inductivity and permittivity in Heaviside's electromagnetic nomenclature.] If not, is it not well to explicitly symbolize them, leaving to the future their true interpretation?[171]

171. Heaviside to Hertz, 8 Dec. 1890, in O'Hara and Pricha, *Hertz and the Maxwellians*, (1987), p. 80.

In his published work, Heaviside emphasized this point again:

> It is possible to so choose the electric and magnetic units that $\mu = 1$, $c = 1$ in ether; then μ and c in all bodies are mere numerics. But although this system (used by Hertz) has some evident recommendations, I do not think its adoption is desirable, at least at present. I do not see how it is possible for any medium to have less than two physical properties effective in the propagation of waves. If this be admitted, I think it may also be admitted to be desirable to explicitly admit their existence and symbolize them (not as mere numerics, but as physical magnitudes in a wider sense), although their precise interpretation may long remain unknown.[172]

Unlike his British counterparts, Hertz did not see Hamilton's principle as the primary principle of rational mechanics. Instead, he based his view on a geometrical principle of least constraint, which contained Newton's first law as a central pillar. In Hertz's system, forces were not primary entities but derived consequences of constraints.[173] Therefore, he did not require Heaviside's explicit view of a medium as the physical bearer of force and displacement. Heaviside, however, forged his solitary way through Maxwell's *Treatise* by stubbornly maintaining the dynamical resemblance between circuits and fields. In this endeavor, the activity formulation of Newton's third law turned out to be the crucial factor. “What I always regard as *the* fundamental principle of dynamics,” he wrote in 1911, “is Newton's celebrated Third Law. If that is not true, the result is Chaos.”[174] In Heaviside's uncompromising force-oriented system, the principle of activity, not an abstract extremum principle, associated all forces with their dynamical effects. Unlike Hertz, therefore, Heaviside simply had to have a tangible picture of the activity in a stressed medium:

> Some do not believe in the materiality of the ether. This view is thoroughly anti-Newtonian, anti-Faradaic and anti-Maxwellian. What mean action and reaction, storage of energy, the transit of force and energy through space &c,. &c., if there is no medium in space?[175]

Thus, if Heaviside's electrodynamics was unique among all other formulations of Maxwell's field theory, it should also be remembered that among the physicists who developed this theory, Heaviside alone came from a back-

172. *Electromagnetic Theory*, Vol. I, pp. 23-4.

173. Heinrich Hertz, *The Principles of Mechanics*, (1899), pp. 27-28.

174. *Electromagnetic Theory*, Vol. III, p. 477.

175. *Ibid.*, p. 479.

ground of practical telegraphy. He never abandoned his interest in circuit theory, and it never took second place to his interest in field theory. In some parts of his work, circuit notions appear as mere means to the end of developing field concepts; in others, field theory is merely a tool serving the development of circuit analysis. The two are quite inseparable in his work when taken as a whole.

Heaviside's commitment to the theory of electric circuits can be immediately perceived by the amount of space he devoted to it. His mathematical work gives further expression to this commitment. As we have seen, Heaviside developed the algebra of vectors as part of his field formulation of Maxwell's theory. His version of the operational calculus, by contrast, was motivated by circuit problems. Heaviside used operational methods as early as the late 1870s, but he really became excited about them in 1886 and 1887, after he found that when properly formulated, they provided a natural language for circuit analysis. With the resistance operator, Heaviside could write down the defining equations of any electrical network containing resistors, inductors, and capacitors using only the operational version of Ohm's law (see appendix 4.2 for more details). What he needed was an operational calculus, namely, a set of algorithms that would enable him to manipulate operational equations, solve them, and turn the solutions back into an algebraic form—a procedure he termed "algebrization." His intense efforts to develop such an operational calculus date back to his discovery of the operationally generalized Ohm's law. In other words, the operational approach seemed to hold the promise of a comprehensive and uniform framework for the analysis as well as solution of networks consisting of any combination of resistors, inductors, and capacitors. A similar promise of comprehensiveness played an important role in Heaviside's decision to adopt and develop the Maxwellian field-dynamical view. Not surprisingly, therefore, he found the operational calculus appealing. Indeed, it was the perfect complement to Maxwell's theory, for it promised a comprehensive *procedural* framework for the analysis and solution of circuit problems for which Maxwell's electrodynamics gave a comprehensive *theoretical* framework.

Also not surprisingly, Heaviside's Maxwellian friends showed little interest in his efforts to develop the operational calculus. The simultaneous differential equations that naturally occur in the study of electrical networks, and to which Heaviside carefully matched his particular methods, are not as common in field investigations. The ease with which the equation of a network is set up with resistance operators does not transform to systems not characterized by Ohm's law. Once again, a field theorist would have to invest a great deal of effort to assimilate a new technique that did not carry immediate promise of enhanced performance, and was beset with problems regarding its mathematical foundations.

The effects of Heaviside's experience as a practical telegraphist extended all the way to his most basic views concerning the scope and purpose of Maxwell's theory. Heaviside always considered Maxwell's theory as an incomplete scheme. He expressed this crisply in 1889 in a letter to Hertz, reflecting a theme that Duhem later turned into a cornerstone of his philosophy of science:

> But I only regard it [Maxwell's theory] as a sort of skeleton-framework; there are plenty of things it does not and cannot account for. Should we revise the theory? I think not much. It would become so cumbrous as to be unworkable. Let the auxiliary facts be tacked on, in the best way that presents itself. Possibly, however, it may require revision to some extent in the framework, but anything of that sort requires to be very cautiously done, on account of the disturbance made.[176]

As usual, Heaviside did not leave such an observation hidden in his private correspondence. Indeed, he had already made the basic observation—of which the above was merely an extension—in 1885:

> Certainly theory must ultimately be made to agree with facts; but when some few facts do not apparently fit into a theory which suits a much greater number of other facts, it becomes a question of balance of advantages whether it would be better to alter theoretical notions, or to leave the facts unexplained for the time, waiting for further information, or for new light on the question of fitting the facts into the theory.[177]

176. Heaviside to Hertz, August 14, 1889, in O'Hara and Pricha, *Hertz and the Maxwellians*, p. 72. Duhem argued that choices in physics cannot be boiled down to purely logical rules, and that an element of "good sense" is always resorted to. One such element of good sense involves sensitivity to the disturbance created in a theoretical framework as a result of modification. See P. Duhem, *The Aim and Structure of Physical Theory*, (1981), p. 217.

177. *Electrical Papers*, Vol. I, p. 418. This too, may be compared with Duhem's later assertion that one compares complex theoretical structures to complex sets of facts rather than single theoretical statements to singular facts. See P. Duhem, *The Aim and Structure of Physical Theory*, (1981), p. 200. Duhem and Heaviside part ways with Duhem's claim that when faced with such discordant situations, scientists make theoretical choices on the basis of "good sense" that cannot be logically justified. Heaviside, on the other hand, would rather suspend judgement until more information is available.

In 1893 he elaborated this point in a paragraph that illuminates the crucial influence of his years as a practical telegraphist:

> Whether a theory can be rightly described as too simple depends materially upon what it professes to be. The phenomena involving electromagnetism may be roughly divided into two classes, primary and secondary. Besides the main primary phenomena, there is a large number of secondary ones, partly or even mainly electromagnetic but also trenching upon other physical sciences. Now the question arises whether it is either practicable or useful to attempt to construct a theory of such comprehensiveness as to include the secondary phenomena, and to call it the theory of electromagnetism. I think not, at least at present. It might perhaps be done if the secondary phenomena were thoroughly known; but their theory is much more debatable than that of the primary phenomena that it would be an injustice to the latter to too closely amalgamate them.... The theory of electromagnetism is then a primary theory, a skeleton framework corresponding to a possible state of things simpler than the real in innumerable details, but suitable for the primary effects, and furnishing a guide to special extensions. From this point of view, the theory cannot be expressed too simply, provided it be a consistent scheme, and be sufficiently comprehensive to serve for a framework.[178]

The unavoidable question that comes to mind is, of course, what did Heaviside mean by primary and secondary phenomena? Nowhere in his work did he supply a direct answer to this question; but it can be gathered quite easily from the phenomena he did discuss, and from those he did not discuss. Throughout his five volumes, comprising nearly 2000 pages of electrical essays, Heaviside never discussed electrochemical phenomena and magneto-optic effects. He briefly discussed thermoelectric effects in the early 1880s, and later outlined an extension of Kelvin's thermoelectric theory, but never developed the subject in detail.[179] A significant part of his work is devoted to the study of electromagnetic waves. The majority of his work, however, revolves around electrical phenomena associated with electrical circuits, whether the discussion is in linear, or field terms. To Heaviside, then, electromagnetic field theory needed first of all to be suitable for the phenomena he encountered as a telegraphist. It extended to electromagnetic waves, but electromagnetic waves were not the main reason he found the theory attractive. Indeed, to a considerable extent, Heaviside pursued the study of electromagnetic waves

178. *Electromagnetic Theory*, Vol. I, Preface.

179. *Electrical Papers*, Vol. I, pp. 303-331; "Extension of Kelvin's Thermoelectric Theory," *Electromagnetic Theory.*, Vol III, pp.183-186.

not because he saw them as the outstanding feature of Maxwell's theory, but because their study enabled him to extend the analysis of circuit elements beyond the scope of linear circuit theory; it provided him with a comprehensive way of regarding an oscillating electric circuit, which reduced to the familiar linear form under the proper conditions.

Field theorists like FitzGerald, or even Joseph Larmor and J.J. Thomson, might have agreed with Heaviside that Maxwell's field theory is remarkably suited to the extended analysis of electrical circuits and their related phenomena. It is highly doubtful, however, that they would have shared Heaviside's research emphasis on these subjects. Most remarkable in this respect is the almost complete absence from Heaviside's work of references to the Hall effect and to the magneto-optic effects of Faraday and Kerr. Throughout the 1880s, the Hall and Faraday effects presented Maxwell's disciples with the most intractable difficulties that faced the field theory of electromagnetism. The intense efforts of FitzGerald, J.J. Thomson, Larmor and Lorentz to resolve these difficulties were closely associated with the most significant development in electromagnetic field theory since Maxwell published his *Treatise*. As Buchwald has shown, between 1894 and 1897 the electric charge and its conservation turned into fundamental principles in a reformulated microscopic field theory, built around the concept of the electron.[180] Statements such as the following from Heaviside (August, 1886) required significant modification:

> The line integral of the magnetic force round a wire measures the current in it, a fact that cannot be too often repeated, until it is impressed upon people that the electric current is a function of the magnetic field, which is in fact what we generally make observations upon, the electricity in motion through the wire being a pure hypothesis. Maxwell made this the universal definition of electric current anywhere.[181]

With the discovery of the electron, however, the electric current became the real motion of real charges through the wire. Rather than the current's exclusive definition, Maxwell's first circuital equation now took the form of a law of correlation between two fundamental, mutually irreducible physical entities: the electromagnetic field, and the electric current that consists of charges in motion. The electric displacement could now justifiably be reinterpreted

180. The role played by the Faraday and Hall effects in the transformation of Maxwell's theory from a macroscopic field theory to a microscopic theory of matter is discussed in J.Z. Buchwald, *From Maxwell to Microphysics*, (1985), pp. 73-98.

181. *Electrical Papers*, Vol. II, pp. 79-80.

as the real electrical polarization of a dielectric medium under the influence of an electric force. The plates of a capacitor could be regarded as storing real electric charges, not merely as surfaces on which lines of displacement terminate. This partially explains why Heaviside's "permittance" never displaced "capacitance," while rational units fulfilling his demand to relegate 4π to its proper place have eventually been adopted along with "impedance," "reluctance," "resistivity" and "inductivity."

Heaviside watched the transformation of Maxwell's theory with great interest, but always from the sidelines. When he commented, it was usually to express measured concern and sometimes displeasure at this turn of events. In 1886 he clearly formulated his doubts about the prospects of a molecular theory of magnetism:

> Although it is generally believed that magnetism is molecular, yet it is well to bear in mind that all our knowledge of magnetism is derived from experiments on masses, not on single molecules, or molecular structures. We may break up a magnet into the smallest pieces, and find that they, too, are little magnets. Still, they are not molecular magnets, but magnets of the same nature as the original; solid bodies showing magnetic properties, or intrinsically magnetised. We are nearly as far away as ever from a molecular magnet. To conclude that molecules are magnets because dividing a magnet always produces fresh magnets, would clearly be unsound reasoning.[182]

Thus, Heaviside did not object to the explicit formulation of microscopic theories because he believed as a matter of principle in continuum theories. However, as long as the physics of molecular processes remained unknown Heaviside saw no reason to pretend that Maxwell's equations were anything but a set of macroscopic relationships:

> The act of transition of elastic induction into intrinsic magnetisation, when a body is exposed to a strong field, cannot be traced in any way by our equations. It is not formulated, and it would naturally be a matter of considerably [sic] difficulty to do it.[183]

At the same time, he believed that Maxwell's theory must eventually be reformulated as a molecular theory. As he wrote to Hertz:

---

182. *Electrical Papers*, Vol. II, p. 39.
183. *Ibid.*, p. 42.

> No doubt ... the el. mag. theory of light is still only in an embryonic stage. It has to be molecular theory as well, and there is the difficulty. Every discovery opens out a fresh field of research, and there is no rest for the wicked.[184]

Heaviside was skeptical of attempts to formulate microscopic theories because he considered them premature in view of the state of knowledge at the time. He considered it possible that molecular physics would involve principles that may significantly differ from those based on studies of macroscopic systems. Given that possibility, he distrusted microscopic theories specifically formed for the purpose of reproducing macroscopic phenomena. Such theories, he felt, could create a false impression of sound knowledge that might actually consist of purely fictitious elements:

> It is the danger of a too special hypothesis, that as, from its definiteness, we can follow up its consequences, if the latter are partially verified experimentally we seem to prove its truth (as if there could be no other explanation), and so rest on the solid ground of nature. The next thing is to predict unobserved or unobservable phenomena whose only reason may be the hypothesis itself, one out of many which, within limits, could explain the same phenomena, though, beyond those limits, of widely diverging natures.[185]

Two things, he wrote, must always be remembered with respect to molecular theories of magnetization:

> First, that the molecular theory of magnetism is a speculation which it is desirable to keep well separated from theoretical embodiments of known facts, apart from hypothesis. And next, that as the act of exposing a solid to magnetising influence is, it is scarcely to be doubted, always accompanied by a changed structure, we should take into account and endeavour to utilise in theoretical reasoning on magnetism which is meant to contain the least amount of hypothesis, the elastic properties of the body, speaking generally, and without knowing the exact connection between them and the magnetic property.[186]

In other words, Heaviside advocated generalized dynamics as the only certain way to formulate sound physical knowledge. His attitude did not change with the emergence of the electron-based electromagnetic field theory. This is not to suggest that Heaviside was indifferent to these developments. Actually, the discovery of the electron had a profound influence on him. In the article on

---

184. Heaviside to Hertz, quoted in O'Hara and Pricha, *Hertz and the Maxwellians*, (1987), p. 68.

185. *Electrical Papers*, Vol. II, p. 41.

186. *Ibid.*, pp. 39-40.

telegraphy that he wrote for the *Encyclopaedia Britannica* in 1902, Heaviside already showed qualified willingness to withdraw from the position that rejects the image of the electric current as charge flowing inside the conductor:

> ... the possibility of a convective explanation of metallic conduction in harmony with Maxwell's theory became obvious when it was established that a moving charge in that theory was magnetically the same thing as a 'current element,' both in itself and under external magnetic force. Only quite lately, however, has it been possible to carry out this notion even tentatively. This has come about by the experimental researches which appear to establish the individuality of electrons of astonishing smallness and mobility. It is now believed by many that the conduction current inside a wire consists of a slow drift of electrons. Naturally, in the present state of ignorance about atoms and molecules, the theory is in an experimental stage. But it does not come into the telegraphic theory sensibly, since the electronic drift is a local phenomenon. It is stationary compared with the wave outside a wire. It may be noted that it is not necessary to consider the electronic drift to be the cause of the electromagnetic wave which has the drift for an after effect.[187]

By 1904 Heaviside converted his fictitious magnetic conductivity into the equally fictitious "magneton," and his notebooks indicate that he eagerly followed any suggestion that such a particle had been discovered.[188] At the same time, he toyed with the notion that the solar system model of a molecule might account for magnetism. This required some way of dealing with the expectation that an electron could not revolve steadily around a positive nucleus owing to a constant loss of energy by radiation:

> Of course, a body at constant temp. is receiving as much radiation as it emits, but that is a matter of averages. So then mag^{sm} a matter of averages too![189]

By and large, however, the electron only intensified Heaviside's feelings that a true molecular theory would bring about far more radical revisions than the one Larmor and Lorentz produced at the turn of the century. If the electron were to be regarded as reflecting an element of physical reality rather than merely as a useful theoretical concept, then searching questions must be asked

---

187. *Electromagnetic Theory*, Vol. III, p. 342.

188. Notebook 3, p. 563, Heaviside Collection, IEE, London. In 1921 he added a note to this page, observing "I find that 'magneton' is generally used to signify a little magnet or a little circular current of electrons. That is quite different from the meaning I attached to magneton as the magnetic analogue of electron."

189. Notebook 17, p. 138, Heaviside Collection, IEE, London.

about its physical properties, from the origin of its inertia to the possible connection between gravitation and electromagnetism:

> There are all sorts of puzzles about the electron. What is its constitution? is it associated with matter? If not, why doesn't it explode? There's a puzzler! Clearly it is not electricity + nothing else. For if electricity only, + only subject to the laws of electricity, it would break up, diffuse itself away ... by its own repulsion. It does not. Something holds it together. So I am inclined to think it is electricity + something else. Is the something else matter or is it ether? If so, why should a definite quantity be capable of free motion through the ether? Evidently there may be coming a revision of theory, as well as discovery of important facts. ... Electrons must have inertia, but have they any weight?[190]

"The mere idea," Heaviside wrote in this connection,

> that electromag. inertia might account for 'mass' occurred to me in my earliest work on moving charges but it seemed so vague & unsupported by evidence that I set it on one side. It explains too much & does not explain enough.[191]

Heaviside could not find any reason why there should be any limitation on the smallness of charge. Consequently, electrons could be composed of smaller parts, but then:

> How account for all moving together? Or what difference will it make on the average if they don't all move together? And if so, what holds them together as a whole? Nuts to crack everywhere.[192]

All of these thoughts and doubts found their way into his published work, where he wrote: "The 'electrical theory of matter,' which has some evidence to support it, is full of difficulties," and proceeded to discuss its various puzzles in the next three pages.[193] It appears, however, that these remarks never went beyond doubts and disjointed speculations. Despite claims to the contrary, no evidence has been found to suggest that Heaviside pursued these thoughts to actually produce the theoretical revisions he considered imminent.[194] He remained a doubtful spectator, watching with considerable fascination the advancing frontier of electromagnetic theory. At the same time, he kept producing additional detailed studies of what he saw as the primary facts,

---

190. Notebook 18, p. 334, Heaviside Collection, IEE, London.

191. *Ibid.*, p. 329.

192. *Ibid.*, p. 335. Notes made in 1902 show that Heaviside was also intrigued by Planck's partition formula and by radioactive effects (p. 343).

193. *Electromagnetic Theory*, Vol. III, pp. 476-78.

namely aspects of electric circuits, using the same dynamical approach he considered best suited for this purpose. Once again then, Heaviside's engineering background came between him and his closest colleagues. It explains why, when Heaviside is regarded through the history of the development of electromagnetic field theory, he appears at one and the same time as "Maxwell's apostle and Maxwellian apostate."[195]

8.2 ...and not at Home with the Engineers

It may seem from all of the above that Heaviside should really be regarded as an electrical engineer, who inadvertently wandered into the territory of Maxwellian field theorists. As our discussion shows, there is much to be said for Sumpner's observation that:

> Heaviside exemplifies a rare case of the combination of great theoretical and mathematical powers with a bias of mind that was strongly practical. ... He found that his work needed mathematics, and he trained himself to be a mathematician. He made himself a physical theorist for the same reason. He was, however, chiefly interested in the practical aspect of signalling problems. He regarded all

194. H.J. Josephs claimed that manuscripts found in Heaviside's home in Torquay contain parts of the unpublished 4th volume of *Electromagnetic Theory*, and that this material goes far beyond the disjointed speculations described above. Josephs described at length his interpretation of these manuscripts as reflecting a fundamentally new view of electromagnetism that Heaviside developed in the early 1900s (see H.J. Josephs, "Some Unpublished Notes of Oliver Heaviside," reprinted in O. Heaviside, *Electromagnetic Theory*, [1971], Vol. III, pp. 523-642, esp, pp. 525-526, 605-639). Unfortunately, there are no references at all in Josephs's lengthy discussion to specific manuscripts, and other historians have failed so far to find any support for his claims in the Heaviside Collection at the IEE. In 1977 B.R. Gossick savagely criticized Josephs's work as wildly speculative and historically unfounded, and chided the Chelsea Publishing Company for including Josephs's work in the 1971 edition of *Electromagnetic Theory* (see B.R. Gossick, "Where is Heaviside's Manuscript for Vol. 4 of His Electromagnetic Theory?" *Annals of Science*, **34** (1977): 601-606). Gossick's criticism is marred by claims that during the last fifteen years of his life Heaviside went mad and could not possibly produce scientific work of any value, while the historical record supplies at best contradictory evidence in support of this claim. Heaviside's fascination with the possibilities opened by the discovery of the electron is well documented. As Josephs himself observed, however, Heaviside's published work contains no hint of the far reaching reforms Josephs claimed on behalf of the missing 4th volume. Without further evidence that Heaviside actually proceeded to systematically develop his scattered speculations, the question should probably be left open.

195. J.Z. Buchwald, "Oliver Heaviside, Maxwell's Apostle and Maxwellian Apostate," *Centaurus*, **28** (1985): 288-330.

> theoretical work as subsidiary. He was a mathematician at one moment, and a physicist at another, but first and last, and all the time, he was a telegraphist.[196]

Indeed, it was within the engineering community that Heaviside found the most loyal support for many of the reforms he wished to bring about. John Perry and W.E. Ayrton proved to be far more sympathetic listeners than Lodge and FitzGerald to his campaign for rational units that properly locate 4π. Lodge in particular found himself the target of a typically caustic Heavisidean castigation in this connection, while John Perry received enthusiastic praise—a rare gift from Oliver Heaviside.[197] While Hertz was the first influential European scientist to adopt the vector formulation of Maxwell's theory, it found its first firm foothold in England among engineering authorities like Perry and Ayrton. To this very day, the vector formulation in terms of the electromagnetic fields is most emphatically used in engineering texts. The potential formulation with the associated exposition of the Lorentz gauge—so crucial to modern field theory—is mentioned in passing, mostly as a convenient vehicle for calculation in certain cases.[198]

A similar pattern emerges with respect to Heaviside's operational calculus. T.J.I'A. Bromwich turned it into a respectable subject of investigation for Cambridge mathematicians. However, the attention of leading engineering authorities such as John Perry, V. Bush, J.R. Carson, E.J. Berg and W.E. Sumpner drastically extended interest in the operational calculus between the mid-1890s and the 1930s.[199] Heaviside himself was particularly pleased with Per-

196. W.E. Sumpner, "The Work of Oliver Heaviside" (Twenty-Third Kelvin Lecture), *Journal of the Institution of Electrical Engineers*, **71** (1932): 837.

197. *Electromagnetic Theory*, Vol. II, pp. 282-285.

198. "As we have seen, time-varying electromagnetic fields are related to each other and to the charge and current sources through the set of differential equations known as Maxwell's equations. It is sometimes convenient to introduce some intermediate functions, known as potential functions, which are directly related to the sources, and from which the electric and magnetic fields may be derived." (S. Ramo, J.R. Whinnery and T. Van Duzer, *Fields and Waves in Communication Electronics*, [1984], p. 156.) Compare this statement to Heaviside's attitude toward the potential formulation, as he expressed it to Hertz: "Speaking of potentials however, I am reminded of your own remarks on the subject. You would abolish them altogether from wave questions. I do not go quite so far myself, because there are occasions when the vector potential is an assistance..." (Heaviside to Hertz, quoted in O'Hara and Pricha, *Hertz and the Maxwellians*, p. 67). Heaviside expressed the same sentiment in his published work: "The clearest course to pursue appears to me to invariably make **E** and **H** the primary objects of attention, and only use potentials when they naturally suggest themselves as labour-saving appliances." (*Electrical Papers*, Vol. II, p. 513)

ry's interest in the operational calculus, because he saw Perry as a "practical physicist":

> It has naturally given me much pleasure to find that the method in question [the operational solution of the heat equation] ... should receive such ready appreciation from a practical physicist. ... It is the fact that he is a practical physicist, without mathematical pretensions, that constitutes the importance of the phenomenon.[200]

The very approach that Heaviside adopted in his exposition of Maxwell's theory beginning in 1882 reflects the affinity between his work and the preferences of an influential engineering eductor such as Ayrton. Heaviside began his exposition of Maxwell's theory with a discussion of the magnetic field, the electric current, and Stokes's theorem, rather than with electrostatics. We have seen him stating that his initial introduction to electricity and magnetism was through the dynamical effects. Compare this to Ayrton's introduction to his own textbook on electromagnetism:

> Readers who have been accustomed only to the ordinary books, commencing with certain chapters on statical electricity, continuing with one or more on magnetism and ending with some on current electricity, will be surprised at the arrangement of the subjects in this book, and will probably be astonished at what they will condemn, at the first reading, as a total want of order. But so far from the various subjects having been thrown together hap-hazard, the order in which they have been arranged has been a matter of the most careful consideration, and has been arrived at by following what appears to me to be the *natural* as distinguished from the *scholastic* method of studying electricity. ... The subject of current is treated first, because in almost all the industries in which electricity is practically made use of, it is the electric current that is employed; secondly, because currents can be compared with one another, and the unit of current (the ampere) defined without any knowledge of potential difference or resistance.[201]

---

199. T.J.I'A. Bromwich, "Normal Coordinates in Dynamical Systems", *Proceedings of the London Mathematical Society*, Series 2, **15** (1916): 401-448; "Examples of Operational Methods in Mathematical Physics, *The Philosophical Magazine*, **37**: 407-419; "Symbolical Methods in the Theory of Conduction of Heat", *Proceedings of the Cambridge Philosophical Society*, **20** (1921): 411-427. John Perry, "On Gamma Function and Heaviside's Operators", *The Electrician*, **34** (Jan. 25, 1895): 375-376; *The Calculus for Engineers*, (1897), esp. pp. 230-242. V. Bush, *Operational Circuit Analysis*, (1929). E.J. Berg, *Heaviside's Operational Calculus as Applied to Engineering and Physics*, (1936). J.R. Carson, *Electric Circuit Theory and the Operational Calculus*, (1926). W.E. Sumpner, "Heaviside's Fractional Differentiator" *Proceedings of the Physical Society of London*, **41** (1929): 404-425.

200. *Electromagnetic Theory*, Vol. II, p. 12.

Having said all that, however, it would be as wrong to turn Heaviside into a mathematically sophisticated electrical engineer as to consider him a theoretical physicist with a penchant for electrical engineering. Half of his published work is directly devoted to the study of Maxwell's field theory, and practically all of the rest is related to it in one way or another. Heaviside was not very interested in the practices of the engineer in the field. This is clearly shown by his collaboration with A.W. Heaviside. A.W. Heaviside designed and built the circuits; Oliver Heaviside supplied the basic design theory to guide and justify the work. Even his publications from the 1870s exhibit this character. He was less interested in specific designs than in basic questions such as the maximum sensitivity of the Wheatstone Bridge, the theory of transmission lines with inductance included, or the theory of electromagnets. In the same way he wanted a general theory of the electric current, and this sustained his interest in Maxwell's *Treatise* not merely as a reference book for various problems in mathematical electromagnetism, but as a starting point for the systematic understanding of artificial electrical systems. Both his interest in circuits and in field theory were expressions of a basic interest in engineering science. Heaviside wanted to understand the basic principles behind the engineering systems he worked with during his six years as a telegraph operator. This desire motivated nearly all of his *Electrical Papers*, and the major part of *Electromagnetic Theory*.

Unlike Heaviside, however, his engineering contemporaries rarely devoted their life to the elucidation of basic engineering theory. The term "engineering science" was only coined in the 1850s,[202] and basic engineering research was in its infancy when Heaviside produced the bulk of his work.[203]

---

201. W.E. Ayrton, *Practical Electricity: A Laboratory and Lecture Course for First Year Students of Electrical Engineering based on the Practical Definitions of the Electrical units*, (1896), p. v.

202. See David F. Channell, "The Harmony of Theory and Practice: The Engineering Science of W.J.M. Rankine," *Technology and Culture*, **23** (1982): 39-52. The explicit realization that telegraph engineering could benefit from a theorist "who was at the same time a practical man" also dates to the same time. It was expressed in 1857 in a letter to Kelvin from a former colleague of his, Lewis Gordon, who was working at the Submarine Telegraph Works at Birkenhead (C. Smith and N. Wise, *Energy & Empire: A Biographical Study of Lord Kelvin*, [1989], p. 666).

203. Basic industrial research marked by a conscious effort to formulate limited-reference theories was emerging more or less concurrently in the realm of electric power engineering. See Ronald Kline, "Science and Engineering Theory in the Invention and Development of the Induction Motor," *Technology and Culture*, **28** (1987): 283-313 (p. 309).

When such research took place it was usually on the basis of specific, momentary need. Formal, institutionalized engineering research appeared only in the first decade of the twentieth century, with the establishment of the first dedicated industrial laboratories. As we have seen, Heaviside's telegraphic background gave his electromagnetic theory a flavor that his theoretical colleagues found strangely remote. At the same time, his uncompromising Maxwellian views, sophisticated mathematical techniques, and interest in basic questions made his work uninteresting to most practicing engineers. The same characteristics made his work inaccessible to the few engineers who did detect the potential interest it contained. As John Perry complained, he admired Heaviside from a distance, but needed someone to write Heaviside down to his own level. Heaviside himself was no pure theoretician; but he had no wish to be a practician either. He expressed a desire to be practical without being a practician:

> The very useful word "practician" has lately come into use. It supplies a want, for it is evident the moment it is mentioned that a practician need not be a practical man; and that, on the other hand, it may happen occasionally that a man who is not a practician may still be quite practical.[204]

8.3 *Alone in the Middle*

In 1863, Maxwell called attention to a gap between pure scientific knowledge and practical knowledge, and outlined what was needed to fill it:

> The progress and extension of the electric telegraph has made a practical knowledge of electric and magnetic phenomena necessary to a large number of persons who are more or less occupied in the construction and working of the lines.... The discoveries of Volta and Galvani, of Oersted, and of Faraday are familiar in the mouths of all who talk of science, while the results of those discoveries are the foundation of branches of industry conducted by many who have perhaps never heard of those illustrious names. Between the student's mere knowledge of the history of discovery and the workman's practical familiarity with particular operations which can only be communicated to others by direct imitation, we are in want of a set of rules, or rather principles, by which the laws remembered in their abstract form can be applied to estimate the forces required to effect any given practical result.[205]

What engineers lacked, Maxwell observed, were systematic rules by which to

204. *Electromagnetic Theory*, Vol. I, p. 2.

guide the application of scientific knowledge to technological ends; but he did not proceed to formulate these required rules. His appendix to the *Report of the British Association on Standards of Electrical Resistance* represents only a first step toward the eventual establishment of such rules, namely, the standardization of electrical measurements.[206] Heaviside, rather than Maxwell, went beyond measurement standardization, and showed with example after painstaking example how such systematic rules are to be formulated. As Maxwell clearly understood, this sort of work belonged neither to the realm of pure science nor to the domain of practical engineering. It occupied a middle ground between them that was largely uncultivated when Heaviside began his work. As Heaviside discovered in the course of his work, the cultivation of this middle ground required its own set of tools. His successful attempts to develop such tools and to apply them to the formulation of basic engineering knowledge were largely responsible for the high regard in which he eventually came to be held by the electrical engineers of the twentieth century. During his own lifetime, however, he found himself preoccupied with the exploration of a rather deserted territory into which both scientists and engineers seldom ventured to wander.

It seems, in the end, that Heaviside spent his life looking for a niche he could not find. In that, he perfectly complemented the constant expressions of incomprehension that accompanied his work and the constant inability of others to agree on whether he should be classified as a mathematician, a physicist, or an engineer. If Heaviside could not find his niche, however, it was to a large extent because the world he lived in did not supply him with one. In the preface to the first volume of *Electromagnetic Theory*, behind a veil of caustic humor, Heaviside actually expressed his dismay at the absence of a category within which he could feel comfortable:

> I had occasion just lately to use the word 'naturalist'. The matter involved here is worthy of parenthetical consideration. Sir William Thomson does not like

205. James C. Maxwell and Fleeming Jenkin (1863), "On the Elementary Relations between Electrical Measurements," *The British Association Reports on Standards of Electrical Resistance*, (Cambridge: at the University Press, 1913), p. 130.

206. "All exact knowledge is founded on the comparison of one quantity with another. In many experimental researches conducted by single individuals, the absolute values of those quantities are of no importance; but whenever many persons are to act together, it is necessary that they should have a common understanding of the measures to be employed. The object of the present treatise is to assist in attaining this common understanding as to electrical measurements." *Ibid.*, p. 131.

> 'physicist', nor, I think, 'scientist' either. It must, however, be noted that the naturalist, as at present generally understood, is a student of living nature only. He has certainly no exclusive right to so excellent a name. On the other hand, the physicist is a student of inanimate nature, in the main, so that he has no exclusive right to the name, either. ... For my part I always admired the old-fashioned term 'natural philosopher'. It was so dignified, and raised up visions of the portraits of Count Rumford, Young, Herschel, Sir H. Davy, &c., usually highly respectable-looking elderly gentlemen, with very large bald heads, and much wrapped up about the throats, sitting in their studies pondering calmly over the secrets of nature revealed to them by their experiments. There are no natural philosophers now-a-days. How is it possible to be a natural philosopher when a Salvation Army band is performing outside; joyously, it may be, but not most melodiously? But I would not disparage their work; it may be far more important than his.[207]

By 1891, the most creative period in Heaviside's scientific career was drawing to a close. His three magnificent volumes of *Electromagnetic Theory* were still ahead of him, but his own observation that for the most part they merely represented development of his original contributions in *Electrical Papers* is basically sound. Furthermore, the character of his work in the *Electrical Papers* and the events of 1886 to 1889 left an indelible mark on *Electromagnetic Theory.* The association between field theory and circuit theory that gave the *Electrical Papers* their unique flavor was further enhanced in *Electromagnetic Theory.* But rather than systematically explaining how engineering concerns shaped his formulation of Maxwell's theory, Heaviside took them for granted and developed his *Electromagnetic Theory* accordingly. His one exposition of these concerns in the form of his disrupted presentation of 1886 and 1887 was never completed. Thus, Heaviside's post-1891 work essentially served to entrench rather than clarify the characteristics that prompted expressions of frustration and incomprehension from his contemporaries.

By the time his *Electrical Papers* were published, Heaviside had earned the recognition and respect of the greatest scientific authorities of his day. But the novel and complex character of his unique interdisciplinary approach to electromagnetism was not properly understood. His work acquired a mysterious aura that was all the more intense for the respect it commanded. With Kelvin's Inaugural Address in 1889, Heaviside successfully emerged from obscurity; but he never dispelled the enigma that surrounded his work.

---

207. *Electromagnetic Theory,* Vol. 1, pp. 4-5.

APPENDIX 3.1: Heaviside's Extended Theorem of Divergence

In the first volume of *Electromagnetic Theory*, Heaviside introduced an extended version of the theorems of Gauss and Stokes. I have not seen these extensions in any other text. Perhaps this is accountable in part because the extensions are not too important from the practical point of view, and in part because Heaviside's texts never enjoyed a wide readership. Yet, the extensions have at least a valuable educational lesson without complicating the basic argument that leads to the more restricted forms of Gauss and Stokes. Only the extended theorem of divergence will be discussed here. The interested reader may find the extended theorem of version (Stokes's) in *Electromagnetic Theory*, Vol. I, pp. 191-193.

In its familiar form, the theorem of divergence states that given a vector function $\mathbf{F}(x,y,z)$, well defined and differentiable throughout space, there exists a scalar function div($\mathbf{F}$) such that for any closed surface S in the vector field $\mathbf{F}$

$$\int_S \mathbf{F} \cdot \mathbf{da} = \int_V \operatorname{div} \mathbf{F} dv. \tag{3.1-1}$$

The first integral is taken over the surface of S, the second one is reckoned throughout the volume V enclosed by the surface. One reckons the direction of the infinitesimal area **da** along the outward pointing normal to S at that point.

The theorem may of course be proved as a purely abstract statement of vector analysis, but this would defy Heaviside's spirit. He would probably suggest that one first consider **F** as representing the three dimensional flow of a material fluid. This would immediately suggest a meaning as well as a physical motivation behind the theorem. The surface integral measures the net amount of flow through the surface S. It adds the amount of fluid existing S at any moment to the amount entering it. If the total is zero, then the two amounts exactly balance each other, with no net flow in or out of the region bounded by S. If positive, the integral measures the excess amount leaving over the amount entering; if negative, it measures the excess fluid entering S over the amount leaving it. Considering that, it would appear that the volume integral on the right sums up the rate at which fluid is taken out of the flow or added to it within the enclosed volume, depending on whether the surface integral is negative or positive. In other words, assuming the fluid is incompressible, the scalar function div(**F**) measures the production or destruction of fluid at a particular point in space. One may say that it expresses the divergence of **F** at that point.

It should be clear that eq. (3.1-1) expresses the physically intuitive notion that if a greater amount of fluid enters the region bounded by S than the amount leaving it, then, being incompressible, the excess fluid must somehow be totally removed from the flow system. From that point of view, the theorem hardly wants further proof. What does require proof is the claim that vectors can meaningfully represent such a system of flow. Thus, in Heaviside's hands eq. (3.1-1) becomes a prior requirement vectors must satisfy if they are to provide a useful language for investigating physical flow systems. He would probably have claimed that vectors were originally invented and designed in just such a manner that they would satisfy it.

Strictly speaking then, his discussion is not a proof of the theorem in the usual sense of the word. He never bothered to boil it down to the rigorous basis of the differential and integral calculus by reckoning limits of ratios between infinitesimal quantities. Perhaps the argument is best described as a reassurance of the intuition. It is an illustration of the vector's fitness to serve as a representative of the flow system rather than a full-fledged proof.

The argument, up to a point, is the one every beginning physics student must have encountered. First, divide S into two (not necessarily equal) parts. Naturally, the two parts share a common surface. Take the integral of **F** over each of the two attached closed surfaces and then add them up. Note that **F** gets integrated over the common surface twice; once for each of the subdivisions. Obviously, **F** is quite unaffected by the imaginary surfaces, so it is reckoned in exactly the same manner in the two integrations over the common surface. However, the direction of each element **da** of the common surface reckoned for the first subdivision is exactly opposed to its direction when reckoned for the second subdivision. Hence, all over the common surface each element $\mathbf{F} \cdot \mathbf{da}$ in the integration over the first subdivision has a countering $-\mathbf{F} \cdot \mathbf{da}$ in the integration over the second subdivision. When the two surface integrals are added, the contributions of the common surface annul each other, leaving only contributions from the original surface S. The same argument will ensure that the sum of all subdivisions of S, regardless of how numerous, will always add up to the surface integral of **F** over the original boundary S.

With the above in mind, the entire volume bounded by S may be subdivided into infinitesimal volume elements, each bounded by an infinitesimal closed surface. The integral of **F** over each such surface may be related to the enclosed volume dv by a number, div(**F**), such that div(**F**)dv is equal to the integral of **F** over dv's bounding surface. Naturally, the value of div(**F**) will vary from volume element to volume element depending on the spatial variation of **F**. In other words, considering the infinitesimal size of the volume elements, div(**F**) varies practically from point to point within the volume bounded by S. We may therefore regard div(**F**) as a well-defined scalar function of space within S. Now, the sum of div(**F**)dv over the entire volume bounded by S is equal to the sum of all the surface integrals over the infinitely many surface subdivisions of S, which by the preceding argument is equal to the integral of **F** over S. This amounts to the statement made by the divergence theorem. There is indeed good intuitive reason to believe that for any well behaved vector function **F** there exists a scalar function div(**F**) that satisfies equation (3.1-1).

Note that unlike many proofs of Gauss's theorem, this one makes no reference to any coordinate system, and does not require an explicit calculable expression for div(**F**). It merely argues that div(**F**) must exist. This distinction between the theorem and the explicit expression of div(**F**) in a specific coordinate system enabled Heaviside to concentrate on a crucial element that makes the theorem possible. It is the fact that the scalar product of **F** and **da** will become its own negative if **da** is turned by 180°. In other words, if **n** is a unit vector in the direction of the outward going normal to S at **da**, we have:

$$\mathbf{F}\cdot\mathbf{da} = (\mathbf{F}\cdot\mathbf{n})\,da = (-\mathbf{F})\cdot(-\mathbf{n})\,da,$$

or:

$$\mathbf{F}\cdot(-\mathbf{n})\,da = -(\mathbf{F}\cdot\mathbf{n})\,da.$$

Indeed, let $G(x,y,z,\mathbf{n})$ be any function, vector or scalar, well defined over any surface characterized by a normal **n**, such that $G(-\mathbf{n}) = -G(\mathbf{n})$. Note that in the specific case of Gauss's theorem, the general function $G(x,y,z,\mathbf{n})$ becomes $\mathbf{F}(x, y, z)\cdot\mathbf{n}$. Under this single condition the integral of $G(\mathbf{n})$ over any closed surface S will be expressible as a volume integral due to the same reasoning that enabled passage from a surface to a volume integral in the case of

Gauss's theorem. That is to say, there will always be some function $H(G)$ such that:

$$\int_S G(\mathbf{n})\,da = \int_V H(G)\,dv.$$

This is Heaviside's extended theorem of divergence, and it is instructive to see what form $H(G)$ takes in Cartesian coordinates. Consider an infinitesimal

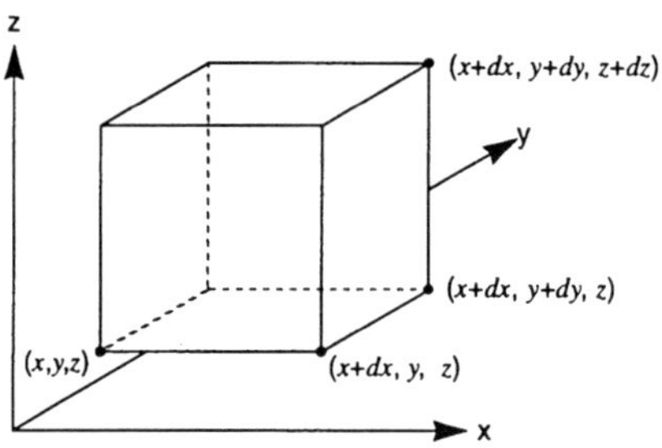

rectangle of sides dx, dy, dz situated somewhere in the range of G. Let its forward facing, lower left corner have the coordinates (x,y,z). The rectangular face closest to us is oriented in the negative y direction, and its area is $dxdz$.[1] Denote the average value of G over this small surface by $G(x,y,z,-\mathbf{j})$, and then the integral of G over the surface is simply:

$$G(x, y, z, -\mathbf{j})\,dxdz = -G(x, y, z, \mathbf{j})\,dxdz.$$

Now consider the opposite face. Its area is once again $dxdz$, but its normal points in the positive y direction. Also, while in reckoning the average value of G over this area we run through the same values of x and z, the value of y changes throughout the surface to $y + dy$. For the integral of G over the positive y surface we may therefore use:

$$G(x, y+dy, z, \mathbf{j})\,dxdz = G(x, y, z, \mathbf{j})\,dxdz + \frac{\partial G}{\partial y}\,dxdydz,$$

1. There is considerable latitude in defining the coordinate system. The above is best envisioned as a left-handed system, in which the upward pointing thumb marks the positive z direction, the forward pointing first finger defines the positive y direction, and the sideways pointing second finger defines the positive x direction.

and the sum of the integral of G over both surfaces becomes:

$$\frac{\partial G(\mathbf{j})}{\partial y} dxdydz = \frac{\partial G(\mathbf{j})}{\partial y} dv.$$

With similar reasoning applied to the surfaces oriented in the x and z directions, the sum of G over the little rectangular surface bounding dv amounts to:

$$\left(\frac{\partial G(\mathbf{i})}{\partial x} + \frac{\partial G(\mathbf{j})}{\partial y} + \frac{\partial G(\mathbf{k})}{\partial z}\right) dv,$$

which implies:

$$H(G) = \frac{\partial G(\mathbf{i})}{\partial x} + \frac{\partial G(\mathbf{j})}{\partial y} + \frac{\partial G(\mathbf{k})}{\partial z}. \tag{3.1-2}$$

In Cartesian coordinates then, Heaviside's extended theorem of divergence acquires the form:

$$\int_S G(\mathbf{n})\, da = \int_V \left(\frac{\partial G(\mathbf{i})}{\partial x} + \frac{\partial G(\mathbf{j})}{\partial y} + \frac{\partial G(\mathbf{k})}{\partial z}\right) dv.$$

In the particular case where $G(\mathbf{n}) = \mathbf{F}(x, y, z) \cdot \mathbf{n}$, this becomes:

$$\begin{aligned} H(G) &= \frac{\partial(\mathbf{F} \cdot \mathbf{i})}{\partial x} + \frac{\partial(\mathbf{F} \cdot \mathbf{j})}{\partial y} + \frac{\partial(\mathbf{F} \cdot \mathbf{k})}{\partial z} \\ &= \frac{\partial F_x}{\partial x} + \frac{\partial F_y}{\partial y} + \frac{\partial F_z}{\partial z}, \end{aligned}$$

which is the well-known expression for div(**F**) in Cartesian coordinates.

Many a student of electricity and magnetism will probably recall two related exercises that were given after the exposition of Gauss's theorem. They involved finding an expression for the surface integral of a scalar function as a volume integral, and then doing the same for the surface integral of the vector product of a vector function with each surface element.[2] Given only the restricted form of the theorem of divergence, the exercise requires conjuring a little "trick." Take a scalar function $P(x,y,z)$. Since the restricted theorem of divergence deals exclusively with the scalar product of vector functions, we

2. See for example, Melvin Schwartz, *Principles of Electrodynamics*, (New York: McGraw-Hill Book Company, 1972), p. 26 (prob. 1.6).

must somehow twist the scalar function for which the volume integral is sought into a vector function, then apply to it the theorem of divergence, then re-extract the scalar function. Letting **A** denote a constant vector, write:

$$\mathbf{A}\cdot\int_S P\mathbf{da} = \int_S (P\mathbf{A})\cdot\mathbf{da}$$

$$= \int_V \operatorname{div}(P\mathbf{A})\,dv \quad \text{(by the theorem of divergence)}$$

$$= \mathbf{A}\cdot\int_V \operatorname{grad}(P)\,dv$$

$$\Rightarrow \int_S P\mathbf{da} = \int_V \operatorname{grad}(P)\,dv.$$

The same principle employed in a more complicated manner holds for transforming the surface integral of $\mathbf{F}\times\mathbf{n}da$ into a volume integral. We take the integral's scalar product with a constant vector function **A**, manipulate it using a vector relationship into a form that involves $(\mathbf{F}\times\mathbf{A})\cdot\mathbf{n}$, then apply to it the theorem of divergence, and finally get rid of the **A** by some more manipulations.

There is much to be said for such exercises, for they illustrate how to work around certain limitations by thinking in terms of a given theorem and by manipulating the formal properties of vector algebra. At the same time, Heaviside's extended theorem possesses a different pedagogical value of equal importance. It demonstrates that thinking *about* a particular theorem is just as beneficial as thinking *in terms* of it. None of the trickery involved in the exercise above is necessary once Heaviside's extension is understood. In the first case, we have $G(x,y,z,\mathbf{n}) = P(x,y,z)\mathbf{n}$ and eq. (3.1-2) immediately implies that $H(G) = \text{grad}(P)$. In the second case, $G = \mathbf{F}\text{x}\mathbf{n}$, which implies almost as quickly that $H(G) = \text{-curl}(\mathbf{F})$.

APPENDIX 3.2: **Unification of Electricity and Magnetism**[1]

Heaviside began innocently enough with a calculation of the relationship between a current distribution **C** and its related vector potential and magnetic field. To follow his reasoning, look first at one element of **C**, say, **c**. Orient a Cartesian coordinate system such that its origin coincides with the point at which **c** is measured, and its z-axis coincides with **c**. In this system $\mathbf{c} = (0,0,c)$. The vector potential **a** of **c** at some point (x,y,z) is:

$$\mathbf{a}(x, y, z) = \left(0, 0, \frac{c}{4\pi r}\right),$$

$$\text{where } r = \sqrt{x^2 + y^2 + z^2}.$$

The magnetic field **b** due to **c** is curl(**a**), namely:

$$\mathbf{b}(\mathbf{r}) = \det\begin{bmatrix} \mathbf{e_x} & \mathbf{e_y} & \mathbf{e_z} \\ \frac{\partial}{\partial x} & \frac{\partial}{\partial y} & \frac{\partial}{\partial z} \\ 0 & 0 & \frac{c}{4\pi r} \end{bmatrix} = \frac{c}{4\pi}\left(\frac{\partial}{\partial y} r^{-1}\mathbf{e_x} - \frac{\partial}{\partial x} r^{-1}\mathbf{e_y}\right). \qquad (3.2\text{-}1)$$

The above is a straight forward application of $\mathbf{B} = \text{curl}(\text{pot}(\mathbf{C}))$, and the vectors $\mathbf{e_x}$, $\mathbf{e_y}$ and $\mathbf{e_z}$ are unit vectors in the x, y, and z directions respectively.[2]

The problem Heaviside really wanted to solve involves the magnetic field of a tiny circular current loop. However, the magnetic field of the straight linear current component just discussed provides all the information necessary for solving for the field due to the current loop. The first thing to note is that curl(pot(**c**)) = pot(curl(**c**)). To find the curl of the current element **c** consider its current as flowing out of a little cylinder of base radius a, and symmetry

1. *Electrical Papers*, Vol. I, pp. 218-223.

2. To stay as close as possible to Heaviside's discussion, I used the same symbols and conventions he used in the original article. It should be noted that **B** above is not the magnetic induction, but the magnetic "force" that Heaviside later symbolized by **H**. It is also safe to assume that were Heaviside to rewrite the article, he would have used his own electromagnetic units so as to eliminate the factor of 4π that keeps recurring throughout the original article.

axis parallel to **c**. The current density through the top and bottom of the cylinder is evidently $c/\pi a^2$ in the z direction. The curl of such a current distribution circulates right handedly around the curved lateral surface of the little cylinder and its strength is $c/\pi a^2$ per unit height of the cylinder. Therefore, using the commutativity of pot and curl, we have:

$$\begin{aligned}\text{pot}\left(\frac{\mathbf{c}}{\pi a^2}\right) &= \text{pot}\,(\text{curl}\,\mathbf{c})\\ &= \text{curl}\,(\text{pot}\,\mathbf{c})\\ &= \mathbf{b},\end{aligned}$$

where **b** is defined by eq. (3.2-1) above. With this in mind, the problem of finding the magnetic field of a small circuit carrying a current C is practically solved. As usual, **B** = curl(pot(**C**)); but now **C** circulates in a small circle just like the curl of the linear element **c** of the previous problem, except that the latter is divided by the area of surface it bounds. Thus, pot(**C**) = **b**dS, where dS is the area enclosed by the path of the current loop **C**, and **b** is defined by eq. (3.2-1) with C substituted for c. Remember, however, that **b** no longer signifies a magnetic field. The quantity **b**dS must be interpreted according to the physics of the new problem, and it now signifies the vector potential of the tiny current loop **C**. To find the magnetic field, we must take the curl of this vector potential. Hence:

$$\begin{aligned}\mathbf{B} = \text{curl}\,(\mathbf{b}dS) &= \frac{CdS}{4\pi}\,\text{curl}\left(\frac{\partial}{\partial y}r^{-1}, -\frac{\partial}{\partial x}r^{-1}, 0\right)\\ &= \frac{CdS}{4\pi}\left(\frac{\partial}{\partial z}\frac{\partial}{\partial x}r^{-1}, \frac{\partial}{\partial z}\frac{\partial}{\partial y}r^{-1}, -\frac{\partial^2}{\partial x^2}r^{-1} - \frac{\partial^2}{\partial y^2}r^{-1}\right).\end{aligned}$$

The z-component of **B** may be written as:

$$B_z = -\left(\frac{\partial^2}{\partial x^2}r^{-1} + \frac{\partial^2}{\partial y^2}r^{-1} + \frac{\partial^2}{\partial z^2}r^{-1}\right) + \frac{\partial^2}{\partial z^2}r^{-1}.$$

The expression in parentheses expresses the convergence of the radial vector field $\mathbf{r}/r^3$. It is the field of a single point charge situated at the origin, and its convergence is everywhere zero except at the origin. Therefore, except for a single point, **B** may be rewritten as:

$$\mathbf{B} = \left(\frac{\partial}{\partial x}, \frac{\partial}{\partial y}, \frac{\partial}{\partial z}\right)\frac{CdS}{4\pi}\frac{\partial}{\partial z}r^{-1}.$$

Thus, $\mathbf{B} = -\mathrm{grad}\Omega$, where:

$$\Omega = -\frac{CdS}{4\pi}\frac{\partial}{\partial z}r^{-1}.$$

The basic relationship between current and magnetic field is $\mathbf{C} = \mathrm{curl}(\mathbf{B})$. In this particular case curl(**B**) is zero everywhere except along the current-loop; but then, wherever the curl of a vector is zero, the vector may be derived as the negative gradient of a scalar function, which we have come to call the scalar potential. Hence, Ω above is the magnetic scalar potential from which **B** may be derived everywhere outside the current loop. Note, however, that the manner in which this result was achieved assumes **C** to be conducted along a geometrical line of no cross section. This means that in practice the result is valid only outside the thin tube within which the current exists. Within the tube, $\mathbf{B} = -\mathrm{grad}\Omega$ is no longer valid, while $\mathbf{B} = \mathrm{curl}(\mathrm{pot}(\mathbf{C}))$ is always true.

Consider now a problem of no apparent connection to the one just solved, namely, expressing the scalar potential of a magnetic dipole. Let there be two parallel surfaces of equal area dS, separated from one another by a small distance dz. Let the bottom surface rest on the x-y plane and coat it with magnetic "charge" of density $-m$. Coat the top surface with magnetic charge $+m$. The magnetic scalar potential due to the negative charge is $P_n(x,y,z) = -mdS/4\pi r$, where r is the distance to the point (x,y,z) at which the potential is calculated. Had the positive charge been at the origin as well, its potential would have been $P_p = mdS/4\pi r$. However, it is moved up a small distance dz, so it may be expressed as:

$$P_p(x, y, z - dz) = \frac{mdS}{4\pi r} - dz\frac{mdS}{4\pi}\frac{\partial}{\partial z}r^{-1}.$$

The total potential is the algebraic sum of P_p and P_n, or in other words:

$$P = -\frac{mdS}{4\pi}\frac{\partial r^{-1}}{\partial z}dz.$$

The comparison of P to Ω makes it immediately evident that they become equal upon setting $C = mdz$. This implies that the magnetic field of the thin

$$-\frac{mdS}{4\pi}\frac{\partial r^{-1}}{\partial z}dz = P(r) = -\frac{CdS}{4\pi}\frac{\partial r^{-1}}{\partial z}$$

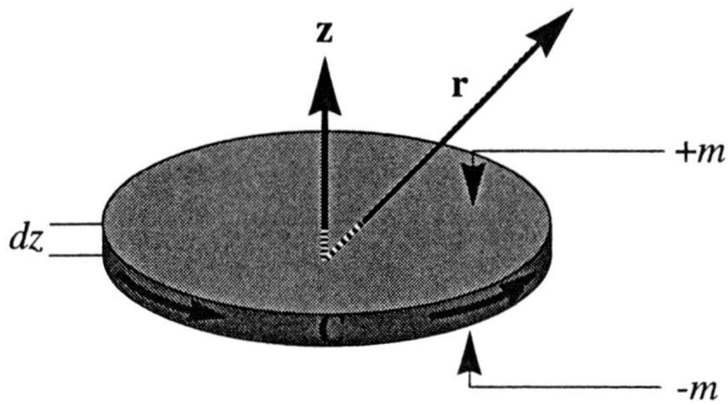

Figure 2.2-1: Comparison of the magnetic scalar potentials of a current loop and a thin magnetic shell.

shell just discussed is identical to the magnetic field produced by a current of strength $C = mdz$ circulating around the shell's boundary. The result is not confined merely to the infinitesimal surfaces and loops used for the derivation. Heaviside had already used what he called "Ampère's ruse" of expressing the current along a closed finite curve as the sum of its infinitesimal subdivisions, each of which carries the same current strength. Likewise, it is clear that any thin magnetic shell of finite surface and constant magnetic dipole moment mdz is the sum of all the sub shells into which it may be divided. Hence, any magnetic shell of finite surface and constant magnetic dipole mdz produces a field identical to the one produced by a current loop round its boundary as long as $mdz = C$. A linear bar magnet may be envisioned as composed of many such shells piled up one on top of the others. Each one of these can be replaced by a current loop, and therefore the whole magnet may be replaced by a current-carrying coil whose loops are wound round the magnet's curved surface. At the same time, "Ampère's ruse" shows that any small part within the magnet may be interpreted as a small current loop, so that instead of envisioning a magnet as composed of magnetic "matter", it may be seen as bulk material within which currents circulate in little loops like so many tiny eddies. In this manner, a theory that begins by identifying the electrical current with the magnetic field's curl, and magnetic "charge" with the field's divergence, demonstrates the reducibility of magnetic "matter" to electrical currents, and unifies electricity and magnetism.

Heaviside, as we have seen, used the argument in a different manner. He demonstrated that *any* vector field whose curl is confined to a closed loop may be derived from the same field's divergence if the latter takes the form of a dipole distribution confined to the surface bounded by the curl. Having demonstrated this result of general vector algebra *regardless* of the existence of currents and magnetic fields, he called attention to Ampère's experimental finding. Putting the two together he concluded that magnetic fields and electrical currents may be described as vector fields obeying the vector-algebraical relationship curl(**B**) = **C**, where **B** is identified with the field and **C** with the current density. With this conclusion he fully mathematicized the physics of electromagnetism, and could now proceed to reason out this physics according to the rules of vector algebra.

APPENDIX 3.3: Note on Heaviside's Derivation of the Mutual Energy of Two Current Systems

Some readers will probably find the intuitive, geometrical picture that Heaviside used to effect the transition from $\Sigma(\mathbf{A}_1 \cdot \mathbf{C}_2)$ to $\Sigma(\mathbf{B}_1 \cdot \mathbf{B}_2)$ quite satisfactory (see chapter III, section *4.9*). Others, however, may feel uncomfortable with it. Looking at a single current tube $C_{2,1}$ of $\mathbf{C}_2$, Heaviside wrote its energy due to the entire current system $\mathbf{C}_1$ as:

$$C_{2,1}\int_{S_{2,1}} \mathbf{B}_1 \cdot \mathbf{da} = \oint_{l_{2,1}} \mathbf{B}_{2,1} \cdot \mathbf{dl}\int_{S_{2,1}} \mathbf{B}_1 \cdot \mathbf{da}, \tag{3.3-1}$$

where the surface integration extends over any equipotential surface of $\mathbf{B}_{2,1}$ and the line integral follows any line of $\mathbf{B}_{2,1}$. Then, in a somewhat vague manner Heaviside suggested that the product of the two integrals on the right may be replaced by a single volume integral of $\mathbf{B}_1 \cdot \mathbf{B}_{2,1}$. From here on the route is clear. Heaviside needed only to point out that the volume integral just derived expresses the mutual energy of $\mathbf{B}_1$ and a single line of $\mathbf{B}_2$. It follows immediately that the total energy is the sum of all such contributions from $\mathbf{B}_2$. This sum obviously amounts to the volume integral of the scalar product $\mathbf{B}_1 \cdot \mathbf{B}_2$, and the transition is complete.

The problematic point then, appears to involve the justification of switching from the product of a line integral and a surface integral to a single volume integral. Heaviside deemed it sufficient to suggest the geometrical picture of a closed line of $\mathbf{B}_{2,1}$ that cuts through each and every one of its own equipotential surfaces at right angles. Since the surface integral is the same for any of these surfaces, one need not stay with one of them, but "move" from one to the next as the line integration is performed without changing the result. This, Heaviside says, amounts to integrating over all of the space permeated by $\mathbf{B}_{2,1}$, and since all other parts of space contribute nothing due to the field being zero there, the volume integral may be extended to all space. Heaviside consciously refused to elaborate beyond that. He merely alluded to the self-energy of a single current system $\mathbf{C}$, then, without proof, extended the comment to the mutual energy of any two current systems:

> As regards the case of a single tube of $\mathbf{C}$, if only the geometrical conditions are pictured in the mind, the division of space into small cubes by the tubes of $\mathbf{B}_1$ cut

across by the equipotential surfaces of $\mathbf{B}_1$, the transformation [to a volume integral] becomes as self-evident as an axiom, and no form of words or sentences is necessary. The less one is cumbered with them the better. And although, the extension to an arbitrary system is less easy, it is still easier to be pictured than logically demonstrated. The transformation might have been seen quite intuitively; it is only when one has to prove it to some one else that clothing the thoughts in words becomes necessary; and, even then, the clothes do not correspond to the original thought, but to those arising in the act of description, and both words and thoughts require to be readjusted, perhaps two or three times, before they will mutually fit with any decency. The Cartesian transformation, breaking up each of the vectors **A**, **B**, **C** into three rectangular components, is short enough, but is gifted with a total absence of visible reason and significance.[1]

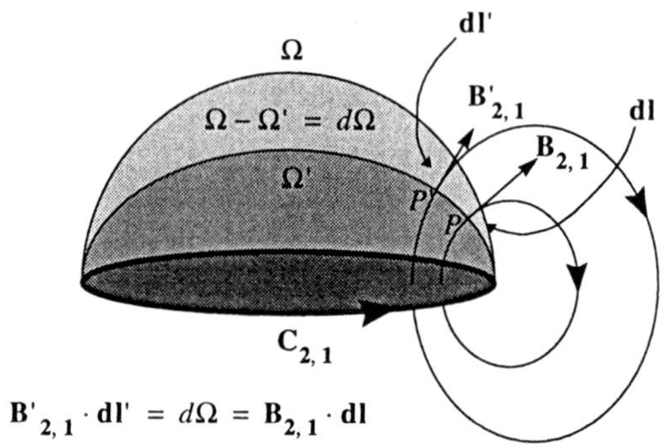

Figure 2.3-1: Illustration of the construction Heaviside used to derive the energy of a current loop immersed in the magnetic field (not drawn) of another current system.

There is much to be said for Heaviside's attitude; yet, one may feel obliged to point out the following. The product of the two integrals in eq. (3.3-1) is executed by taking a component $\mathbf{B}_{2,1} \cdot \mathbf{dl}$ of the line integral, multiplying it by the surface integral of $\mathbf{B}_1$ over the entire equipotential surface that intersects the line of $\mathbf{B}_{2,1}$ where **dl** is reckoned, then moving on to the next section of the line integral until the entire loop has been completed. It is true, as Heaviside suggests, that at the point of intersection **dl** and **da** describe a small volume element. However, this is only an infinitesimal part of the surface in-

1. *Electrical Papers*, Vol. I, p. 246.

tegral. At all other points the product of $\mathbf{B}_{2,1} \cdot \mathbf{dl}$ with $\mathbf{B}_1 \cdot \mathbf{da}$ is taken at a different place than the point of intersection. To fix ideas, let the point of intersection be P, and denote by P' any other point on the equipotential surface cut perpendicularly at P by the line of $\mathbf{B}_{2,1}$. The transition to a volume integral will hold if:

$$(\mathbf{B}_{2,1} \cdot \mathbf{dl})\,(\mathbf{B}'_1 \cdot \mathbf{da}') = (\mathbf{B}'_{2,1} \cdot \mathbf{B}'_1)\,dv' \qquad (3.3\text{-}2)$$

Once this is shown, the rest follows easily: the equality above allows replacing the product of $\mathbf{B}_{2,1} \cdot \mathbf{dl}$ and the surface integral of $\mathbf{B}_1$ at a single point P by a single volume integral of $\mathbf{B}_{2,1} \cdot \mathbf{B}_1$ throughout the thin shell bounded by two equipotential surfaces separated by $\mathbf{dl}$ at P and by $\mathbf{dl}'$ at P'. To effect this transition, consider first that throughout the construction depicted in fig. 2.3-1, the product $\mathbf{B}_{2,1} \cdot \mathbf{dl}$ yields the constant difference $d\Omega$ between the equipotential surfaces Ω and Ω'. Note also that the normal to $\mathbf{da}'$ coincides with $\mathbf{dl}'$ and with $\mathbf{B}'_{2,1}$, since $\mathbf{da}'$ is part of an equipotential surface of $\mathbf{B}_{2,1}$ which must, therefore, be cut perpendicularly by the field line of intensity $B'_{2,1}$ at P'. Hence, we may express a unit vector in the direction of $\mathbf{da}'$ as $\mathbf{B}'_{2,1}/B'_{2,1}$. Putting all this together, we obtain the required proof of equation (3.3-2):

$$\begin{aligned}(\mathbf{B}_{2,1} \cdot \mathbf{dl})\,(\mathbf{B}'_1 \cdot \mathbf{da}') &= d\Omega\left(\mathbf{B}'_1 \cdot \frac{\mathbf{B}'_{2,1}}{B'_{2,1}}\,da'\right) \\ &= B'_{2,1}dl'\left(\mathbf{B}'_1 \cdot \frac{\mathbf{B}'_{2,1}}{B'_{2,1}}\,da'\right) \\ &= (\mathbf{B}'_1 \cdot \mathbf{B}'_{2,1})\,dv'.\end{aligned}$$

Thus, the geometrical picture Heaviside sketched provides a direct and fully justified argument for the transition to the volume integral.

The transition may also be effected in three short lines using a vector identity and Gauss's theorem.[2] But while this proof is much easier, being short and mechanical, it merely glances over the geometrical relationship between a current tube, its related field, and the latter's scalar potential. Heaviside's goal in this exercise was not merely to derive an already well-known result. He wanted to teach field thinking, and one can hardly doubt that his geometrical ap-

2. See for example, Melvin Schwartz, *Principles of Electrodynamics*, (New York: McGraw-Hill Book Company, 1972), p. 163.

proach to the derivation of the mutual energy accomplishes that task far better than the short, mechanical proof.

APPENDIX 4.1: The KR Law and the Distortionless Condition

To properly understand the theoretical background of the dispute between Heaviside and Preece, one must clearly establish the relationship between Heaviside's distortionless condition and Preece's notorious "KR law." Consider first the problem of telegraphic communication by an imaginary submarine cable. Using Heaviside's symbols, let the cable be of length l and have a resistance R per mile, capacitance S per mile, leakance K per mile, and inductance L per mile. Then [K] = Sl and [R] = Rl are respectively the *total* capacitance and *total* resistance of the entire cable, and [K][R] = RSl^2. As Kelvin well knew, the theoretical analysis of telegraphy by a submarine cable is quite accurately served under the assumption of negligible leakage and inductance. Preece started out from a well-known result of Kelvin's analysis, according to which the time it takes for a signal's intensity to reach a given percentage of its final strength at the receiving end is proportional to the product of the line's total capacitance and total resistance, namely, to [K][R]. This is the basis of what Preece referred to as the "KR law." To preserve a semblance of similarity to the original language, the law will be referred to here as the [K][R] law, where [K] stands for the *total* capacitance of the cable, and [R] for its *total* resistance. As the distance of the receiving point from the sending end increases, so does the value of [K][R] and accordingly the signals rise and decay more slowly. This means that if two originally sharp signals are transmitted with only a short interval between them, there will be a point far enough along the line at which the rising part of the second signal will overlap the more developed part of the first in such a manner that the two will no longer be distinguishable. As a very rough first approximation, a telephone signal may be regarded as a series of closely following pulses in the line (it seems almost certain that Preece never realized just how rough this approximation really is). In view of the above it seems clear that at distances in excess of a certain value of [K][R], telephone conversations will become completely unintelligible over a given line. This suggests a statistical determination of the highest value of [K][R] that still allows for intelligible telephony over operating telephone lines of known properties,. As Heaviside noted, the data in his brother's paper fixed this value at about 0.015.[1] Having determined that, it necessarily follows that since the capacitance and resistance of wires are always finite quantities, there exists an absolute limit to the possible range of communications.

Within this theoretical limit the engineer can extend practical ranges by decreasing as far as possible the resistance and capacitance of transmission lines. The most important point to keep in mind, however, is that these conclusions stem from, and hence are applicable to, the particular class of submarine cables, characterized by negligible inductance and leakance.

Now consider the above in light of Heaviside's distortionless condition. It requires that $R/L = K/S$ be satisfied for distortionless transmission. In a submarine cable, L is much smaller than R, and K is much smaller than S. Hence the ratio R/L is far greater than the ratio K/S. There are four possible remedies and combinations thereof: decrease R, increase L, increase K and decrease S. It will immediately be seen that if one chooses to neglect the inductance and leakance, the [K][R] law's recommendations are a subset of Heaviside's. In general, however, the two are not necessarily in agreement. Beginning with a properly balanced cable in accordance with the distortionless condition, any exclusive decrease in resistance will introduce unwanted distortion—a state of affairs that the [K][R] law cannot justify. Furthermore, one can have a cable with a huge value of [K][R], and still have perfectly distortionless telephony by properly adjusting the leakance and inductance. In particular, given the situation in a submarine cable, keeping S and R constant, increase the inductance L until $R/L = K/S$. This will enable vastly increased ranges for telephone communications. Under these circumstances the overall length l may be increased at will while S and R are kept constant. Therefore $[K][R] = SRl^2$ can be increased to any arbitrary value, well in excess of, say 0.015, without any disturbing effect on the clarity of conversation. This is precisely what Heaviside alluded to when he promised that he could make any value of [K][R] amenable

1. Preece, it should be noted, always presented the KR law as purely empirical. Confined to the statistical average that he correlated with maximum intelligible telephony, the claim is quite justified. The conclusions Preece wished to base on this empirically determined number, however, are solidly grounded in the hypothesis that all telephone lines are adequately described within the limited framework of Kelvin's theory. Thus, the conclusion that intelligible telephony is not possible in excess of KR = 0.015, is quite empirical when limited to the sample of lines used to determine this value. The extrapolation that telephony in excess of this number will never be possible, is thoroughly theory dependent. This is because only with Kelvin's theory in mind does KR become a meaningful predictor of telephony ranges. Without a theoretical significance, it is no more than an arbitrary measure reflecting a momentary state of affairs. Thus, while Preece's KR value could be used as an acceptable guide to the state of telephony at the time, it simply could not provide a sound engineering guide for future designs.

to good telephony, provided he be allowed to depart from the strict class of submarine cables.

This last point proved to be the insurmountable obstacle for Preece. For him, both electrostatic inductance, otherwise referred to as capacitance, and magnetic inductance, or simply the inductance, were harmful to telephony. Both cause retardation, therefore Preece concluded that both should be minimized. Beyond this qualitative association of inductance and capacitance, Preece was quite unable to understand the fine points of Kelvin's theory. Thus, he treated inductance as having the same effect as [K] in the [K][R] law, and all of Heaviside's developments were quite beyond his reach. What he could clearly understand was that Heaviside suggested to increase the self-inductance, while Preece's unjustifiably extended [K][R] law told him the opposite should be done. Naturally, he could not understand why Heaviside insisted that copper wires made for good telephony because of their relatively high self-inductance.

It should be noted that Preece's error is not an unreasonable one. Indeed, it seems that far more sophisticated observers such as S.P. Thompson made the same mistake. The unreasonable part of Preece's behavior was that he absolutely refused to examine Heaviside's pinpointed explanation of the root of the error, while others, physicists *and* engineers, were able to comprehend it.

APPENDIX 4.2: Notes on Heaviside's Operational Calculus

The description of a transmission line as a linear circuit involves two basic relationships between current and voltage:

$$\left.\begin{aligned} -\frac{\partial V}{\partial x} &= RC + L\frac{\partial C}{\partial t} \\ -\frac{\partial C}{\partial x} &= KV + S\frac{\partial V}{\partial t} \end{aligned}\right\} \tag{4.2-1}$$

From these we get the telegraph equation in its usual differential form:

$$\frac{\partial^2 V}{\partial x^2} = LS\frac{\partial^2 V}{\partial t^2} + (RS + LK)\frac{\partial V}{\partial t} + RKV. \tag{4.2-2}$$

Consider a wire of length l with a constant electromotive force V_0 impressed at one end while the other end is put to ground. If we wish to describe the condition of the wire after the current in it has become stable, we must set all time variations to zero, which transforms the telegraph equation into the following simple form:

$$\left.\begin{aligned} &\frac{d^2 V}{dx^2} = F^2 V, \text{ where } F = \sqrt{RK}, \\ &\text{subject to: } V(0) = V_0; \quad V(l) = 0. \end{aligned}\right\} \tag{4.2-3}$$

the general solution is:

$$V = Ae^{Fx} + Be^{-Fx}.$$

The integration constants are specified by the two boundary conditions in (4.2-3) so that the steady-state solution for the particular problem above becomes:

$$V = V_0 \frac{e^{F(l-x)} - e^{-F(l-x)}}{e^{Fl} - e^{-Fl}}. \tag{4.2-4}$$

Heaviside obtained this steady state solution ten years before he started his series “On the Self-Induction of Wires.” It is a useful form because it simply relates the voltage at any point along the line to the impressed electromotive force V_0 at its beginning. This solution is also instructive because it clearly shows how closely Heaviside fashioned his operational approach to harmonize with the needs of solving transmission-line problems. Proceeding to the discussion in "On the Self-Induction of Wires," we find Heaviside substituting p for d/dt, and rewriting the telegraph equation (4.2-2) as:

$$\frac{d^2V}{dx^2} = [LSp^2 + (RS+LK)\,p + RK]\,V,$$

or:

$$\frac{d^2V}{dx^2} = F^2V;\ \text{where: } V(0,t) = V_0(t);\ V(l,t) = 0. \qquad (4.2\text{-}5)$$

Formally identical to eq. (4.2-3), equation (4.2-5) is substantially different because F is not a constant but a function of the time derivative d/dt, and because V_0 is not a constant, but any function of time. In other words, this is not an equation for the asymptotic steady state, but one that describes the line at any time under the effects of an arbitrarily varying impressed electromotive force at $x = 0$. These differences between eqs. (4.2-3) and (4.2-5) taken together with their formal similarity, immediately suggest how eq. (4.2-4) should be adapted to a formal solution of (4.2-5). The two integration constants, now pertaining to a time-dependent problem must be allowed to vary with time, while e^{Fx} must now be taken as denoting an operator function of the time derivative. In other words, the general solution to (4.2-5) should look like:

$$V(x,t) = e^{Fx}A(t) + e^{-Fx}B(t),$$

where the two exponential functions should be understood as representing a differential operation on the time-dependent functions $A(t)$ and $B(t)$. Proceeding now exactly as in the steady state case, $A(t)$ and $B(t)$ are defined by the two boundary conditions in eq. (4.2-5), which makes the specific solution an exact operational analogue of (4.2-4):

$$V(x,t) = \frac{e^{F(l-x)} - e^{-F(l-x)}}{e^{Fl} - e^{-Fl}}\,V_0(t). \qquad (4.2\text{-}6)$$

Once again, we have what appears as a simple, direct way of relating the voltage at any point along the line to the impressed force at its beginning. However, unlike the steady-state case, the ratio between the voltages is no longer a function of *F*, the square root of *RK*, but a function of the operator:

$$F = \sqrt{LS\frac{\partial^2}{\partial t^2} + (RS + LK)\frac{\partial}{\partial t} + RK} \tag{4.2-7}$$

The problem outlined above requires the solution of a partial-differential equation with constant coefficients, and to a considerable extent, Heaviside shaped his operational calculus with this requirement in mind. From the point of view of applied mathematics, the solution of such equations is *the* purpose of operational techniques, and most historical reviews of the development of these techniques naturally emphasize this purpose. Heaviside, however, did not develop his operational methods out of a general interest in differential equations *per se*, but as a result of his interest in transmission-line analysis. The operational formulation of the transmission-line problem held for him an additional attraction beyond its usefulness as an equation-solving tool. To see this, substitute p for time differentiation in equations (4.2-1), divide the first equation by the second, and multiply through by V to yield:

$$\frac{\partial V}{\partial x}V = \left(\frac{R+Lp}{K+Sp}\right)\frac{\partial C}{\partial x}C,$$

or, integrating with respect to x:

$$V^2 = \left(\frac{R+Lp}{K+Sp}\right)C^2,$$

$$\therefore V = \sqrt{\frac{R+Lp}{K+Sp}}C. \tag{4.2-8}$$

In other words, the relationship between voltage and current throughout the transmission line may be expressed as a generalized Ohm's law, in which an operational impedance replaces the traditional resistance. Since the impedance operator contains all of the electrical information regarding the line, much useful knowledge may be gathered from its analysis even without obtaining explicit expressions for the current and voltage. As in the solution of the telegraph equation, a complete description of the line still requires that we establish procedures for applying the impedance operator to known functions.[1]

Returning to the telegraph equation, the quadratic function under the radical in eq. (4.2-7) may suggest at first sight that eq. (4.2-6) has no more than a symbolic value, and that in order to obtain a workable solution of eq. (4.2-5) we have to revert back to the standard methods of solving partial-differential equations with constant coefficients. That, however, is not the case. The seemingly impenetrable eq. (4.2-6) may be expanded in a power series of p. Under various conditions, the series may be truncated to yield useful approximations. Having obtained such approximated expressions, involving only positive and negative powers of p, one merely needs to differentiate and integrate $V(0,t)$ accordingly, to obtain the voltage anywhere along the line. Obviously, the difficulty invariably resides in the manipulation and interpretation of the resulting series expansions of expressions like eq. (4.2-6). Investigations of such series stand at the heart of Heaviside's most difficult papers. He produced most of this work from 1892 on, but its original basic motivation is clearly discernible in his transmission line work in "On the Self-Induction of Wires," and in "On Resistance and Conductance Operators," which he published in *The Philosophical Magazine* in December of 1887.[2] Indeed, the special, highly innovative flavor of Heaviside's operational calculus stems from the nature of the transmission-line problem. An important aspect of transmission-line analysis consists of tracing the effects of a particular voltage source turned on at $t = 0$ at a certain point on the line (usually its beginning). However complicated the time dependence of the signals impressed by the source, they are always enveloped by a simple step function; zero before a certain point in time, one afterwards. In the following pages this step function will be denoted by a boldface **1**, such that:

$$\mathbf{1}(t) = \begin{cases} 0, & (t \le 0)\,; \\ 1, & (t > 0)\,. \end{cases}$$

1. The ease with which the impedance operator has been derived here is somewhat deceptive. In the form (4.2-8) it applies only to a semi-infinite line with $V(0,t)$ at the origin. This may be verified by substituting (4.2-6) with infinite l into the first of eqs. (4.2-1) to extract the impedance operator. Heaviside obtained both the general and semi-infinite line's impedance operators in a more involved, though highly instructive analysis that makes explicit use of the line's electrical properties (see *Electrical Papers*, Vol. II, pp. 367-69.

2. They are reproduced under the same titles in Heaviside's *Electrical Papers*, Vol. II.

or, more generally:

$$\mathbf{1}(t-t_0) = \begin{cases} 0, & (t \leq t_0); \\ 1, & (t > t_0). \end{cases}$$

Thus, a harmonically oscillating signal coupled to a transmission line at $t = 0$, is symbolized by $\mathbf{1}\sin(nt)$. The product ensures that the voltage is zero before $t = 0$, and $\sin(nt)$ afterward. Heaviside quickly discovered that the interpretation of his operational expressions can usually be reduced to the specification of their operation on the unit function. Accordingly, much of his operational work concentrates on attempts to discern the meaning of $F(p)\mathbf{1}$ where $F(p)$ represents an operational function.[3]

In modern texts on differential equations, the unit function $\mathbf{1}$ is sometimes designated as $H(t)$ and referred to as "the Heaviside unit function." Heaviside used it extensively in his work. It forms the basis of his own version of the operational calculus, and very clearly distinguishes it from other approaches to this branch of mathematics.[4] In much of his work, Heaviside also used the

3. A particularly lucid example of such investigations may be followed in Heaviside's *Electromagnetic Theory*, Vol. II, pp. 286-301.

4. It may be pointed out that the mere manipulation of d/dt as an algebraic quantity in certain differential equations does not exhaust Heaviside's contribution to the operational calculus. Such symbolical approaches to differential equations existed long before Heaviside, and it was with justice that some individuals rebelled against the practice of regarding Heaviside as *the* inventor of *the* operational calculus. Having said that, however, one feels compelled to accept Lutzen's observation that Heaviside's contribution to the field was highly original (J. Lutzen, "Heaviside's Operational Calculus and the Attempts to Rigorize It," *Archive for History of the Exact Sciences*, **21** (1979): 161-200). There are currently at least five other operational calculi, due to Bromwich, Doetsch, Van Der Pol, Mikusinski and Moore. Heaviside's version predates all of them, and differs from them in its basic reliance on the unit function. Moore's is the closest in spirit to Heaviside's in its conscious attempt to formulate a natural language for dealing with impulsive dynamical systems. I have found no convincing evidence to show that the particular marks of Heaviside's approach, based on the unit function and its derivative, have been anticipated by previous workers, but see S.S. Petrova, "Heaviside and the Development of the Symbolic Calculus," *Archive for History of the Exact Sciences*, **37** (1987): 1-23.

derivative of the unit function, whose basic property is easily discernible from what we have done so far:

$$\int_{-\infty}^{t} \frac{d}{dt}\mathbf{1}f(t)\,dt = \mathbf{1}f(t)\,\Big|_{-\infty}^{t} = f(t)$$

$$= \int_{-\infty}^{t} \mathbf{1}'f(t)\,dt + \int_{0}^{t} f'(t)\,dt$$

$$= \int_{-\infty}^{t} \mathbf{1}'f(t)\,dt + f(t) - f(0)$$

$$\Rightarrow \int_{-\infty}^{t} \mathbf{1}'f(t)\,dt = f(0)\,.$$

The above property is easily generalized to any point t_0 contained within the integration region by a simple transformation of variables. Thus, in general, we have for $a < t_0 < b$:

$$\int_{a}^{b} \mathbf{1}'(t - t_0)f(t)\,dt = f(t_0)\,. \tag{4.2-9}$$

Heaviside often used this property, treated $\mathbf{1}'(t\text{-}t_0)$ as an impulse of infinitesimal width and infinite height situated at $t = t_0$, and knew how to represent it as a Fourier series.[5] This may tempt one to think that had Heaviside given this function a proper name, we might well call it today the "Heaviside Impulse" rather than the "Dirac Delta". On further thought, it will be seen that the temptation should be resisted. Neither Dirac nor Heaviside were the first to use the delta function; but no one had built a mathematical scheme around the delta function as Dirac did in his formulation of quantum mechanics. Similarly, several mathematicians devised symbolic methods for solving differential equations before Heaviside; but none of them built an operational calculus around the unit step function as Heaviside did.

Insufficient sensitivity to the role of the unit function in Heaviside's work led to interpretations of his operational calculus which make it appear to suffer from more mathematical flaws than it really does. In 1927, Harold Jeffreys,

5. *Electromagnetic Theory*, Vol. II, pp. 55, 99-100.

later Plumian Professor of Astronomy at the University of Cambridge, published a short treatise entitled *Operational Methods in Mathematical Physics.*[6] Jeffreys began his exposition of the operational calculus by treating definite integration from zero to t as the basic operator, which he denoted by Q. He noted that Heaviside treated integration as the inverse of the differential operator p, so that $pp^{-1} = p^{-1}p = 1$, and that this relationship enabled Heaviside to treat differential equations as algebraic equations. Jeffreys then showed by an elementary consideration that in fact differentiation and integration from zero to t do not commute as the above relationship requires:

$$\frac{d}{dt}\int_0^t f(t)\,dt = f(t),$$

but:

$$\int_0^t \frac{d}{dt} f(t)\,dt = f(t) - f(0).$$

The difficulty may be avoided by restricting operations to functions that vanish for values of t less than or equal to zero, and indeed, such functions characterize most of the electrical problems Heaviside investigated. Jeffreys, however, proceeded to show that the problem is more intractable than that. The relationship:

$$\begin{aligned} f(t+h) &= \left(1 + h\frac{d}{dt} + \frac{h^2}{2!}\frac{d^2}{dt^2} + \frac{h^3}{3!}\frac{d^3}{dt^3} + \dots\right) f(t) \\ &= \left[1 + hp + \frac{1}{2!}(hp)^2 + \frac{1}{3!}(hp)^3 + \dots\right] f(t) \\ &= e^{hp} f(t), \end{aligned}$$

defines e^{hp} as the "shifting operator" that shifts a function $f(t)$ by $-h$ units along the t-axis. This operator, which Heaviside used extensively, does not commute with p^{-1} even when operating on the unit function, which clearly sat-

6. H. Jeffreys, *Operational Methods in Mathematical Physics*, (Cambridge: At the University Press), 1927. The title itself may imply a response to Heaviside's controversial "On Operators in Physical Mathematics."

isfies the requirement of vanishing for negative values of t. The argument is best seen graphically:

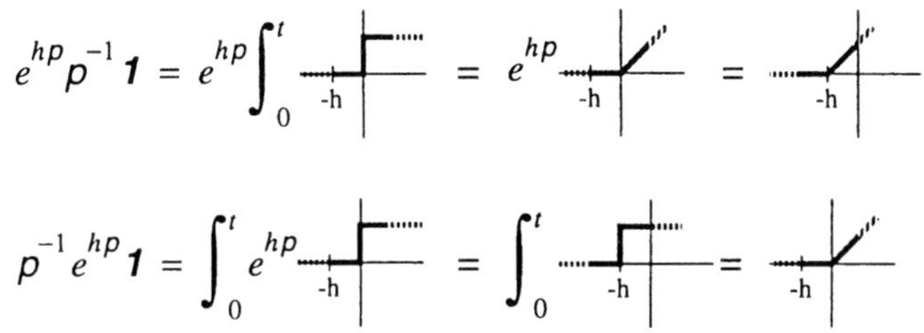

Jeffreys repeatedly invoked these two demonstrations of commutative failure as indications of the real source of difficulty with Heaviside's operational calculus. Thus, he wrote in 1950:

> Heaviside was a largely self-taught genius and found out much for himself that some of the greatest mathematicians might have been glad to discover. But he had not enough mathematical background to know that many of his difficulties were old ones and could be got round by known methods. The attacks on him for using divergent series, and the absence of attacks for commuting non-commutative operations, showed only that the contemporary pure mathematicians had neglected parts of their own subject.[7]

Again, in 1966:

> Unfortunately, though Heaviside noticed that the operators of differentiation and integration combine with constants without restriction, he did not notice that they do not commute with each other.... Heaviside obtained a considerable number of wrong results through interchanging the order of differentiation and integration, and their explanation in terms of this non-commutative property was first given by H. Jeffreys.[8]

To support his claim of "a considerable number of wrong results," Jeffreys referred to his own 1927 work. The alleged errors he illustrated in this earlier work involve various cases of illegal commutation. In the same 1927 analysis, however, Jeffreys did not claim that Heaviside actually committed these errors; indeed, Jeffreys noted that Heaviside *avoided* errors by refraining from

7. H. Jeffreys, "Heaviside's Pure Mathematics," *The Heaviside Centenary Volume*, (London: 1950), p. 91.

8. H. Jeffreys and B.S. Jeffreys, *Methods of Mathematical Physics*, (Cambridge: At the University Press, 1966), p. 229.

altering the order of integration and differentiation.[9] It may be, however, that Heaviside managed to avoid errors because his operational manipulations were guided by use of the unit function that differs from Jeffreys's. In fact, it seems that the source of Jeffreys's difficulties is in his own definition of p^{-1} as representing integration from zero to t. Recalling that Heaviside represented all initial conditions as $\mathbf{1}f(t)$, consider first the commutativity of p and p^{-1} when the latter is taken to represent integration from $-\infty$ to $t > 0$:

$$\frac{d}{dt}\int_{-\infty}^{t} \mathbf{1}f(t)\,dt = f(t) = \int_{-\infty}^{t} \frac{d}{dt}\mathbf{1}f(t)\,dt. \tag{4.2-10}$$

Note that the presence of the unit function is essential for the equality to hold, because without it, the integral on the left would be $f(t)$ while the one on the right would be $f(t) - f(-\infty)$. Thus, it is the presence of the unit function that makes integration the exact inverse of differentiation. In other words, under these conditions integration can always be expressed simply as p^{-1}, and manipulated as an algebraic quantity relative to p, so that $p^{-1}p = pp^{-1} = 1$. Now consider again the case of the shifting operator e^{hp} under the same definition of p^{-1}:

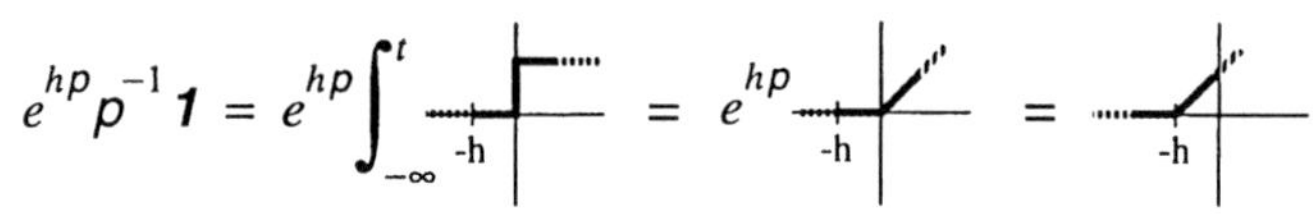

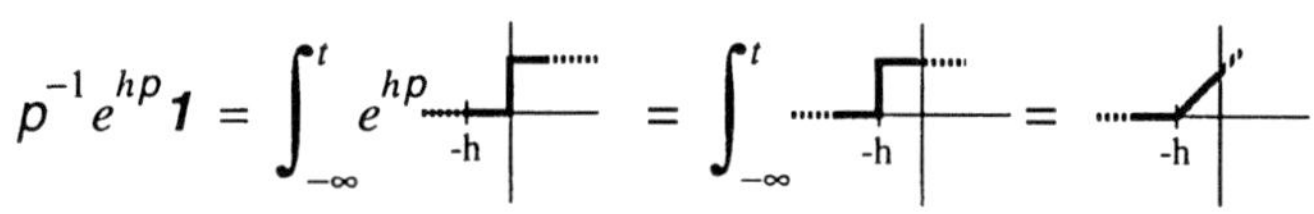

Thus, both difficulties disappear once Heaviside's work is given a different interpretation than the one Jeffreys imposed on it. In both cases, due consideration of the unit function removes the difficulties. Jeffreys ran into these problems because he manipulated the operational relations involved without proper attention to their associated operands. This does not properly represent Heaviside's procedures, in which an operand is always implied, even when, as in the case of the unit function, it is not always explicitly marked.[10] Indeed,

9. H. Jeffreys, *Operational Methods in Mathematical Physics*, (Cambridge: Cambridge University Press, 1927), p. 16. See also, O. Heaviside, *Electromagnetic Theory*, Vol. II, p. 298.

the specific example Jeffreys cited to demonstrate his claim that Heaviside avoided errors because he refrained from altering the order of operation with p and p^{-1}, demonstrates Jeffreys's careless reading of Heaviside. The paragraph Jeffreys referred to actually demonstrates that Heaviside did not refrain from commuting p and p^{-1}. It shows, rather, that Heaviside made the two operators commute by proper consideration of integrating the impulse $p1$:

> Thus $pp^{-1}1 = pt = 1$, but $p^{-1}p1 = p0 = 0$, unless we say $p^{-1}p1 = p^{-1}\frac{t^{-1}}{(-1)!} = \frac{t^{0}}{0!} = 1$. This property has to be remembered sometimes.[11]

The basic property Heaviside tried to express here by extending the usual integration rule to negative powers, is that the time derivative of the unit function yields an impulse function concentrated at $t = 0$, and as we have already seen, he knew the result of integrating it quite well. He described this basic property in plain words:

> We have to note that if Q is any function of the time, then pQ is its rate of increase. If, then,... Q is zero before and constant after $t = 0$, pQ is zero except when $t = 0$. It is then infinite. But its total amount is Q. That is to say, $p1$ means a function of t which is wholly concentrated at the moment $t = 0$, of total amount 1. It is an impulsive function, so to speak.[12]

In fact, the first quotation above is a straightforward application of the second to yield eq. (4.2-10) in direct contradiction of Jeffreys's claims.

All of this, however, is not to say that everything is fine with Heaviside's mathematics and that Jeffreys's criticism merely reflects a sort of academic dogmatism and a refusal to read a text sympathetically because it does not conform to certain stylistic requirements. Actually, Jeffreys was by no means an enemy of Heaviside and his general appreciation of Heaviside's operational calculus was undoubtedly favorable. The title of his 1927 treatise is *Operational Methods in Mathematical Physics*, which may well be a tribute to Heaviside's controversial "On Operators in Physical Mathematics." In the preface to his 1927 monograph, Jeffreys wrote:

---

10. This may be contrasted with Jan Mikusinski's operational calculus, in which the distinction between operator and operand is obliterated, and the concept of "operator" becomes a generalized number that includes numbers, functions and differential and integral operations. Under such circumstances it becomes possible to deal with self-contained relations among operators. See Jan Mikusinski, *Operational Calculus*, 2nd Edition, (Oxford: Pergamon Press, 1983), Vol. I, pp. 12-37.

11. *Electromagnetic Theory*, Vol. II, p. 298. There is an error in the text. It should have been: $p^{-1}p1 = p^{-1}0 = 0$.

12. *Electromagnetic Theory*, Vol. II, p. 55.

> Heaviside's own work is not systematically arranged, and in places its meaning is not very clear.... [A]s a matter of practical convenience there can be no doubt that the operational method is far the best for dealing with the class of problems concerned. It is often said that it will solve no problem that cannot be solved otherwise. Whether this is true would be difficult to say; but it is certain that in a very large class of cases the operational method will give the answer in a page when ordinary methods take five pages, and also that it gives the correct answer when the ordinary methods, through human fallibility, are liable to give a wrong one.[13]

In view of this, it would be imprudent to dismiss Jeffreys's misdirected criticism as a reflection of narrow-minded rigorism. Indeed, to the extent that the preceding discussion mitigates Jeffreys's objections, it also reveals the problematic nature of Heaviside's work. As we have seen (p. 315), Heaviside used the well-accepted convention of equating 1/(-1)! with 0. This enabled him to give the impulse function at t = 0 a simple mathematical expression. This is customarily justified as follows. Using Euler's Gamma function,

$$\Gamma(x) = \int_0^\infty t^{x-1} e^{-t} dt, \qquad (4.2\text{-}11)$$

the factorial operation may be extended to include both whole and fractional positive numbers, so that $x\Gamma(x) = \Gamma(x+1)$ in general, and in particular for integral values of x, $\Gamma(x) = (x-1)!$. The integral that defines the Gamma function diverges for negative values of x. However, $x!$ for negative values of x may still be evaluated by a recursive procedure. We evaluate $\Gamma(x)$ for $-1 < x < 0$ as $(1/x)\Gamma(x+1)$, then use these values to calculate $\Gamma(x)$ for $-2 < x < -1$ and so on. From this it turns out that $1/\Gamma(x)$ for $x < 0$ is a decaying oscillating function of unit wavelength, which cuts the x-axis at each negative integral value of x.[14] This justifies Heaviside's conventional use of $t^{-1}/(-1)!$ to represent an impulse at $t = 0$. His manner of justifying the convention, however, would probably make many mathematicians cringe:

> Let the operand be $t^n/n!$ where $n!$ is the factorial function 1.2.3....n. Then
>
> $$p\frac{t^n}{n!} = \frac{t^{n-1}}{(n-1)!}, \quad p^n\frac{t^n}{n!} = 1, \quad p^{-n}1 = \frac{t^n}{n!} \qquad (11)$$

13. Harold Jeffreys, *Operational Methods in Mathematical Physics*, (1927), p. v.

14. See e.g. D.B. Scott and S.R. Tims, *Mathematical Analysis, An Introduction*, (Cambridge: At the University Press, 1966), pp. 406-408.

> [...] Now the fundamental property of $n!$ is
>
> $$n! = n(n-1)!$$
>
> with the addition that its value must be fixed for any one value of n, for instance, $1! = 1$. It follows that $0! = 1$ also, and that $n!$ is ∞ for all negative integral values of n. Consequently (11) ... [is] also valid when n is negative. For example, $p1 = 0$, provided t is positive. It is really an impulse at the moment $t = 0$. Also $p1 = t^{-1}/(-1)!$, and this is zero, unless t is also zero.[15]

The striking simplicity by which Heaviside arrived at the conclusion that $n!=\infty$ for negative integers is undoubtedly tempting: after all, $0! = 1 = 0.(-1)!$ (by "the fundamental property of $n!$") which implies that $(-1)!$ is infinite. On second thought, however, the conclusion involves an ambiguity of sign, because by this reasoning $(-1)!$ may equally well be $-\infty$. Regarded from the vantage point of the Gamma function, it will be seen that $\Gamma(x)$ tends to $-\infty$ when x approaches -1 from above, and to $+\infty$ when x approaches -1 from below. While this may prove important in certain situations, it is quite irrelevant to Heaviside's purpose because $1/(-1)!$ is zero either way, and this is the result that he put to use. Note that Heaviside's procedure of evaluating $(-1)!$ is fully analogous to the one based on the Gamma function. He established an algorithm for calculating $n!$ for positive integers ($n! = 1.2.3. \dots .n$) in conjunction with a basic property of $n!$, namely, $n! = n(n-1)!$. Then, using the algorithm-calculated value for $1!$, he extended the evaluation of n! to zero and negative integers by recursively employing the basic property. The same procedure is applied to the Gamma function. Here, the integral in equation (4.2-11) plays the role of the calculating algorithm for any positive number, while the basic relationship is generalized to $x\Gamma(x) = \Gamma(x+1)$. Beginning with the calculated values for $0 < x < 1$, we recursively employ the generalized basic relationship to obtain values of $\Gamma(x)$ for $x < 0$. The difference between the two procedures is that the latter resolves the sign ambiguity contained in Heaviside's. As it stands, then, Heaviside's evaluation of $n!$ for negative values of n is incomplete, and while it leads to no errors in his actual applications, it raises a problem that goes beyond a question of mere formalities.

As in the particular case above, the problem with Heaviside's operational methods is not one that involves strict errors; rather, it involves the use of terms and procedures that are seldom fully defined.[16] This point was clearly expressed in 1926 by one of Heaviside's more capable and sympathetic read-

15. *Electromagnetic Theory*, Vol. II, p. 289.

ers, John R. Carson, with respect to Heaviside's use of divergent series. Carson first noted that even from the practical point of view, the state of the theory of divergent series solutions to operational equations was as yet unsatisfactory because it lacked a sure general rule of application. He then continued:

> Furthermore, the precise sense in which the expansion asymptotically represents the solution cannot be stated in general, but requires an independent investigation in the case of each individual problem. On the other hand, when an asymptotic expansion is known to exist, the Heaviside Rule finds this expansion with incomparable directness and simplicity, the problem of justifying the expansion being a purely mathematical one, which usually need not trouble the physicist. Furthermore, on the purely mathematical side, the Heaviside Rule is of large interest and should lead to interesting developments in the theory of asymptotic expansions.[17]

The fact that Heaviside used his own conventions without proper introduction naturally made matters worse. Thus, the use of the unit function is implied throughout his operational work, but he never bothered to symbolize it clearly, and treated it as analogous to the algebraic 1 that implicitly multiplies any algebraic expression. One has to go through hundreds of pages of operational exercises in *Electromagnetic Theory*, Vol. II, before this is explained as an afterthought, in a parenthetical remark:

> (If an operand is always understood to be at the end, the unit operand may be omitted in general, just as in arithmetical and algebraical operations, and it is sometimes an advantage to omit it.)[18]

As we have seen, Jeffreys's objections do not hold with respect to Heaviside's work. Heaviside's subsequent difficulties with the Royal Society were not founded on such a basic level. In other words, neither an absence of rigor that led to demonstrable errors on Heaviside's part, nor a purely stylistic demand for rigor on the part of his critics underscored the objections raised by the Royal Society.[19] In his quest for interpretation of various operational expressions, Heaviside encountered and made rather creative use of a host of divergent series. Jeffreys's observations notwithstanding (see p. 313), it seems

16. This is made particularly evident by Heaviside's implicit use of rules of limited applicability, which yield correct results with often surprising effectiveness as long as they are used judiciously. See J.R. Carson, *Electric Circuit Theory and the Operational Calculus*, (New York: McGraw-Hill Book Company, Inc., 1926), pp. 62-78, esp. pp. 67-71.

17. *Ibid.*, p. 78.

18. *Electromagnetic Theory*, Vol. II, p. 294

that Heaviside's theory (or lack of it) of divergent series was largely responsible for the critical reaction of certain Royal Society mathematicians. At the heart of his work with divergent series stood a notion of "equivalence" between series, of which Heaviside wrote:

> The reader should be cautioned against concluding that equivalence as of a divergent and a convergent formula, means identity. The fact that they are different shows that they are not alike in *all* aspects, and cannot be interchanged under *all* circumstances. I am inclined to think that this is true even when the equivalence exists between two convergent formulae of different types, in fact, what rigorous mathematicians call an identity. Or there may be equivalence when the argument is real, but not when it is imaginary or even negative. The extent to which equivalence persists is an interesting matter, but is better observed in the practical concrete examples than theorised about upon incomplete data. Experience and experiment must precede theory.[20]

It is hardly surprising that such statements, and mathematical procedures reflecting such statements, were received with grave suspicions by the mathematicians of the Royal Society. While Heaviside's unrigorous procedures did not lead to explicit errors, their critique clearly involved more than the esthetic

---

19. For an account of the historical roots of Heaviside's operational calculus, its special character, and the challenges it presented to other mathematicians, see Lutzen's excellent analysis in "Heaviside's Operational Calculus and the Attempts to Rigorize it;" *Archive for History of the Exact Sciences*, **21**(1979): 161-200. For an attempt to justify the Royal Society's censorship, see J.L.B. Cooper, "Heaviside and the Operational Calculus," *The Mathematical Gazette*, **36** (1952): 5-19. For a different point of view, see B.J. Hunt, "Rigorous Discipline: Oliver Heaviside versus the Mathematicians," in Peter Dear, (ed.), *The Literary Structure of Scientific Argument: Historical Studies*, (1991), pp. 72-95. Other accounts of the affair may be found in P.J. Nahin, *Oliver Heaviside: Sage in Solitude*, (1988), pp. 222-227; E.T. Bell, *The Development of Mathematics*, 2nd Ed. (1945), pp. 413-415; E.T. Whittaker, "Oliver Heaviside," reprinted in O. Heaviside, *Electromagnetic Theory*, Vol. I, pp. xxix-xxx. Most accounts, Cooper's notwithstanding, seem to agree that while the operational calculus was by no means Heaviside's invention, his version of it was innovative, and that he successfully applied it to a range of problems which were not previously tackled with its aid. The controversy seems to have revolved especially around Heaviside's use of divergent series. He used them in highly unrigorous ways that give an impression of mathematical tinkering rather than sound mathematical work. This must have seemed particularly undesirable at a time when rapid progress in the theory of divergent series was being made by Poincaré and others. However, it is difficult to tell to what extent Heaviside incurred the wrath of his mathematical critics because they were aware of more dependable methods that Heaviside ignored. In fact, while Heaviside's methods were mathematically problematic, they were not nearly the sort of trial and error guesswork they appear to be at first sight. Throughout, he was guided by constant reference to physical theory, as Lutzen has shown in his careful analysis of how Heaviside "proved" the expansion theorem.

20. *Electromagnetic Theory*, Vol. II, p. 250.

sensibilities and territorial jealousy of a professional community. Equally plain, however, is that the severity of the punishment far exceeded the severity of the "crime."

Bibliography

Appleyard, Rollo, *Pioneers of Electrical Communication*, (London: Macmillan and Co., Limited, 1930)

______ q.v. "Heaviside, Oliver." *The Dictionary of National Biography*, 1922-1930.

Ayrton, W.E., *Practical Electricity: A Laboratory and Lecture Course for First Year Students of Electrical Engineering based on the Practical Definitions of the Electrical units*, (London: Cassell and Company, Limited, 1896, first published -1887).

Ayrton, W.E., and Perry, John, "Modes of Measuring the Co-Efficients of Self and Mutual Induction," *The Electrician*, **19** (May 13 - June 3, 1887): 17-21; 39-41; 58-60; 83-85.

Baker, E.C., *Sir William Henry Preece, F.R.S., Victorian Engineer Extraordinary*, (London: Hutchinson, 1976).

Bell, E.T., *The Development of Mathematics*, 2nd Ed. (New York: McGraw-Hill Book Company, Inc., 1945).

Berg, E.J., *Heaviside's Operational Calculus as Applied to Engineering and Physics*, (New York: McGraw-Hill Book Company, Inc., 1936).

Berkson, William, *Fields of Force: The Development of a World View from Faraday to Einstein*, (London: Routledge and Kegan Paul, 1974).

Bewley, L.V., *Two-Dimensional Fields in Electrical Engineering*, (New York: The MacMillan Company, 1948).

Bowers, Brian, *Sir Charles Wheatstone*, (London, 1975).

Brittain, James E., "The Introduction of the Loading Coil: George A. Campbell and Michael I. Pupin," *Technology and Culture*, **11** (1972): 36-57.

Bromwich, T.J.I'A., "Examples of Operational Methods in Mathematical Physics," *The Philosophical Magazine*, 37: 407-19.

______ "Normal Coordinates in Dynamical Systems," *Proceedings of the London Mathematical Society*, Series 2, **15** (1916): 401-48.

______ "Normal Coordinates in Dynamical Systems," *Proceedings of the Londond Mathematical Society*, **15** (1916) : 401-448.

______ "Symbolical Methods in the Theory of Conduction of Heat," *Proceedings of the Cambridge Philosophical Society*, **20** (1921): 411-27.

Buchwald, J.Z., *From Maxwell to Microphysics: Aspects of Electromagnetic Theory in the Last Quarter of the Nineteenth Century*, (Chicago: The University of Chicago Press, 1985).

______ "Oliver Heaviside, Maxwell's Apostle and Maxwellian Apostate," *Centaurus*, **28** (1985): 288-330.

Bush, V., *Operational Circuit Analysis*, (New York: John Wiley & Sons, 1929).

Carson, J.R., *Electric Circuit Theory and the Operational Calculus*, (New York: McGraw- Hill Book Company, Inc., 1926).

Channell, David F., "The Harmony of Theory and Practice: The Engineering Science of W.J.M. Rankine," *Technology and Culture*, **23** (1982): 39-52.

Cooper, J.L.B., "Heaviside and the Operational Calculus," *The Mathematical Gazette*, **36** (1952): 5-19.

Crowe, Michael J., *A History of Vector Analysis: The Evolution of the Idea of a Vectorial System*, (New York: Dover Publication Inc., 1967).

Culley, R.S., *A Handbook of Practical Telegraphy*, Adopted by the Post Office and by the Department of Telegraphs in India, 5th edition, revised and enlarged, (London: Longmans, Green, Reader, and Dyer, 1871).

______ *A Handbook of Practical Telegraphy*, 6th edition, (London: Longmans, Green, Reader, and Dyer, 1874)

Doetch, G., *Theorie und Anwendung der Laplace-Transformation*, (Berlin: Springer Verlag, 1937).

Duhem, Pierre, *The Aim and Structure of Physical Theory*, (New York: Atheneum, 1981).

Escher, M.C., *Escher on Escher: Exploring the Infinite*, (New York: Harry N. Abrams, Inc., Publishers, 1989).

Fabre, Jean Henri, *Life of the Fly*, Tr. by A.T. de Mattos, (London: Hodder and Stoughton, 1919).

Feynman, R.P., and Hibbs, A.R., *Quantum Mechanics and Path Integrals*, (New York: McGraw-Hill Book Company, 1965).

FitzGerald, G.F., "Heaviside's Electrical Papers," *The Scientific Writings of the Late George Francis FitzGerald*, edited by Joseph Larmor, (Dublin, 1902).

Fleming, J.A., *Fifty Years of Electricity: The Memories of an Electrical Engineer*, (London: The Wireless Press, LTD, 1921).

Gossick, B.R., "Where is Heaviside's Manuscript for Volume 4 of his 'Electromagnetic Theory'?," *Annals of Science*, **34** (1977): 601-606.

Hartog, Philip Joseph, q.v. "Ayrton," *The Dictionary of National Biography*, Supplement, January 1901 - December 1911, p. 73.

Heaviside, Oliver, *Electrical Papers*, 2 vols., 2nd Ed., (New York: Chelsea Publishing Comapny, 1970).

______ *Electromagnetic Theory*, 3 vols., 3rd Ed., (New York: Chelsea Publishing Company, 1971).

Hertz, Heinrich, *The Principles of Mechanics*, Translated by D.E. Jones and J.T. Walley, (London: Macmillan and Co., LTD, 1899).

Hofstadter, Douglas R., *Gödel, Escher, Bach*, (New York: Vintage Books, 1980).

Hughes, David E., "The Self-Induction of an Electric Current in Relation to the Nature and Form of its Conductor," *The Electrician*, **16** (Feb. 5, 1886): 255-258.

Hunt, B.J., "Rigorous Discipline: Oliver Heaviside versus the Mathematicians," In Peter Dear, (ed.), *The Literary Structure of Scientific Argument: Historical Studies*, (Philadelphia: University of Pennsylvania Press, 1991), pp. 72- 95.

______ "The Origins of the FitzGerald Contraction," *British Journal for the History of Science*, **21** (1988): 67-76.

______ *The Maxwellians*, (Ithaca: Cornell University Press, 1991).

Jeffreys, H., and Jeffreys, B.S., *Methods of Mathematical Physics*, (Cambridge: At the University Press, 1966)

Jeffreys, H., *Operational Methods in Mathematical Physics*, (Cambridge: At the University Press), 1927.

Jordan, D.W., "D.E. Hughes, Self-Induction and the Skin-Effect," *Centaurus*,

26 (1982): 123-153.

______ "The Adoption of Self-Induction by Telephony, 1886-1889," *Annals of Science*, **39** (1982): 433-461.

Josephs, H.J., "Some Unpublished Notes of Oliver Heaviside," reprinted in Heaviside, O., *Electromagnetic Theory*, (New York: Chelsea Publishing Company, 1971), Vol. 3, pp. 523-642.

Kargon, R., and P. Achinstein, (eds.), *Kelvin's Baltimore Lectures and Modern Theoretcial Physics*, (Cambridge, Mass.: MIT Press, 1987).

Keele, Kenneth D., qv. "Leonardo da Vinci, Life, Scientific Methods and Anatomical Works," *Dictionary of Scientific Biography.*

Kelvin (William Thomson), "On the Theory of the Electric Telegraph," in *Mathematical and Physical Papers*, 6 vols., (Cambridge: Cambridge University Press, 1882-1911).

______ "On Vortex Motion," in *Mathematical and Physical Papers*, Vol. IV, (Cambridge: At the University Press, 1910).

Kennelly, A.E., *Artificial Electric Lines: Their Theory, Mode of Construction and Uses*, (New York: McGraw-Hill Book Company, Inc., 1917).

Klein, Felix, "On the Mathematical Character of Space Intuition and the Relation of Pure Mathematics to the Applied Sciences," in *Lectures On Mathematics*, (New York: McMillan and Co. 1894).

Kline, Ronald R., *Steinmetz: Engineer and Socialist*, (Baltimore: The Johns Hopkins University Press, 1992).

______ "Science and Engineering Theory in the Invention and Development of the Induction Motor," *Technology and Culture*, **28** (1987): 283-313.

Kragh, Helge, "Ludvig Lorenz and the Early Theory of Long-distance Telephony," *Centaurus*, 35 (1992): 305-324.

Lamb, Horace, "On the Induction of Electric Currents in Cylindrical and Spherical conductors," *Proceedings of the London Mathematical Society*, **15** (1884): 139-149.

Larmor, Joseph, "Electromagnetic Induction in Conducting Sheets and Solid Bodies," The Philosophical Magazine, 5th Series, 18 (Jan. 1884): 1- 23.

Lee, Sir George, "Oliver Heaviside - The Man," *The Heaviside Centenary Volume*, (London: The Institution of Electrical Engineers, 1950), pp. 10-17.

Lodge, Oliver, "Oliver Heaviside, F.R.S.," *Electrical World*, (Feb. 21, 1925).

Lodge, Oliver, *Lightning Conductors and Lightning Guards*, (London: Whittaker & Co., 1892)

Lutzen, J., "Heaviside's Operational Calculus and the Attempts to Rigorize it," *Archive for History of the Exact Sciences*, **21** (1979): 161-200.

Maxwell, James C., and Jenkin, Fleeming, "On the Elementary Relations between Electrical Measurements," *Reports of the Committee on Electrical Standards, Appointed by the British Association for the Advancement of Science*, (Cambridge: at the University Press, 1913), pp. 86-140.

Maxwell, James Clerk, *A Dynamical Theory of the Electromagnetic Field*, (Edinburgh: Scottish Academic Press, 1982).

______ *A Treatise on Electricity and Magnetism*, 1st Edition, (Oxford: At the Clarendon Press, 1873).

______ *A Treatise on Electricity and Magnetism*, 3st Edition, (New York: Dover Publications, Inc., 1954).

Mikusinski, Jan, *Operational Calculus*, 2nd Edition, (Oxford: Pergamon Press, 1983)

Nahin, P.J., *Oliver Heaviside: Sage in Solitude*, (New York: IEEE Press, 1988).

Newton, Sir Isaac, *Principia*, Motte's translation, revised by Cajory, (Berkeley: University of California Press, 1962).

O'Hara, J.G. & Pricha, W., *Hertz and the Maxwellians*, (London: Peter Peregrinus Ltd., 1987)

Olesko, Kathryn M., *Physics as a Calling: Discipline and Practice in the Koenigsberg Seminar for Physics*, (Ithaca: Cornell University Press, 1991).

Pais, Abraham, *Subtle is the Lord: The science and Life of Albert Einstein*, (Oxford: Oxford University Press, 1982).

Perry, John, "On Gamma Function and Heaviside's Operators," *The Electrician*, **34** (Jan. 25, 1895): 375-76.

______ *Applied Mechanics: A Treatise for the use of Students who have Time to Work Experimental, Numerical, and Graphical Exercises Illustrating the Subject*, 2nd ed., (London: Cassell and Company, Limited, 1898).

______ *The Calculus for Engineers*, (London: Edward Arnold, 1897).

Petrova, S.S., "Heaviside and the Development of the Symbolic Calculus,"

Archive for History of the Exance Sciences, **37** (1987): 1-23.

Pol, B. Van Der, and Bremmer, H., *Operational Calculus, Based on the Two-Sided Laplace Integral*, (Cambridge: At the University Press, 1950).

Preece, W.H., & Sivewright, J., *Telegraphy*, (New York: D. Appleton & Co., 1876).

Preece, W.H., "Fast Speed Telegraphy," *The Electrician*, **19** (Sept. 23, 1887): 423-426.

______ "On the Limiting Distance of Speech by Telephone," *The Electrician*, **18** (Mar. 11, 1887): 395-397.

______ "On Copper Wire," *The Electrician*, **19** (Sept. 9, 1887): 372-374.

______ "Multiple Telegraphy," *The Electrician*, **3** (June 7, 1879): 34-36.

Purcell, E.M., *Electricity and Magnetism*, (New York, McGraw-Hill Book Company, 1963).

Ramo, S., Whinnery, J.R., and Duzer, T. van, *Fields and Waves in Communication Electronics*, 2nd edition, (New York: John Wiley & Sons, 1984).

Rayleigh (Robert John Strutt), *The Life of Sir J.J. Thomson*, (Cambridge: At The University Press, 1942).

______ "Acoustical Observations - IV," *The Philosophical Magazine*, **8** (May, 1882): 340-47.

______ "On the Duration of Free Electric Currents in an Infinite Conducting Cylinder," *Report of the British Association for the Advancement of Science - 1882*, pp. 446-47.

______ "On the Self Induction and Resistance of Straight Conductors," *The Philosophical Magazine*, 5th Series, **21** (May, 1886): 381-94.

Russel, A., "Mr. Oliver Heaviside, F.R.S.," *Nature*, **115** (1925): 237-238.

Schey, H.M., *Div, Grad, Curl, And All That*, (New York: W.W. Norton & Company, 1973).

Schwartz, Melvin, *Principles of Electrodynamics*, (New York: McGraw-Hill Book Company, 1972).

Scott, D.B., & Tims, S.R., *Mathematical Analysis, An Introduction*, (Cambridge: At the University Press, 1966)

Searle, G.F.C., *Oliver Heaviside, The Man*, edited by Ivor Catt, (St. Albans: C.A.M. Publishing, 1987).

Smith, C., and Wise, M.N., *Energy and Empire: A Bioraphical Study of Lord Kelvin*, (Cambridge: Cambridge University Press, 1989).

Staff of Research and Education Association, Dr. M. Fogiel, Director, *The Electromagnetic Problem Solver*, (New York: Research and Education Association, 1983).

Sumpner, W.E., "Heaviside's Fractional Differentiator," *Proceedings of the Physical Society of London*, **41** (1929): 404-25.

______ "The Measurement of Self-Induction, and Capacity," *The Electrician*, **19** (June 17 - July 22, 1887): 127-128; 149-150; 170-2; 189-190; 212-213; 231-232.

______ "The Work of Oliver Heaviside," *Journal of the Institution of Electrical Engineers*, **71** (1932): 837-851.

Thompson, Silvanus P., *Life of William Thomson, Baron Kelvin of Largs*, 2 Vols. (London: MacMillan and Co., Limited, 1910).

Tyndall, John, *Heat: A Mode of Motion*, 4th edition, (London: Longmans, Green, and Co. 1870).

Valkenburg, M.E. van, *Network Analysis*, (Englewood Cliffs, New Jersey: Prentice-Hall, 1974).

Vincenti, Walter G., *What Engineers Know and How They Know It*, (Baltimore: The Johns Hopkins University Press, 1990).

Warwick, A., "On the Role of the FitzGerald-Lorentz Contraction Hypothesis in the Development of Joseph Larmor's Electronic Theory of Matter," *Archive for History of the Exact Sciences*, **43** (1991): 29-91.

Webb, F.C., "Momentary Currents in Wires," *The Electrician*, **2** (Apr 26 - May 10, 1879): 275; 286-287; 297.

Weber, Ernst, "Oliver Heaviside," preface to O. Heaviside, *Electromagnetic Theory*, (London: E.&F.N. Spon, LTD, 1951).

Weinberg, Steven, *Gravitation and Cosmology: Principles and Applications of the General Theory of Relativity*, (New York: John Wiley & Sons, 1972).

Whitehead, Alfred North, *Science and the Modern World*, (New York: Mentor Books, 12th printing, 1963).

Whittaker, Edmund T., "Oliver Heaviside," *The Bulletin of the Calcutta Mathematical Society*, **20** (1928-1929): 199-220. Reprinted in O. Heaviside,

Electromagnetic Theory, Vol. I, pp. xiii-xxxiv.

Wise M.N., and C. Smith, "The Practical Imperative: Kelvin Challenges the Maxwellians," in R. Kargon and P. Achinstein (eds.), *Kelvin's Baltimore Lectures and Modern Theoretical Physics*, (Cambridge, Mass.: MIT Press, 1987), pp. 323-348.

Yavetz, I., "Oliver Heaviside and the Significance of the British Electrical Debate," *Annals of Science*, **50** (1993): 135-173.

Index

Page numbers in *italics* indicate entries in footnotes.

CPSIA information can be obtained at www.ICGtesting.com
Printed in the USA
LVOW080412030712

288636LV00002B/37/P

9 783034 801768